PELICAN BOOKS

MODERN COSMOLOGY

Dr Jagjit Singh, General Manager of the South Eastern Railway, Calcutta, was born at Amritsar in 1912. After receiving his M.A. in Mathematics from the Government College, Lahore (1933), he dedicated himself to the study of mathematical statistics, especially in the field of engineering. As president of the Operational Society of India, fellow of the Royal Statistical Society, London, and member of the Indian Statistical Institute, Calcutta, Dr Singh has contributed greatly to operational research. He was adviser on operational research to the Government of India in 1965. He was also responsible for the operations-research movement on the South Eastern Railway, which resulted in a spectacular increase of profits. The author of a large number of papers and books, he has written: *Mathematical Ideas – Their Nature and Use, Great Ideas in Information Theory, Language and Cybernetics, Great Ideas of Operations Research, Some Eminent Indian Scientists,* and *Statistical Aids to Railway Operation.* Several of these have been translated into foreign languages. Dr Singh was the first Asian recipient of the UNESCO Kalinga Prize for his works in popularization of science. He was awarded the honorary degree of Doctor of Science by the Roorkee University in 1968.

JAGJIT SINGH

Modern Cosmology

REVISED EDITION

Penguin Books

Penguin Books Ltd, Harmondsworth, Middlesex, England
Penguin Books Australia Ltd, Ringwood, Victoria, Australia

—

First published under the title
Great Ideas and Theories of Modern Cosmology by Constable 1961
Revised edition published by Penguin Books 1970

—

Copyright © Jagjit Singh, 1961, 1970

—

Made and printed in Great Britain by
Hazell Watson & Viney Ltd,
Aylesbury, Bucks
Set in Monotype Times

TO
Jawaharlal Nehru

Contents

Acknowledgements 9

Preface to the Pelican Edition 11

INTRODUCTION

1 Digging Up Eternity 15

PART I: ASTROPHYSICAL THEORIES

2 The Physics of Celestial Fires: An Astrophysical Preamble 27

3 How the Stars Live and Die 51

4 Stars in the Making 75

5 The Birth of Galaxies 92

6 The Birth of Galaxies: A Turbulent Alternative 103

PART II: COSMOLOGICAL THEORIES

7 Space, Time and Gravitation or Relativistic Cosmology 123

8 A Flight of Rationalist Fancy 168

9 Cosmology and Continuous Creation 192

10 A New Cosmological Principle 219

11 Cosmology *a priori* 239

12 Quasars and Cosmology 254

13 Nuclear Physics and New Cosmology 267

14 The Last Dusk of Reckoning 280

PART III: ORIGINS

15 The Origin of the Elements 305

16 The Origin of the Planetary Worlds 323

CONTENTS

PART IV: LIFE IN THE UNIVERSE

17 Life in the Universe 363

18 God and Cosmology 387

Appendix: On How We See the Universe 397

Name Index 409

Subject Index 414

Acknowledgements

It is a great pleasure to acknowledge with gratitude my debt to National Professor S. N. Bose, F.R.S., Professor E. C. G. Sudarshan, Mr G. H. Keswani, Dr S. M. Chitre and Mr Shibdas Burman, who read the revised manuscript and made several valuable suggestions for improvement. With the exception of Plates 1, 11, 12 and 13 all photographic illustrations were supplied by the Mount Wilson and Palomar Observatories, to whom grateful acknowledgement is made. All other sources are acknowledged beneath the relevant Plate or Figure.

Preface to the Pelican Edition

THE increasing sophistication of astronomical observation and physical theory has radically altered the cosmological perspective during the past ten years since the early version of this book was written. Astronomical observations have revealed the existence of celestial objects as well as events that were not even suspected then. Cases in point are quasars, radio galaxies, blackbody 3·2 cm. microwave background radiation, the putative relic of the 'creation' of our universe and the like. But the most exciting and recent of these discoveries is that of pulsars which are by no means rare even though so far only forty odd have been observed. Since February 1968, when the discovery of the pulsar was first announced, there has appeared in *Nature* a veritable spate of articles and letters on them in addition to those on cosmological topics. Indeed, recent years, to use *Nature*'s graphic phrase, have been a 'vintage era' for cosmology and astronomy.

Similar advances in physical theory, in particular Einstein's relativity that launched cosmology some fifty years ago, have not only yielded many new ideas but also extensions of Einstein's relativity itself. Some of the new developments in physical theory are the gravitational collapse phenomenon, the effect of magnetic turbulence and cosmical electrodynamics, the role of nuclear physics in understanding stellar evolution, supernovae explosions, etc. The emergence of these new explanations and interpretations of cosmic phenomena is not only changing cosmology. It bids fair to change basic physics derived from our terrestrial experience as well.

Inevitably therefore the book has had to be extensively revised for the new Pelican edition. Most of the old chapters have been re-written and two entirely new ones have been added. My first reaction was to compensate the addition of these two chapters by deletion of an equal number – Chapters 8 and 11 dealing respectively with the cosmologies of Milne and Eddington. These are points of view which, though very much in vogue some two to three decades ago, are now way off the main stream of cosmological thought. However, I finally decided to retain them in deference to advice that I normally respect. But any reader not particularly interested in them may omit them altogether.

In a subject that is growing even faster than the 'exponential' rate normal for other fields of science and technology, it is impossible to be up-to-date. But one may hope that the new Pelican edition will survive the next few years of cosmological research sufficiently to remain at any rate in essentials a reasonably valid appraisal of the current situation in cosmology.

Calcutta   JAGJIT SINGH
November 1968

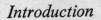

Introduction

CHAPTER 1

Digging Up Eternity

ONE may say of cosmology what has been said of Wisdom in Ecclesiastes: 'The first man knew him not perfectly, no more shall the last find him out.' For it is the one branch of knowledge where our deepest plumbings will fail to reach bottom for a long time to come, perhaps forever. An awareness of this bottomless complexity is of course only a recent occurrence. Yet far from diminishing our interest, it has on the contrary further stimulated that vague but powerful longing for cosmic speculation which has dominated human thinking since its very dawn. So great is this human itch to know how the universe began and how it is going to end that the problem of cosmic beginnings is verily the beginning of human speculation. That is why some sort of primitive cosmology was developed in all ancient civilizations as part of their religion or folklore, even before there was any astronomy. Naturally the earlier cosmological speculators were incredibly naïve. Mostly they were content to see the origin of the cosmos in the churning of the oceans by Vishnu; the marriage of the Heavens to Chaos; the fishing exploits of a sea-god like Tiki; the slaughter by Marduk of Tiamat, the primeval sea, and the lifting up of half her body to form the sky; the emergence of a Spirit on the face of the waters; or simply in the fiat of an almighty God. But the amazing success of rationalism in the various fields of scientific endeavour during the past three centuries has not only undermined the appeal of such a theological-cum-mythical approach but has also encouraged the hope that by a rational approach in this field, too, we are likely to obtain more satisfactory answers to the riddle of the universe.

Nevertheless, even now in spite of all the stupendous advances of science at which we marvel, the scientist who essays cosmology is fully aware that this ambitious field of study is beset with great difficulties if not dangers. For if the universe is the whole of creation, embracing everything from the tiny electron to the most distant galaxies and quasars, we get into deep waters the moment we ask about the origin of the most inclusive totality that can be imagined. If we were a little less ambitious and were to inquire about the origin of something not quite as all-inclusive as the entire universe, we might hope to get a satisfactory answer. Suppose, for instance, we wanted to inquire into the origin of life. A good enough answer might be that it originated from the virus or some still more atomic or electronic form of protoplasm. It is quite otherwise when we begin to talk about the origin of an all-embracing entity such as the entire universe. Suppose I were to say that the present state of the universe originated from some earlier state. Immediately the question would arise from what state this earlier state itself arose, and so on indefinitely.

There are only three ways of escape from this bottomless regress – or rather I should say two because the third is really a dodge. To take the dodge first, we might be able to arrange the past states through which our universe has evolved in some sort of sequence. Having done so, we might be able to discover one which could in some way be said to be like zero in the series of whole numbers. Here we blink over the issue as to what preceded the zero state itself. The second way of escape – and the first real one – depends on luck. It may happen, by chance, that the universe passes through a sequence of evolutionary states very much as the hand of a clock passes over the figures of the dial. If so, this cycle of evolutionary states repeats itself endlessly. There is then no beginning or end and the universe moves like an endless wavy curve. The final way of escape simply postulates that the universe has always been and will always be like it is now, so that there is no beginning or evolution at all.

All three of these ways of escape have actually been adopted by various speculators. The first way that was mentioned leads to theories which start the universe with a bang from a zero state, somewhat akin to the 'darkness on the face of the deep' from which, the Bible says, the universe emerged at God's command. This may not appear satisfactory to some as it seems to smack of a throwback to a fiat origin of the cosmos.

But the other escapes have their drawbacks too. The second way, the way of endless repetition, reminds us of Dante's futile running perpetually after a whirling standard in the Vestibule of Hell, while the third way requires a sort of *deus ex machina* on the job all the time to prevent the universe from changing.

But to continue the enumeration of the difficulties in this field of study: our second difficulty is due to the fact that observational knowledge of the actual nature of the universe and its contents is still very limited. The greatest depth to which we have been able to see with the aid of the world's most powerful telescopes, both optical as well as radio, is a distance of about 8,000 million light years, that is, the distance which light travelling at the rate of 186,000 miles per second would travel in 8,000 million years, or about 53,000 million million million miles. This depth seems incredibly great. Nevertheless, it is virtually nothing compared to the total depth of space if the universe is infinite. Even if we accept the proposition that the universe is not infinite (as some experts tell us), 8,000 million light years is only a fraction of the 'greatest' distance attributed by these scientists to the universe. A theory of evolution of the universe as a whole, based on such meagre data, would be like a study of biological evolution that took account of, say, only the Primates and ignored all other forms of life.

Our third difficulty is that the evolution of the universe takes an infinitely longer time than the span of human life or even the span of human civilizations. It is as though a fruit fly, having had a glimpse of a collection of human beings of all ages during its brief life of a few hours, were to try to figure out the relationship between adults and infants.

Attempts have been made to overcome these difficulties inherent to cosmology by appeal to other sciences. That is why cosmology has had to become a rendezvous of astronomy, physics, philosophy, and even pure mathematics. Cosmology has to utilize astronomy and physics because physical laws discovered by the study of down-to-earth phenomena have been extrapolated to the scale of the heavens. For example, the falling-apple law has been extrapolated to apply to the motion of stars in the interstellar void. It has thus been possible to piece together a large body of astronomical data (obtained by direct observation) into some sort of coherent cosmological story. But the procedure is not free from danger. The most fundamental danger is that these so-called laws may not apply to the universe as a whole even though they have been found to hold in the tiny corner of the universe that we have explored so far. As P. W. Bridgman has remarked, cosmological extrapolations at the base of the story are truly 'hair-raising'. To justify them, cosmology has been obliged to examine the 'foundations' or inner rationality and logicality of the physical principles used. In other words, recourse is had to philosophy in an endeavour to deduce the observed physical laws, such as those of gravitation, electrodynamics, and the like, from some plausible world principles or axioms assumed to be true *a priori*. But philosophic attempts to deduce scientific laws from *a priori* conceptions have been unsuccessful in the past. The Greeks who initiated this idea by the development of deductive geometry were nevertheless restrained by their delight in sensuous observation. The Middle Ages inherited the idea without the Greek restraint. The result was a disaster – a widespread choking of progressive thought by the weeds of scholastic quibbling and hair-splitting. Voltaire did not exaggerate when, in his story *Zadig*, he ridiculed the scholastic physician Hermes who wrote a book to 'prove' that the wound in Zadig's eye was incurable even after it had healed itself.

The truth of the matter is that we depend for our knowledge of the universe on experience and observation, on the one hand, and thought and deductive reasoning on the other. That is why every observation is shot through and through with theory, and vice versa. Consider, for instance, such a purely observational routine as the determination of the position of a star. It involves almost the whole of optical theory besides necessitating a series of corrections and adjustments that require a good deal of theoretical background. For even the correction of errors of observation cannot be undertaken without delving deeply into the heart of statistical theory. On the other hand, the all-embracing *theory* of gravitation required a whole host of observations of the planets which took Tycho Brahe and Kepler nearly a century to complete.

In spite of the interpenetration of theory and observation at all levels, it is nevertheless possible, starting from a set of axioms, to formulate certain types of knowledge by pure deduction. But what is deduced depends inevitably on what the chosen axioms contain. If they embody a large number of principles derived from observation, the deduced laws have considerable material content and are significant – but are built on the very foundation which the proof is designed to avoid. If, on the other hand, they incorporate only a few facts of observation, the deduced laws become vacuous – of some heuristic value perhaps, but nothing more. There is no escape from this dilemma. That is why *a priori* 'proofs' of fundamental physical laws are likely to be mere manipulations of preconceived concordances, if not outright special pleadings. Take, for instance, Eddington's deduction of the observed speed of the recession* of the galaxies from what he considered to be a set of *a priori* principles. Since the observed speed has now been greatly reduced by several subsequent revisions of the galactic distances by Walter Baade, Allan Sandage and others, one wonders if we should set much store by such arguments.

However, quite apart from the danger that a seemingly infallible *a priori* proof of a physical law may be debunked later, the main trouble in the use of an *a priori* approach in cosmology is that different sets of what seem to be equally plausible world axioms lead to different kinds of universes and laws that govern them. How is one to choose between them? This is the stage where the pure mathematician steps in to cut the Gordian knot with the suggestion that cosmologists should be concerned only with building up all possible model universes, in the same way a geometer creates all manner of possible geometries without any regard to what is the actual geometry of the space around us. The construction of model universes thus becomes a branch of pure mathematics, a free creation of the mind, a *jeu d'esprit*, and the identification of any of the created models with our actual universe becomes a minor if not an irrelevant detail.

We need not take too seriously the pure mathematician's refusal to make his choice. At worst it is only an instance of what Alfred North Whitehead has called the extreme timidity of expert scholarship in facing the complex problems of reality – a tendency that is becoming widespread in many fields. But a more relevant

* See Chapter 5.

consideration in this context is whether there is any real choice at all. As Herbert Dingle has asked, since there is one and only one universe, is its uniqueness, so to speak, essential or accidental? In other words, can we conceive of a whole hierarchy or series of possible universes out of which a Creator could make his choice when setting out to create one, or is any conceivable alternative to the actual universe necessarily impossible? This question must be answered before we can decide the validity of the pure mathematician's approach to the universe as a whole. Dingle seems to doubt the validity of this approach because he feels that we have no assurance that the existence of any universe other than the one around us *is* possible. Apparently the present situation in cosmology is no different from that in geometry prior to the invention of non-Euclidean geometries. No one doubted then that there could be one and only one type of space – the type whose properties are embodied in the Euclidean geometry we still learn at school. Philosophers like Kant even 'proved' that there was only one space of which what we call 'spaces' are parts, *not* instances. But thanks to the ingenuity of pure mathematicians, we now have a rich abundance of several kinds of spaces, of which Euclidean space happens to be just an *instance*. It therefore seems that there is no valid reason for denying to the mathematician the conceptual possibility of constructing model universes so long as we do not evade the problem of identifying some one of them with the actual universe around us and showing, moreover, why it is the actual one that came into existence. To demand a formal prior proof of the existence of such universes is to block a possible line of advance.

The upshot of the above discussion is that we can approach the cosmological problem from three different points of view. First, we may build up some sort of coherent cosmological story by blending physical theory with astronomical observation. Secondly, we may construct a series of model universes on the basis of some fundamental laws like those of relativity and thermodynamics and then try to identify one of them with the actual universe in which we live. Thirdly, we may try to figure out the origin of the cosmos and its possible evolution by rational deduction from a number of primitive cosmological axioms or principles chosen on essentially philosophic grounds. One may imagine that such a plurality of approaches would be an advantage in that it would enable their outcomes to be cross-checked. Actually it is an embarrassment. For the conclusions to which the different approaches lead are in flat contradiction to one another and, unfortunately, there is no way yet in sight of reconciling or choosing among them.

The reason it is difficult either to reconcile them or to decide among them is twofold. First, the gaps in our observational knowledge of the universe are still very wide. They let pass too many conflicting ideas. The situation is somewhat analogous to that of trying to find a unique solution of an equation in two unknowns. There can be none unless we find another equation. This difficulty of solving the equivalent of indeterminate algebraic equations is multiplied n-fold in cosmology. While the extensive series of observational programmes now under way the world over in the great observatories equipped with giant telescopes is designed to bridge these gaps and provide the counterparts of additional 'equations', the new 'equations' often involve new unknowns which require yet newer 'equations' and so on *ad infinitum*. It is therefore no wonder that new observations often raise almost as many new problems as they solve. Thus the extension of our ability to see the universe with radiation other than visible light all the way from long radio waves to short X-rays has raised new cosmological enigmas. A case in point is the discovery of an entirely new species of celestial objects called quasars* discovered by means of giant radio telescopes, or neutron† stars suspected to be the source of X-ray radiation recently detected by rocket-borne instruments. Obviously there can be no end to astronomical-cum-cosmological observations any more than there is an end to the pursuit of one's horizon. This leads us to the second difficulty hinted at earlier – the immense length of time required to collect an adequate basis of observational facts to strengthen the existing foundation of shifting quicksands under the cosmological theories. Moreover, observational tests designed to decide between the rival theories lead in many cases to such minute observational differences as to require

* See Chapters 2 and 5.   † See Chapter 3.

centuries and even millennia to accumulate sufficiently to come within our ken.

Cosmologists are therefore obliged to select those solutions that appeal to them out of the many possible ones for other than purely scientific reasons – aesthetic, emotional, moral and even metaphysical reasons. Thus to an Eddington the conception of a universe cyclically repeating itself endlessly seems from a moral standpoint wholly 'retrograde'. He would prefer to see the universe whimper to a heat death such as the second law of thermodynamics* seems to predict than submit it to the futility of a cyclic whirl perpetually repeating itself. On the other hand, to a Tolman the idea of the beginning and end of the universe is so repugnant as to drive him to resort to this very 'morally retrograde' course of endless cycles as the only way of keeping the universe alive forever. More recently, the discovery of quasars has proved such a cosmological enigma that it has been invoked to support a variety of diametrically opposite speculations according to the predilections of its sponsor so that cosmology is quite literally now in its hundred-flowers bloom. After all, the universe is very large and enough is not known about it. Given sufficient ingenuity almost anything can be explained or explained away. This is why a wide diversity of ideas from various fields like particle physics, chemistry, nucleogenesis, geology, astrophysics, radio astronomy, etc., may be borrowed to plant almost any exotic bud in the cosmological garden. For example, Dirac and Jordan reared a cosmology on the novel idea that universal constants of nature like the constant of gravitation are not true constants but vary with the 'age' of the universe. Likewise, Bondi and Lyttleton founded a cosmology on the assumption of a very slight departure from the usual belief that the electron and proton charge magnitudes are equal. R. O. Kapp built up yet another cosmology on a further extension of Hoyle's hypothesis of continuous creation by postulating that matter is not only being continuously created out of nowhere† but is also continuously disappearing into nowhere by extinction. By a blend of the twin ideas of continuous creation and continuous extinction he claimed that gravitation was not the 'signature tune of matter'

*See Chapter 14.   †See Chapter 9.

as supposed by tradition but only its 'swan song' at the moment of its extinction.

It may be regretted that modern cosmological research drives our feelings in such contrary directions as to make impossible the unity of scientific outlook which characterized the writings of earlier cosmologists like La Mettrie and Haeckel. Nevertheless, when a cosmologist has done all he can *not* to let his scientific theorizing be influenced by his hopes and desires, how can one blame him for making his choice out of the alternatives that even a strict striving after the knowledge of truth may allow? This is why cosmology is likely to remain for long full of controversy and conflict. Nor is there any immediate prospect of resolving the question and reaching the ultimate cosmological truth or anything like it. But what we do have is the excitement of a quest

> To follow knowledge like a sinking star,
> Beyond the utmost bound of human thought.

PART I

Astrophysical Theories

The Physics of Celestial Fires:
An Astrophysical Preamble

IN Part I we propose to give an account of what may be called astrophysical theories of the origin and evolution of the universe derived by piecing together a large body of observational data by means of astrophysical and mathematical theory. Before we begin theorizing, let us pull the veil of the heavens aside to see what there is in the universe. Looking out into the depths of space from our earthly abode, we see the moon a little more than a second away, and the sun about eight minutes. Here, as is customary in such cases, we reckon distances not in miles but in terms of the time their light, moving at the rate of 186,000 miles a second, would take to reach us. Farther away, within distance ranges varying from a few minutes to a few hours, we may spot some of those knights-errant of the sky, the planets of our solar system, whose remotest member, Pluto, is about five hours away.

Beyond the solar system, there are the fixed stars – 'fixed' because they are so remote from us that their own motion does not affect their apparent position in the sky as seen by us. These stars are bodies more or less resembling our own sun but immensely remote from us and from one another. In the neighbourhood of our sun they are, on the average, about five light years distant from their nearest neighbours; but as one goes in towards the centre of the galaxy the stars are much more crowded together and may, indeed, be separated from each other by no more than a fraction of a light year.

Most of the stars we see seem to be concentrated in a bright band of light called the Milky Way. The reason for this concentration of stars in one particular region of the sky is that we are located in a huge group of stars that is shaped like a gigantic wheel or disc as shown in Figure 1. The diameter of this disc is about 100,000 light years but it is only 3,000 light years thick near its outer rim, in our neighbourhood, though it is about six

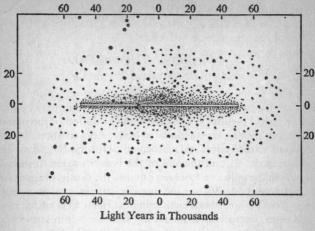

Light Years in Thousands

Figure 1.

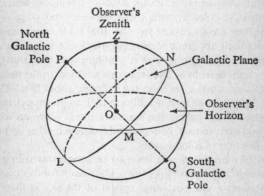

Figure 2. The sphere is the observer's sky. The line *LMN* is the projection on the sky of the Milky Way line or the galactic plane. Points *P* and *Q* are the galactic poles. The galactic plane is the counterpart of the terrestrial equator. It can be used to define the position of any celestial object on the sky by means of two numbers, its galactic longitude and latitude.

times thicker at its centre. Hence when we look along the radius of the disc, that is, along the galactic plane, we are looking in a direction that is studded with stars along a line thirty-three times as long as when we look in the perpendicular direction, that is, in the direction of its thickness or the galactic poles of the Milky Way (see Figure 2). Surrounding the disc and extending to distances about three times its radius is a great diffuse halo of stars. This huge system is known as the galaxy.

The galaxy, however, is no mere collection of stars sprinkled over immense distances. New techniques of observation like those of radio astronomy have now shown conclusively that it is pervaded by an extremely thin substratum of invisible interstellar matter in the form of cosmic dust and gas of exceedingly low density. Earlier astronomers, who observed the Milky Way during the first thirty years of the present century, had no means of detecting it. This is why they were haunted by this ghost of a material, which, while it could not be directly observed, nevertheless made its presence felt in many indirect ways. We now have, thanks to radio astronomy, quite powerful means of directly observing it. Thus its major constituent – neutral hydrogen – shows itself by the emission of radio noise emitted at 1,420 megacycles per second frequency or 21 cm. wavelength. In fact, observation of 21 cm. 'radio' radiation from the Milky Way has been so successful in revealing the presence of neutral hydrogen in interstellar space that its observation coupled with Oort's theory of galactic rotation has given us a complete picture of the distribution of neutral atomic hydrogen in our Milky Way system. It is now believed that the interstellar matter consists on the average of 1 per cent dust to 99 per cent gas, mostly neutral hydrogen atoms. The total mass of interstellar matter in the galaxy is only 2 per cent of its total mass of about a hundred billion suns. This amounts to an average density of one atom per cubic centimetre, although there are many regions of denser concentrations where it may be ten to twenty or even a few thousand times greater.

Even these regions of comparatively high concentrations of interstellar gas are rarer than the most rarefied vacuum we can create in a laboratory. But on the cosmic scale these vacua –

regions of high concentrations – make their presence known quite readily in the form of dark or luminous nebulae. The former appear as dark patches interrupting the rich star fields that can be observed in the Milky Way even with a small telescope. A case in point is the Coal Sack Nebula near the Southern Cross shown in Plate 1. This fact, together with the appearance among them of both bright and dark patches and still darker globules, shows that they are real dark clouds rather than starless gaps in heaven. Indeed astronomers have now devised ways of even measuring the absorptions, distances, and thickness of these dark clouds, the dark nebulae, by measuring the apparent distances of stars both through a cloud and in a neighbouring field outside the cloud. Thus the Coal Sack Nebula of Plate 1 is known by these means to be a region of scattered dust and gas particles thirty light years across situated about forty light years away from us.

Luminous nebulae too are easily distinguished. For while the stars always appear as more or less bright points both to the telescope and the naked eye, luminous nebulae look like diffuse nebulosities even on photographs taken with the largest telescopes. These luminous nebulosities – the galactic nebulae – are of two main types. First, there are the irregular, diffuse nebulae which are sometimes drawn out in long filaments like the Network Nebula in Cygnus (see Plate 2). Secondly there are small disc-shaped symmetrical formations with well-defined edges similar to the images of the planets such as Uranus and Neptune of our solar system (see Plate 3). But they are definitely not planets because they remain as fixed on the vault of the heavens as any of the fixed stars. Nevertheless, their superficial resemblance to the planets has earned them the misleading name of planetary nebulae.

The galaxy with its stars, globules, nebulae, dust and gas clouds is by no means the end of what we can observe in the heavens. It is in fact merely the end of a new beginning. For there are in the sky hundreds of billions of telescopic objects which also look like diffuse nebulae but which are situated far beyond the limits of the galaxy. Indeed, they are themselves galaxies, that is, immense conglomerations of stars and nebulae very similar to the galaxy in which our solar system is located.

This completes the list of actors that according to the current school of cosmologists are concerned in the drama of the cosmos. But it now seems likely that the cosmic play they have been trying to produce may well be very much like *Hamlet* without the Prince of Denmark. For they could not take account of new species of celestial objects that the large radio telescopes in England and Australia have begun to reveal only in the early sixties. They are some ten thousand discrete sources of radio noise. While some of them have been identified with nearby objects in our own galaxy such as the galactic nebulae, there are many which are way beyond it. Approximately one hundred of these *extragalactic* radio sources of the latter kind have been associated with visible galaxies about 100,000 light years wide. They are the so-called 'radio' galaxies, like M 82 shown in Plate 4. But what upset the cosmological apple cart in the early sixties was the further refinement of radio telescopes which in 1960 enabled astronomers for the first time to identify a strong radio source with what appeared to be a single star instead of a galaxy. This was indeed astounding because radio astronomy, despite all its refinements, could hardly be expected to reach a stage of pinpointing a source as small as a star. Yet in the next two years as radio data volleyed back and forth between the British and Australian radio astronomers, on the one hand, and the American optical astronomers at Mount Palomar on the other, three small, strange, *star-like* objects were identified. Plate 5 shows one of them, the quasi-stellar radio source 3C 273.

Although so far about 200 such quasi-stellar radio sources, or quasars for short, have been discovered, we still do not understand what they are. Thus while they are much smaller in size than even a midget galaxy, they are a hundred million times as heavy as the sun and they radiate energy at a rate 300 billion times that of the sun so that their power output exceeds that of a giant galaxy. They are presumably among the remotest objects in the universe. All in all their behaviour is so curious that no theory is able to account for it. In particular, no known process of energy generation can sustain their prodigious outpourings of radiation into space.

Even more puzzling is the recent discovery by Allan Sandage

of a still newer species of quasars, called quasi-stellar galaxies. Like the earlier sources, the newly identified objects are small (and therefore quasi-stellar); but they are immensely powerful emitters of blue and ultra-violet light. They differ from the normal quasars in two ways. First, they are *not* noticeable sources of radio noise. They are the quiet ones. Second, there are many more quiet quasars than radio quasars – about 500 times as many according to Sandage's original estimate. It is likely that quasars are in fact quasi-stellar galaxies going through a temporary stage of intense radio emission as Sandage has suggested.

Further, while the cosmologists of the past decade failed to see the quasar Prince, they were unable to notice the King's ghost – those extremely short X- and γ-ray* background radiations at the far end of the electromagnetic spectrum, which the rocket-borne instruments launched in outer space have started detecting only within the last few years. Obviously the implications of both quasars and short background radiations for cosmology are serious. We shall describe them more fully in later chapters.† Meanwhile we may pause to examine how we may classify the stars and galaxies in some sort of sequence.

To begin with the stars in our own galaxy, we may do so on the basis of some of their attributes very much as the fruit fly mentioned in Chapter 1 might try to arrange a group of human beings of which it had a momentary glimpse according to their height, hair length, weight, colour, and the like. We realize, however, that some of these categories (for instance, skin colour and hair length) would give our tiny sage no clue whatever regarding the evolution of an infant into an adult. We also appreciate that by arranging them according to their height he might erroneously consider an overgrown child to be older than an aged dwarf. Despite such possible pitfalls the height classification might with a stroke of luck lead him to surmise that the taller individuals of the group were in general older than the shorter ones. In a somewhat similar search for evolutionary clues, we have to select some attributes of stars in order to classify them.

If we study the huge collection of stars in our galaxy, we find a very wide diversity of conditions. Some stars, like the companion

* See Appendix. † See Chapters 6 and 12.

of Sirius, are extremely dense and compact – so dense that a matchbox full of such material would weigh a few tons. At the other extreme there are stars like the red giant Betelgeuse, which are so diffuse that they are more tenuous than the most perfect vacuum we can create in our laboratories. Some give out a million times as much heat and light as our sun. Some have a surface temperature averaging 28,000° to 35,000°C; others have a temperature as low as 2,000°C. Some stars, like Antares, are so large that their diameter is several hundred times that of most others. Some contain more hydrogen, others more helium. Some, like our modern dictators, want to shine all alone; others do not mind a few companions, while still others prefer to flock together in huge herds that astronomers call star clusters and associations. The only characteristic that is fairly uniform is their mass, or the amount of matter that constitutes them. The heaviest star would not be more than fifty times as massive as the sun and the lightest not less than one-tenth or -twentieth as massive, if a few exceptions, known to be over one hundred times as massive as the sun, were disregarded.

Among the various attributes that we could use for classifying stars, the most important are:

1 Mass.
2 Radius.
3 Luminosity or brightness.
4 Spectral class.
5 Chemical composition.

We assume that the reader understands what is meant by the mass and radius of a star.

As far as luminosity or brightness is concerned, we must be particularly on our guard against appearances. Here all that glitters is not gold. Our sun appears to us a prince among the luminaries only because it is, relatively speaking, so close to us. If it were removed as far away from us as, say, Sirius, so that its light took about eight years to come to us instead of only eight minutes, it would appear about twenty-five times fainter than Sirius. In other words, this lord of the heavenly luminaries would sink down to the level of an ordinary star rather faintly visible to the naked eye. The reason is that the luminosity of the star diminishes inversely as the square of the distance from which it is observed. We eliminate this dilution effect of the distance variable by converting the apparent luminosity of a star into its intrinsic or *absolute* luminosity. We do this by calculating the luminosity that it would exhibit when placed at a standard distance of about thirty-three light years from us.* The calculated luminosity is usually measured in 'magnitude' units on the basis that a star of any magnitude x is 100 times *brighter* than a star of magnitude $x + 5$. An increase of magnitude by one unit therefore corresponds to a *diminution* of luminosity by a factor of

$$100^{1/5} = 2 \cdot 512.$$

A star, however, has three different kinds of magnitudes depending on whether it is 'seen' by the human eye, a photographic plate or a photocell.

All these three types of detectors are not equally sensitive to all the different colours of light radiated by the star. The human eye has a maximum sensitivity in the yellow-green, that is, in the region of wavelength $\lambda 5,700\text{Å}$.† On the other hand, the common photographic emulsion is sensitive in the blue-violet, in the region of $\lambda 4,000\text{Å}$. As a result a blue star which radiates mostly in the violet and very little in the yellow region will appear brighter to the photographic emulsion than to the eye. The opposite will be the case for a yellow star. Consequently visual magnitude (V) of a star is different from its photographic magnitude (B). Both will differ from the magnitude measured by a photocell or a

---

* In actual practice the procedure is reversed. In most cases the stars are too remote to show any measurable parallax, that is, any apparent shift in their position in the sky owing to the earth's orbital motion around the sun, so that their distances cannot be measured by a straight extension of the surveyor's method of triangulation to celestial objects. These distances are therefore determined indirectly by comparing apparent and absolute magnitudes. The former is observed directly but the latter is inferred indirectly from certain empirically derived relations between absolute magnitude and spectral behaviour or, if the star is a Cepheid variable, between the period of its light variation and absolute magnitude. The distance of the star is then proportional to the square root of the ratio of the absolute brightness to the apparent brightness.

† Wavelength here is measured in Ångström units (10^{-8} cm.).

thermocouple, because these are equally sensitive to all radiations. The magnitude measured by such non-selective detectors is called bolometric.

For greater precision the magnitude of a star is actually measured in some preselected colour such as blue, yellow or any other by means of an appropriate filter. It is usual to measure stellar brightness in three colours – ultra-violet, blue and yellow. Indeed such triplets of stellar magnitudes form the basis of the system of three-colour photometry. If U, B, V are the magnitudes of a star in ultra-violet, blue and yellow light, the latter two, namely B and V, are respectively photographic and visual magnitudes. Their difference $(B - V)$ is called the colour index of the star. The difference $U - B$, the so-called ultra-violet excess, is also computed. It is a measure of the difference in a star's radiation at short wavelengths *vis-à-vis* its radiation at longer but neighbouring wavelengths.

We next come to the fourth attribute, namely the spectral class of a star. When we pass light from a star through a glass prism or grating, it is decomposed into its composite colours just as sunlight is broken up into rainbow colours when viewed through a prism. What we see after it has passed through a prism is known as a spectrum. Now although a spectrum is only a band of bright or dark lines on a dark or coloured background, its lines are a sort of hieroglyphics which an expert can decipher. When deciphered they tell us many things about the stars. A stellar spectrum may, for instance, give us a lot of information about the chemical composition of the atmosphere of the star, its radial velocity in the line of sight, its surface temperature, and the like.

The stellar spectrum can yield this information because the continuous spectrum of a ray of light that has passed through a relatively cool gas (the atmosphere of a star) will be crossed by dark lines that are the imprint of the chemical composition of the gas. The location of a spectral line relative to the colour panorama of its background yields what is in effect its identifying or registration mark, usually known as its wavelength. By studying the wavelengths of lines produced here on the earth by glowing gases of known chemical composition and by comparing them with those produced by celestial sources we can tell which of the elements are present in the celestial objects. It is indeed remarkable that this method has also revealed the existence of elements hitherto unknown. It was in this way that helium was first discovered in the sun before it was isolated here on earth.

To understand the messages the spectral lines write for us, we have to learn the syntax of spectral language, that is, the selection rules devised by spectroscopists. The hard core of these syntactical rules is that profound amalgam of atomic theory and empirical observation called quantum mechanics. We cannot go into it and its vast ramifications here except to remark that the syntax of the spectral hieroglyphics is based on the idea that spectral lines are emitted when the planetary electrons,* circulating around the atomic nucleus, pass from one excited state to another. In the old Bohr theory of the atom this was pictured in terms of electrons in orbits. According to this picture, which we may use as a helpful guide, a bright line is emitted when the electron jumps from an exterior orbit to one nearer the nucleus, and a dark line in the reverse case.

The syntax of spectral lines permits us to envisage what types of lines are likely to arise in spectra because of the presence of various kinds of elements in various states of ionization. While our home-made terrestrial spectra rigidly obey the selection rules and show only such lines as are permitted by the syntactical rules of spectral hieroglyphics, the celestial spectra exhibit some 'forbidden' lines, that is, lines not allowed by the spectroscopists' selection rules. There was a time when this anomaly was interpreted as an indication of the existence of some unknown new elements. Two of them, nebulium and coronium, were even christened before their existence was proved. The chemists, however, reported 'full house', as there was no room in their periodic table of the elements for any new light atom. It is now realized that these 'forbidden' lines are produced by ordinary elements like nitrogen and oxygen but under such peculiar conditions that they can occur only in strange celestial environments which cannot be duplicated in the laboratory. In fact, the identification of 'forbidden' lines is one of the principal modes of knowing what these celestial environments are.

* See also Chapter 13.

Now the same spectral line of any particular registration mark or wavelength may etch itself on different spectra with varying intensity of brightness (if it is a bright line) or shades of darkness (if it is a dark line), just as the same letter of the alphabet may be written in inks of varying degrees of visibility. This is fortunate, for the intensities of these lines depend upon such factors as the chemical abundances, the temperature, the pressure in the atmosphere, the state of ionization and the like; and although the relationships involved are rather complex, astronomers have shown us how to derive a great deal of information from an analysis of the spectral lines. Nor are we limited to the spectral lines alone for our information; the radiation that gives rise to the continuous spectrum is also very useful, for we can, by means of Planck's radiation formula, calculate the temperature of the surface from which this radiation emanates. Thus, if we measure the radiation intensity of a glowing tungsten filament of an incandescent lamp, we can employ a modified Planck's formula* to determine its temperature. An extension of this technique to stellar light enables us to calculate the surface temperature of the radiating star.

However, we must be careful to exercise caution in deciphering the messages of spectral lines. For while the position of spectral lines (or what comes to the same thing, their identifying wavelengths) is a clear index of the presence of their sponsoring elements, there is the possibility that *all* of them may be shifted to the *same* extent against the colour background of their spectra, either towards the violet or red end, by an entirely fortuitous circumstance. This is the well-known Doppler shift of spectral lines resulting from the relative motion of the source in the direc-

---

* Planck's radiation formula yielding energy E_λ in ergs per cm.^{2} per sec. per wavelength interval radiated at wavelength λ is the equation

$$E_\lambda = \frac{C_1}{\lambda^5(e^{C_2/\lambda T}-1)},$$

where T is the absolute temperature, that is, centigrade temperature plus 273°, e is the base of the Napierian logarithm, a number equal to 2·718 and C_1 and C_2 are two constants whose values are $3·70 \times 10^{-5}$ and $1·4323$ respectively. Since the values of C_1 and C_2 are known, a knowledge of E_λ for any given λ is sufficient to determine T.

tion of the observer's line of sight. If the source is receding, the lines are shifted towards the red end, but if it is approaching they move towards the violet end. In either case the magnitude of the shift is a measure of the velocity of the source. This again has become grist for the cosmological mill. For it enables us to tell whether a celestial object is receding from or approaching us, and, if so, at what rate. It even reveals the rotation of the object; if the spectral lines of a celestial object show a shift of both kinds (and hence a broadening of the lines for stars), as is the case, for instance, with certain stars and some of the nearer galaxies, it means that while one end of its limb is receding, the other is approaching. This can happen only if it is in rotation. A case in point is the dual spectral shift of the spectral lines of the Andromeda galaxy, which shows that it is revolving around its centre in about 150 million years.

On the basis of information furnished by stellar spectra we could classify stars in several ways. We could group together stars whose spectra indicate that they have the same or very nearly the same surface temperature. In this way, ignoring unstable stars with peculiar behaviour, we obtain eight main types, referred to as O, B, A, F, G, K, M and N. The very hottest stars that we know have surface temperatures averaging about 30,000° to 35,000°K. Their spectra are called type O. The next type, called B, belongs to a somewhat cooler class of stars with a surface temperature of about 25,000°K. The third class of spectra is designated class A, and corresponds to stars with a surface temperature of about 11,000°K. Next in the series are spectra of classes F, G, K, M and N, with stars of progressively lower surface temperatures, the average temperature ranging from 7,500° for F class to 3,500° for M and 3,000° to 2,000° for the last class, although the use of infra-red methods has very recently revealed sets of ultra cool stars in the centre of the galaxy with surface temperature as low as 800°K.

Since the energy radiated by a star at any particular wavelength or colour is strictly linked with its surface temperature in accordance with Planck's formula, we should expect a distinct correlation between the 'colour' of radiation, and the temperature at which it is emitted. The most familiar instance of this kind is the successive changes in colour assumed by a piece of metal as it is heated. It assumes at first a red colour becoming white hot as temperature increases. But as it cools, the colour changes from white to white-yellow, from yellow to orange and from orange to red. It is the same with stellar surface temperature. The hotter stars radiate mainly in the violet and blue and appear to us of a whitish blue colour and the cooler ones appear red. The 'colour index', defined earlier as the difference $(B - V)$ between photographic (B) and visual (V) magnitudes, is an objective measure of the stellar colour and hence indirectly of its surface temperature and spectral class. Indeed the correlation between spectral class, colour index and surface temperature is very close as shown in the table below:

TABLE 1

Spectral class	Colour	Colour index	Surface temperature (absolute)
O	blue	−0·30	>30,000°K.
B	blue-white	−0·15	30,000 to 15,000°K.
A	white	0·00	15,000 to 10,000°K.
F	white-yellow	+0·40	10,000 to 7,000°K.
G	yellow	+0·80	7,000 to 5,000°K.
K	orange	+1·20	5,000 to 4,000°K.
M	red	+1·80	4,000 to 3,000°K.
N	infra-red	?	<3,000°K.

We shall now take up the question of chemical composition. The earlier cosmologists believed that stellar material, or at least its inner core, consisted of heavier materials like iron and silicon. But the assumption did not agree with theory. What theory? This leads us to a slight digression on the theory of stellar structure.

When we look upon a large mass such as, say, the mass of water in the Pacific Ocean, we generally do not realize to what terrific pressure its lower depths must be subject – unless we happen to be in a submerged submarine. This pressure is simply due to the weight of overlying water. Every thirty-three feet of its layer creates a pressure of about fourteen pounds per square inch. At a depth of five miles, where we may find Gray's 'gems of purest ray serene', the pressure is over 10,000 lb. per square inch. These gems are hard enough to withstand this pressure. But it is a good deal more difficult for materials in the very bowels of our good earth. Those at its centre have to bear the strain of not merely the overlying layers of oceanic waters but also the solid rocks, which are much heavier. It is estimated that the material at the centre of the earth is subject to a pressure of fifty million pounds per square inch, which is 5,000 times greater than that at the bottom of the Pacific. This is colossal, but our terrestrial materials seem tough enough to withstand it. However, the sun is 340,000 times more massive than the earth, and the solar central core has to withstand a pressure 20,000 times greater than any within the earth. No terrestrial material could endure so great a compressional force.

If we could imagine a giant rolling up about 340,000 earths to make up a mass of terrestrial material equal to the solar mass enclosed within a sphere of solar radius, we should see it change explosively into a hot gaseous sphere of matter just like the sun. The reason is that the weight of all this overlying material would bring about a sudden collapse which in turn would result in a rapid increase in the temperature. Indeed, the temperature would continue to rise until the material had all vaporized and the atoms and molecules were moving about fast enough to support, by means of this rapid motion (it is this motion that gives rise to the pressure in a gas), the weight of the material pressing down.

Now the average velocity with which the atoms and molecules of a material move is also a measure of its temperature. Thus the molecules of air at 20°C. move on the average at about 1,000 miles per hour. To prevent material at the core of our imaginary aggregate of 340,000 earths from simply caving in, we should have to endow its particles with such terrific velocities that when we convert their mean velocity in terms of degrees of temperature we find it simply supercolossal. It happens that this temperature at the centre of our imaginary aggregate turns out to be 13,000,000°C.

We may remind you that the temperature of an electric furnace is a paltry 3,000°C. We can now understand why the mere act of pooling up 340,000 earths to form one huge mass the size of our sun would result in an instantaneous blaze. But such a blaze would be rocked by an internal 'seething turmoil' between the pressure and the weight of overlying layers that have not yet had time to reach a state of balance. It would therefore look as if this ball of fire were breathing 'in fast thick pants' a mighty fountain of fire into a lifeless void. However, the panting phase is only a temporary affair and lasts a mere twinkle. For very soon the internal seething turmoil turns into a balanced tug of war wherein both parties pull the rope with equal might. When this happens the aggregate attains equilibrium and remains in that condition for aeons. This is how our own sun continues to shine and remain in a steady state.

It would thus seem that fire and light lie grovelling in all matter, to be kindled by that cosmic incendiary gravitation when it manages to herd together a sufficient quantity of matter within the ambit of its sweep. This is why one may say, with greater literal truth and less poetic licence, that if life and consciousness are the fever of matter, as Thomas Mann once remarked, then stars and galaxies are its flame.

However, stellar flames are not ignited as simply as the foregoing account of how a star blazes forth from a sufficiently large aggregation of dark matter might seem to suggest. The question is a little more complicated than finding the mean motion of the particles in the interior of a large mass of matter the size of a star so that it will be able to withstand the weight of overlying layers. For the blaze calls into play a new ally to aid the gas pressure of the aggregated material now completely vaporized in its contest with gravitation (or weight of overlying layers) to attain equilibrium. This ally is the radiation pressure of radiant light.

That a beam of light exerts pressure is difficult to appreciate in everyday life because no sultan's turret has ever been squeezed or dented by a shaft of light. Nevertheless, at higher temperatures it is indeed a power to reckon with. The reason is that it increases with the fourth power of temperature whereas ordinary gas pressure increases only in simple direct proportion. Thus while a tenfold rise in temperature means a paltry tenfold increase in gas pressure, it catapults radiation pressure 10^4, or 10,000 times.

However, in spite of this rapid growth of radiation pressure with increasing temperature, it is still the gas pressure that is most responsible for maintaining the dynamical equilibrium conditions inside stars like our sun. It is only when we are dealing with the very massive stars, such as the red supergiants, that the radiation contributes more than a small percentage to the total pressure. In the case of the sun and other stars like it, the radiation pressure is about 3 per cent of the total pressure, but for a star that is ten times more massive than the sun the radiation pressure may be as much as 40 per cent of the total. Thus radiation pressure may be neglected in considering the internal structure of stars like the sun, but must be taken into account for the very hot luminous stars like Rigel.

As mentioned before, the older cosmologists began with the guess that stellar material, at any rate in a star's interior, consisted of heavy atoms like iron, silicon and so forth. This led to trouble. The observed luminosities of stars did not at all conform to the theoretical formula; they were far too low. It was discovered that observation would fit the formula if the stars were to consist of either 30 per cent hydrogen or over 99 per cent, assuming that no helium was present. Until about thirty years ago the astrophysicists believed that stars consisted primarily of heavy atoms and that at most hydrogen was present to the extent of about 30 per cent. These ideas, which were due to Elis Strömgren, were based on the assumption that there was a negligible amount of helium in the stars. Strömgren, however, using the no-helium assumption, arrived at two possible solutions for the internal structure of a star, one of which indicated a hydrogen content of 30 per cent and the other a hydrogen content of 99 per cent. It was clear from this that one could not neglect the helium content if one was to obtain an unambiguous solution to the problem of the structure of stellar interiors.

The fact that neither of these models obtained by Strömgren could apply to actual stars was clear from the investigations by Russell, Menzel, Pannekoek, Unsold and others who were able to show that the stellar atmospheres of stars like the sun contain more than 80 per cent hydrogen and large quantities of helium. They calculated the abundances of elements as a function of temperature and pressure of stellar atmospheres by means of Saha's theory of atomic ionization which describes the behaviour of ions, that is, atoms stripped of one or more of their satellite electrons. These studies showed that hydrogen and helium constitute more than 97 per cent of stellar atmospheres. All the recent investigations made since the end of the war into the internal constitution of stars have taken into account both the hydrogen and the helium content, and it has been definitely demonstrated that hydrogen and helium together account for more than 96 per cent of the material in stars like the sun.

Having settled the question of chemical composition, it now remained to discover the mode of energy generation in stellar interiors. Until recently it was a major astrophysical enigma, although it has been recognized since antiquity that the sun and stars would need fresh supplies to replenish the continuous discharge of the precious flame that they continually pour out into space. Thus Lucretius said in *On the Nature of Things*: 'We must believe that sun, moon and stars emit light from fresh and ever fresh supplies rising up.' But how the fresh supplies came into being neither Lucretius nor even most of his scientific successors down to our own day could ever conceive. For before the recent discovery of nuclear energy we knew of only two sources of heat and fire, both of which we use in our homes. One is chemical combustion and the other gravitation, that is, fall of materials under their own self-attraction. We resort to the former when we warm ourselves by burning coal or gas, and we tap the latter when we turn falling torrents like Niagara into electricity. Neither of these would enable the sun to shine at its present rate for more than a mere twinkle.

To take the latter first, a simple calculation shows that even if the sun contracted from infinity to its present dimensions, the total gravitational energy released thereby would not last more than twenty million years, whereas the sun is known to have been radiating energy at its present rate at least since the emergence of life on earth some 500 million years ago. The combustion source has an even lesser staying power. If this were all the sun could rely on for its energy output, it would be bankrupt in mere millennia. To make the sun draw its daily sustenance from either of them, or even both together, is worse than setting Baron Munchausen's bears on the sultan's solitary bee for the latter's honey.

This difficulty about the source of solar energy was not resolved till H. A. Bethe's discovery in 1939 of two chains of nuclear reactions known as the proton-proton chain and the carbon-hydrogen cycle. In both these reactions four protons of hydrogen nuclei are fused into an α-particle or helium nucleus with release of prodigious amounts of energy. The reason is that although barely 0·7 per cent of the hydrogen mass is lost in the fusion process (the balance reappearing as helium), the lost mass manifests itself as energy in accordance with Einstein's mass-energy relation

$$E = mc^2,$$

where E is the energy released in ergs, m the lost mass in grams, and c the velocity of light in centimetres per second.

Since c is $3 \cdot 10^{10}$ cm./sec., the loss of a single gram of hydrogen yields 9×10^{20} ergs, or twenty-five million kilowatt-hours of energy. At this rate of conversion the loss of only 1 per cent of solar mass will suffice to keep the sun shining at its present rate for more than 1,000 million years. If the sun remained in its present steady state, even with an expenditure of 10 per cent of its initial stock of hydrogen, it could maintain itself in this state for 10,000 million years. The nuclear or atomic source thus turns out to be over 500 times more plentiful than gravitational attraction considered earlier. Nevertheless, as we shall see presently, gravitational attraction does play a role both in the initial and later stages of the life of a star, that is, when it is condensing out of interstellar gas, and later when it becomes unstable after having exhausted nuclear sources of energy.

If we assume that the stellar material is practically all hydrogen and that the energy that it radiates is produced by one or the other of these two nuclear processes, the known mass enables us to calculate all its physical features such as size or radius, brightness or luminosity, internal pressure, and central and surface temperatures, provided we adopt a specific 'stellar model'.

The model means postulating some further conditions regarding the generation of energy in stellar interiors, the mechanism of its escape to the surface, chemical composition and the extent of inhomogeneity of its material. Thus the transformation of hydrogen into helium with the release of energy may occur throughout a fairly large region of the star, as in the case of the sun, or in a small concentrated core near the centre, as in the case of the hot blue-white giants, or yet again in a hot shell surrounding a central core, as in the case of the red supergiants like Betelgeuse and Antares. Whether the first, second, or third of these processes occurs depends on how hot the interior of the star is and how old it is. For the very bright young stars (blue-white giants like Rigel) the central temperatures must be so high that the point-source model applies, whereas for the very old supergiants with a helium core the transformation of hydrogen into helium will take place in the thin shell around the core. In a medium-aged star like the sun, in which not very much hydrogen has changed into helium and in which the central temperature is not greater than fifteen million degrees, the release of energy will occur throughout a fairly large region extending out from the centre.

Further, energy generated in the stellar interior may be transported either by radiation, like the propagation of warmth from a chimney fire, or by large-scale motions of the heated gases, that is, by convection currents, like the flow of heat from the Tropic of Cancer to the shores of Spain by the Gulf Stream, or by conduction. Which of these three modes of transport will prevail in the star depends on temperature differences in the region considered and the density and pressure of the material. If the temperature difference required for radiative transport is too high, the layers will be unstable and the slightest disturbance, as Cowling showed, will cause large-scale convection currents to be established. Consequently, in models where the mode of energy generation requires high temperature, as in the case of the carbon-hydrogen cycle, convection currents occur in the core of the star. If the proton-proton reaction supplies the energy, the convection core may be very small or even non-existent. Because of this dependence of the mode of energy transport on temperature difference, it may happen that while convection occurs at one place, energy flows by radiation in another. Conduction occurs only in the white dwarfs where the gases are extremely dense and where there are many free electrons moving about.

Besides the mechanism of energy transport, the model must also specify the opacity of the material in order to determine the extent of obstruction it offers to the outward flux of radiant energy from its energy-generating core. Fortunately this also is known from quantum theory. It is given by Kramers's law as a simple function of density, temperature, and chemical composition, though different models employ somewhat different variants of it.

If the stellar model prescribes the rate of energy generation in the interior, the mode of its transport to the stellar surface, and the extent of obstruction it encounters on the way, the problem of energy flow in the model is in essence no different from that of fluid flow through a porous slab. The rate of fluid seepage is simply proportional to the pressure difference between the two faces of the slab and inversely to the resistance of the slab to its escape. In exactly the same way the rate at which energy oozes out at the surface of a star, that is, its absolute luminosity L, is determined by the radiation-pressure gradient and the opacity of its material. As we saw before, radiation pressure is proportional to the fourth power of temperature, and opacity is given by Kramers's law. It is thus possible to evaluate L and all other attributes, such as surface temperature T and radius R of the star, given its mass M.

That these calculations are no mere abracadabra and do really work is no longer in doubt. We have observational verification of the predicted results. We may mention two of them here: the mass-luminosity and luminosity-temperature laws. Calculation shows that for certain stellar models

$$L = \frac{\mu^7 M^5}{\varepsilon^{1/40} k} \text{ and } L = \frac{\varepsilon^{1/7} k^{1/2}}{\mu^{4/3}} T^{5 \cdot 5}, \qquad (1)$$

where k is the coefficient of opacity and ε that of energy generation of the material, and μ the molecular weight of the 'gas' particles of the material.

It is true that uncertainties in the estimated values of these constants do affect the predicted values of L and T. But owing to a lucky concordance of circumstances the derived formulae are robust enough not to be greatly perturbed by these uncertainties. Take first ε. Its relatively low exponent in the two formulae is the reason the luminosity L is insensitive to any uncertainty in its estimate. Even if it were in error by a factor of 100 – a very wide margin indeed, as it has been determined within a factor of 2 by nuclear experiments – the end product L of the calculations would not be affected by more than a factor of $100^{1/40}$, that is, by not more than 12 per cent. Further, the mean molecular weight μ of the gas particles depends only to a very minor extent on the chemical composition as the known abundance of hydrogen and helium in stellar interiors largely determines it, and the small extent of heavier elements within any likely limits does not affect the mean molecular weight of the material as a whole. The reason is that even a heavy atom – for example, iron of atomic weight fifty-six times that of a hydrogen atom – will barely retain three of its twenty-six planetary electrons under conditions prevailing in stellar interiors, so that the entire weight of the atom is averaged over twenty-four particles (twenty-three disrupted electrons plus one nucleus). The result is that the mean molecular weight of the material contributed by iron is only 56/24 = 2·3 times that of a hydrogen atom. Finally, the opacity coefficient k is also known within small limits of error from quantum mechanics.

Using the known values of these constants, we can cast our two equations in a numerical form and verify to what extent the observed values of L, M, and T, in the case of real stars satisfy them. We find that within the limits allowed by uncertainties of chemical composition the observed values do conform to the theoretical equations (1) in a vast majority of cases. But there are many exceptions, too.

One reason exceptions occur is that our idealized stellar model, on which the derivation of the equations is based, is assumed to be homogeneous. But obviously this is an unrealistic assumption. For even if our star is initially homogeneous – practically all hydrogen – during the course of its evolution, the process of energy generation (if it is confined to some specific region of its interior) is likely to lead to some inhomogeneity by the exhaustion of hydrogen in that zone. For unless it is in very rapid rotation, there appears to be no means whereby a star can keep its material well mixed outside convective zones; within the convective zones the prevalence of convection currents continually stirs the material. Therefore the theory of stellar models must also take account of inhomogeneity. This can easily be done because once a homogeneous stellar model has been constructed it is possible to proceed to a more realistic state of affairs by grafting on to it a measure of heterogeneity, provided we prescribe some particular pattern for it. Several such patterns may be visualized.

First, a rapidly rotating star may remain nearly homogeneous, for it may manage to stir up its material so well as to smooth out the differences in composition caused by nuclear transmutations. In such a case the evolutionary changes in its life are slow and not very significant. Secondly, in the absence of any significant rotation or convection currents in the interior, the proportion of helium may gradually increase with age from its initial value at the periphery to a maximum at the centre. The greatest effect of such a creeping advance of inhomogeneity will be felt when the central helium core attains sufficient size. As Strömgren has shown, the structure of the star then begins to change gradually and its radius increases. Thirdly, strong convection currents may prevail in the inner central region, to which nuclear transmutation of hydrogen into helium is confined. As a result, a homogeneous energy-generating convective core with diminishing hydrogen content surrounded by a homogeneous envelope of higher hydrogen content may develop.

What course of evolution the star will then follow has been studied by Schönberg and Chandrasekhar. They show that the radius of the star will again increase with age. Later, when the hydrogen in the inner convective core is completely transmuted by nuclear fusion into helium and this hydrogen-free core grows to a sufficient extent, the Schönberg-Chandrasekhar model too breaks down. The further course of events is then described by another model – that constructed by Sandage and Schwarzschild. They show that in the case of such inhomogeneous stars with hydrogen-exhausted cores, the radius again increases with age but the increase is much more pronounced. Finally, we may have an inhomogeneous central core surrounded by a convective envelope such as that recently computed by Schwarzschild, Howard and Härm to account for the observed large abundances of heavy elements in the sun.

We thus observe that stellar models are of several types. This plurality of stellar models is necessary to take account of the immense diversity of types of stars in the universe. Complicated as these stellar models seem to be, there is no doubt that they are not complicated enough. For they greatly oversimplify a very complex state of affairs by their neglect of a number of features that are important in many, if not all, cases, such as the existence of turbulence and shock-waves, magnetism, rotation, presence or absence of accretion from near-by clouds of interstellar gas, and so on.

Thus, for example, H. W. Babcock has discovered the presence of strong magnetic fields around certain fast-rotating stars of spectral types O and B as well as stars having periodically variable magnetic fields. Although we do not yet know the origin of these magnetic fields, it is certain that their existence must profoundly modify the behaviour of stellar matter. Consequently, the construction of stellar models must be based on magneto-hydrodynamical laws instead of the purely hydrodynamical laws on which they are based at present. But unfortunately, the laws of this new science of magneto-hydrodynamics have to be formulated before they can be applied to stellar models. Nevertheless, in spite of the neglect of magnetic fields and other features in the computation of stellar models, it is indeed remarkable that in many cases the calculations made do enable us to divine with reasonable accuracy the physical state of stellar interiors and to calculate main features of stars, such as luminosity L, radius R, surface and internal temperatures, and pressure, given the mass as well as the course of their evolution.

As we shall see later, we may actually observe stellar systems evolving along lines at least qualitatively in agreement with the predictions of stellar horoscopes cast on the aforementioned astrophysical principles. Meanwhile, we may merely note here that in the case of our own sun, where energy is generated by the proton-proton chain, the calculated values of L and R agree surprisingly well with the accurately known observational values. These results certainly confirm our faith in the theory of stellar interiors, at any rate as a first approximation, and in the proton-proton chain as the source of solar energy.

CHAPTER 3

How the Stars Live and Die

WITH the completion of the astrophysical preliminaries in Chapter 2, it is now time to take up one of the main themes of this book, the life history of a star. Here again we may do well to consider what our old friend the fruit fly might do. If he were as wise as we think we are, he might attempt to correlate the weight and height of the individuals under study. That could be done easily enough. He could draw on graph paper two lines at right angles to each other and represent weight along one line and height along the other. Each individual could thus be represented by a dot on the graph paper. The scatter of these dots would then show that, in general, greater weight went hand in hand with greater height.

Now we can do this sort of thing with our stars by taking any two of their attributes: we can take their mass and luminosity, which will give us the mass-luminosity diagram; or we can take mass and radius; or luminosity and surface temperature or, what is the same thing, its spectral class. All these diagrams have been drawn by astronomers and studied in detail. But for our present purpose it will suffice if we confine our attention to the last-mentioned diagram, the luminosity/spectral class or Hertzsprung-Russell (H-R) diagram. Here, it may be repeated, we denote a star on graph paper by means of a dot whose ordinate is its absolute luminosity, and abscissa its spectral class, spacing equally the eight classes *O*, *B*, *A*, *F*, *G*, *K*, *M* and *N* or, what amounts to the same thing, its colour index, or the logarithm of its surface temperature, because of the link between all the three embodied in Table 1 of the last chapter.

We find that a majority of the dots cluster around one main line *PQ* as shown in Figure 3. But there are plenty of exceptions, most of them concentrated towards the top right of the diagram. This does not mean that the left hand exceptions are few even though our diagram shows only three; they happen to be too faint to be

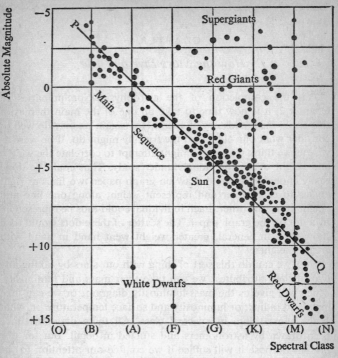

Figure 3.

seen unless they are very close to us. The exceptions to the right are known as red giants because they are so luminous and large that they could easily accommodate within their bosom the whole of our terrestrial orbit, including the sun, Mercury, Venus and earth, and yet have room to spare. In spite of their bulk they are practically hollow inside. In fact, they are so diffuse that their density is lower than that of a good terrestrial vacuum.

In the opposite camp are the three left deviationists located towards the lower left corner of our diagram. Because of their small luminosity and high surface temperature, they are called the white dwarfs. They are packed to the brim with material, a matchbox of which would easily weigh several tons. Earlier we mentioned one such star, the companion of Sirius.

These left deviationists are said to be in a 'degenerate' state. This is a technical term devised by astrophysicists to denote a peculiar condition of matter in its last stage of condensation. When we compress anything (say, an ordinary gas), we soon reach the limit of its compression because its atoms refuse to interpenetrate one another no matter how hard we press it. But what cannot be done on earth is not too difficult inside stars. For at pressures and temperatures such as prevail there, the atoms are stripped bare of their electrons and a good deal of interpenetration takes place. But in the white dwarfs even the separate identities of individual atomic nuclei are destroyed by further pressure and the nuclei and electrons are packed cheek by jowl so that they cannot come any closer. Such a gas is said to be degenerate. Degenerate matter therefore is the ultimate in condensation, or at least the penultimate now that matter inside the recently discovered quasars or the newly postulated neutron stars is believed to be in an even denser state of concentration. That is why the estimate of its density quoted above – several tons per cubic inch – is no Munchausen story. It is the outcome of one of the most precise calculations modern science can make. That is also why the stuff of which white dwarfs are made has its own peculiar laws and its peculiar appellation. Even energy transport in its interior takes place neither by radiation nor by convection but by conduction, as in a metallic bar with one end in a firebox.

The precise link between the main-sequence, red giant, white dwarf and other kinds of stars shown in Figure 3 is best understood by following the life history of a star. As we shall explain more fully in the next chapter, a typical star begins its life as a bloated relatively cool condensation of interstellar gas and dust held together by its own gravitation. As a condensation of this sort – the embryo star – contracts under its self-gravitation, its temperature gradually rises. The period of contraction before its temperature becomes high enough to make it luminous depends on the mass of the protostar. Table 2 below gives the duration of the contraction phase as a function of its mass before its emergence as a main-sequence luminous star:

TABLE 2

Mass of the protostar	Duration of gravitational contraction
20 suns	3×10^4 years
3 ,,	2×10^6 ,,
1 sun	5×10^7 ,,
0·6 ,,	2×10^8 ,,
0·2 ,,	10^9 ,,

The mass of the protostar also determines the point on the main sequence where it will emerge as a luminous star. Having emerged it remains for a while in this state of what is called radiative or convective equilibrium. In other words, the star maintains itself in a steady state without having to shrink or expand, as the outflow of radiation from its outer surface keeps pace with energy generation in its interior by nuclear transformation of hydrogen into helium.

As we mentioned earlier on page 44, these nuclear transformations in the interior of stars are of two main types. In the case of stars with masses equal to and lower than that of our own sun the energy they radiate is derived from the proton-proton chain, which leads to the building of helium nuclei through direct interaction between protons, with the consequent mass-loss released in the form of nuclear energy.

In stars with masses in excess of three solar masses the star derives its energy from the carbon-hydrogen cycle wherein carbon nuclei act as catalysts to produce the same net result of transforming protons into helium nuclei. In stars with masses intermediate between that of the sun and three times as great, both processes may operate simultaneously – the carbon cycle in the central parts and proton-proton chain in the outer shells surrounding the stars' interior core. In either case the star remains in its stable main-sequence stage for a period of time which may last from a few million years in the case of very massive stars to a few billion years in the case of light ones. Table 3 below gives the duration of the stable main-sequence stage for a few typical values of stellar masses:

TABLE 3

Stellar mass	Duration of main sequence stage
20 suns	8×10^6 years
3 ,,	5×10^8 ,,
1 sun	10^{10} ,,
0·6 ,,	10^{11} ,,
0·2 ,,	10^{12} ,,

Although these time intervals are very uncertain, they are all about 100 to 1,000 times longer than the first stage of gravitational contraction shown in Table 2. But in both stages of its life, namely as a contracting globule of gas and as a main-sequence star, the more massive the star, the briefer its life span. Thus a star of one solar mass has sufficient store of hydrogen to enable it to shine in a steady state for ten billion years, while stars with a significantly larger stock of hydrogen have much shorter lives. The reason is that a more massive star squanders its talents – hydrogen – like a true prodigal. A star ten times as massive as the sun, for example, will squander its hydrogen a thousandfold faster. Its fast life has an inevitable nemesis. It passes away in a paltry ten million years instead of the normal ten billion years it might have lived. In human terms, it barely lives three weeks instead of the ripe threescore and ten.

Whatever its life span, a star usually remains on the main sequence in a steady state for about 90 per cent of its total life. The residual 10 per cent is spent in various stages as a red giant, supergiant, etc., till its extinction or terminal state as a white dwarf. Naturally the precise course of its evolution depends on various parameters, most important of which are its mass and chemical composition. Thanks to recent advances in computer technology and in our understanding of the physics of the atomic nucleus and of elementary particles, we can now calculate quite precisely the luminosity and surface temperature (or spectral class) of an evolving star of any given mass and chemical composition at all stages of its evolution. If we plot the computed values on an H-R diagram, we obtain its evolutionary track. Figure 4 depicts three such typical evolutionary paths of stars with the same chemical composition pattern as the sun but with

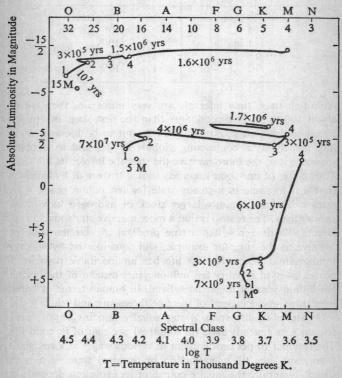

Figure 4. Luminosity/spectral-class diagram showing the evolutionary tracks of three typical stars with masses 15, 5 and 1 times the solar mass. Numbers shown on top are the absolute values of temperature in thousand degrees K. (Adapted from Icko Iben, Jr, 'Stellar Evolution: Comparison of Theory with Observation', *Science*, Vol. 155, 17 February 1967, pp. 785–96. Copyright © 1967 by the American Association for the Advancement of Science.)

masses equal to 15, 5 and 1 times the solar mass. The tracks are punctuated with numbered points all the way from 1 to 4 to indicate the various phases of its evolution from its main-sequence stage onwards. During the stage from point 1 to 2 the star remains a stable main-sequence star. Between point 2 to 3 it is in a state of transition from main sequence to red giant and beyond three it enters the red-giant phase. The time spent in the phases is shown in Table 4 below:

TABLE 4

Mass of the star in units of solar mass	Time spent in years during the phase between			
	point 1 and 2	point 2 and 3	point 3 and 4	beyond point 4
15	10^7	3×10^5	1.5×10^6	1.6×10^6
5	7×10^7	4×10^6	3×10^5	1.7×10^6
1	7×10^9	3×10^9	6×10^8	—

By computing the evolutionary tracks for different masses but for a given pattern of chemical composition we can predict the course of stellar evolution and confront the prediction with observation. When we do so we may actually see it in progress on the vault of the heavens for we find in our galaxy confederations of stars whose close association in groups shows that they are all of the same age, even though one group as a whole may be older than another. If we spread, in our H-R diagram, the stellar population of these groups, such as the young double cluster in Perseus and the middle-aged Pleiades, we find that while the stars of both the systems congregate around the main sequence towards the lower right-hand end of the diagram, the distribution of the stars in the upper half of the diagram is widely different for the two groups (see Figure 5). These upper-end deviations from the standard main sequence are in conformity with the theoretical evolutionary tracks predicted by Schonberg-Chandrasekhar stellar models.

The diagram also shows that some supergiants of the

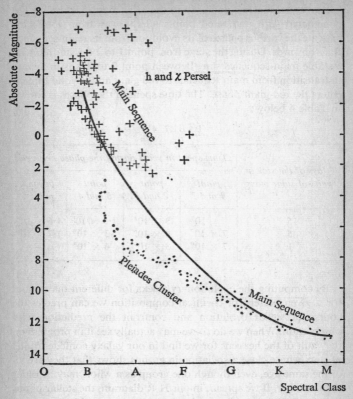

Figure 5. Luminosity/spectral-class diagram for the nuclear regions of *h* and *χ* Persei as well as Pleiades. The solid line is the standard main sequence.

Perseus cluster are far away from the main sequence. Their location too agrees with the synthetic evolutionary tracks yielded by Sandage-Schwarzschild stellar models. What is true of these two clusters has now been verified for nine more clusters. Figure 6 exhibits similar evolutionary differences between the stars of ten galactic clusters and one globular cluster including Perseus and Pleiades in a schematic form. The dotted lines in the diagram

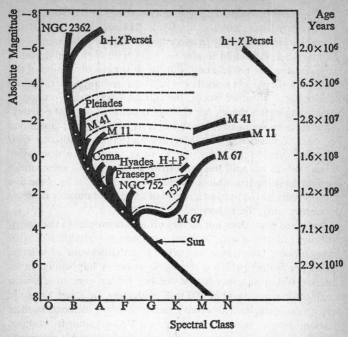

Figure 6. Luminosity/spectral-class diagram. (From Allan Sandage, 'Observational Approach to Evolution', *Astrophysical Journal*, Vol. 125, No. 2, March 1957).

are the fate lines of each cluster prescribed by the theory of stellar evolution. The actual scatter of stars of each system in close proximity to its own fate line is a confirmation of the theory underlying the evolutionary tracks. These confirmations of the theory of stellar evolution give us great confidence that the course of evolution we are about to describe is no mere mathematical fantasy.

As a main-sequence star burns its hydrogen in the hottest central region, a core of helium ash is deposited at the centre and gradually grows in size. When it accumulates to about 10 per cent of the mass of the star, its configuration ceases to be stable.

For the core shrinks, and the gravitational energy which the shrinkage releases heats it up, while the outer envelope of the star expands enormously in accordance with the behaviour of the Sandage-Schwarzschild model for stars with hydrogen-depleted cores referred to in Chapter 2. But the enormous distension of the outer envelope cools it so that it begins to radiate cooler, that is, redder, light. In other words, the star now becomes a red giant. According to this theory our sun will become a red giant one day. Then as it grows in size, it will begin to swallow the planets one by one, commencing with Mercury and ending with our own earth or even Mars. If so, the Koranic vision of a doomsday when the sun will grill the earth from a distance of a spear and a half might come true unless we choose to advance it by some five billion years by the massed ignition of a sufficient number of those miniature suns, the H-bombs.

But a red giant does not merely bring catastrophe to the neighbouring planets, if any. It itself runs into a veritable storm of catastrophes, cataclysms, and crises in an endeavour to balance its energy budget, which is repeatedly upset by happenings in its interior. What happens is that the hot helium core inside, at a computed temperature of 100 million degrees, becomes a cosmic witch's cauldron brewing atomic nuclei of lighter elements such as carbon, oxygen, neon and possibly even magnesium from those of helium. Although even after deviating from the main sequence, the bulk of the star still consists of hydrogen and the bulk of the energy production still flows from the hydrogen-helium conversion in a thin shell outside the core, the atomic consolidation now under way in the core also contributes its share of nuclear energy in the core to keep it at its temperature of 100 million degrees.

As the core continues to contract and the temperature exceeds 100 million degrees, the alpha particles (that is, the helium) interact with the neon nuclei to form magnesium, and then with these nuclei in turn to form the ordinary isotopic forms of the nuclei of all the elements up to iron whose atomic weights are multiples of 4. After all the helium has thus been used up, the core will begin to contract again with a release of gravitational energy, and this new contraction sparks the core temperature to still dizzier heights enabling fusion of still heavier nuclei to take place, with release of nuclear energy. The fusions triggered are of two main kinds depending on the type of star. For, as we shall explain more fully in the next chapter,* the stars that we see in our own Milky Way, as also those in other galaxies, evolved in two distinct stages corresponding to the two distinct types of stars populating it. Thus the stars of the Milky Way exhibited in the H-R diagram of Figure 3 are stars in the region near the sun. They are known as population I and are quite different from those of population II located in its more distant central regions. Because population II stars are believed to have been formed earlier, probably all at the same time at a fairly early stage of cosmic evolution when the primordial gas clouds from which they condensed had relatively greater abundance of hydrogen, they contain a very small amount of helium, and other heavy elements like carbon, nitrogen, oxygen, etc., all of which are grouped by astrophysicists under the omnibus title 'metals'. Population I stars like our own sun, on the other hand, seem to be second generation stars which were formed later. For they have a considerable amount of heavy-element abundance, thus indicating that they condensed out of hydrogen clouds that were impregnated with greater quantities of 'metals' forged in the interiors of massive population II stars which are presumed to have since blown themselves away in supernova explosions. It is also very likely that the heavier elements were spewed out by them in periodic outbursts before their final extinction. In any case, population I and II categories of stars have condensed out of two distinct types of hydrogen clouds – population I having arisen from those more richly impregnated with metals than population II and therefore comparatively younger source material than the comparatively purer primeval hydrogen cloud.

In the case of a population II star, formed out of primeval hydrogen clouds, the nuclear reactions occurring in its core towards the end of its red-giant stage produce heavier nuclei such as carbon, oxygen and neon, to produce additional quantities of multiple-four type nuclei right up to iron 56. In the case of population I stars, which are formed, not out of pure hydrogen

* See page 81.

cloud but out of material containing small proportions of the multiple-four type nuclei (such as carbon 12, oxygen 16 and so forth), the interactions of these nuclei with the protons give rise to the nuclei that are not multiples of four such as nitrogen 14, sodium 23 and so on. In both cases the nuclear reactions going on within the stars of different types at different stages of their evolution are so complex as to require a separate chapter for their exposition. We shall dwell on them more fully in Chapter 15. Meanwhile it will suffice for our present purpose to observe that as a star exhausts its supply of one kind of nuclear fuel it begins to burn the 'ash' of the previous burning process. Thus, as we saw at the outset, hydrogen in the core burns to produce helium 'ash'. With the exhaustion of hydrogen, helium burns into carbon, carbon to oxygen, neon and magnesium, magnesium to sulphur and sulphur to iron. The duration of the various stages of burning may be as long as ten billion years as in the case of hydrogen into helium or as short as one year as in that of sulphur to iron. As each stage nears its end, a period of gravitational collapse ensues in order to obtain the energy required to raise the core temperature to bring about the next stage of nuclear synthesis. While the period of such successive gravitational collapses may last from 100 to 10,000 years depending on the mass of the star, the temperatures shoot up prodigiously from a paltry ten to twenty million degrees in the first stage of the fusion of hydrogen into helium to five billion degrees for the conversion of sulphur to iron. This is why when a star has burnt its stock of helium 'ash' it departs from its state of red giantism into a state of instability in trying to obtain the energy it radiates by switching from one type of nuclear fuel to another. The instability manifests itself in a variety of ways depending on conditions not yet fully understood. For example, the star may be thrown into violent but rhythmic pulsation of its spherical surface causing periodic changes in its luminosity like those we actually observe in variable stars like R.R. Lyrae or the Cepheids. It appears, however, that the Cepheid instability is not structural. For a recent calculation by N. Baker seems to show that such rhythmic pulsations are a temporary aberration in the life of some stars when their evolutionary track happens to cross the so-called 'instability strip', a region on the H-R diagram defined by two close-ranged and almost parallel vertical lines.

However, the ageing star cannot escape structural instability if only for the reason that prodigious increases in its central temperature caused by successive bouts of gravitational collapse due to failing supply of its nuclear fuels raise its radiation pressure even more steeply. For as the central temperature (T) rises sharply with age its radiation pressure, which as we noted on page 41 varies as T^4, increases even more rapidly. Thus an increase of central temperature from twenty million degrees to, say, two billion degrees escalates radiation pressure 100 million-fold. Such escalation of radiation pressure strains the hydrostatic balance between gravitation and pressure to the limit of its endurance so that the smallest disturbance – no matter what its immediate cause – sparks catastrophic effects. Consequently the evolving star enters a period of structural instability which manifests itself in the star flaring up as a nova, that is, increasing its brightness thirty to 100,000-fold, possibly not once but several times, and ejecting considerable quantities of material. About twenty to thirty novae explosions are observed to occur every year in the Milky Way and there is reason to believe that fits of novaitis are likely to recur, as a number of repetitive novae have been discovered.

There is, however, some indication that many novae-like stars and ordinary novae are peculiar binary systems, which has led Robert P. Kraft to put forward a purely dynamical alternative explanation of nova outbursts. Thus let us imagine a binary star system originally consisting of two main-sequence stars, one a little more massive and considerably less luminous than the sun, orbiting around each other in, say, twelve hours. In course of time the more massive star evolves into a red giant. During this transition it is possible under certain conditions for material in the outer envelope of the red giant to pass partly into the gravitational orbit of the smaller star which then appears as a nova. Because of the depletion of hydrogen-rich material the evolution of the more massive star is greatly speeded up, while accretion of material by the secondary star causes its evolution to be slowed down. Eventually the more massive component becomes a white dwarf and stops losing mass. Later the lighter component evolves into a red giant and begins losing matter to the white dwarf. This is the stage of the dwarf nova we observe.

Unfortunately this 'dog-eat-dog' hypothesis sheds no direct light on the source of the outbursts. Further, it relies for the passage of material from the outer envelope of the expanding red giant into the smaller companion (nova) on certain limiting solutions of the famous three-body problem – the problem of finding the motion of a tiny body such as an atom of gas under the influence of two massive bodies rotating round their common centre. As is well known, there is no unique solution and the dynamical theory of nova outbursts leans rather heavily on the possibility that the expanding outer envelope of the red giant reaches the so-called critical zero-velocity surface. This surface may be regarded as the boundary around a binary within which a particle would be permanently retained by the system. A particle on the surface may drift either into the gravitational ambit of the smaller companion or escape into free space, but any beyond the surface will be irretrievably lost.

Whether nova outbursts arise because of structural changes in stellar interiors or from purely dynamical effects may be a moot point. But there is no doubt that the ageing star's career of violent distensions, pulsations, ejections and explosions comes to a close when it produces iron 56 by nuclear transmutation in the core. The reason is that it can no longer burn iron 'ash' to obtain the energy it needs, because further fusion or for that matter even fission of the iron group of elements *absorbs* energy instead of releasing it. The star therefore passes for its energy diet from a regime of nuclear transformation to that of gravitational contraction *alone*. As it contracts, it converts its gravitational potential energy into heat which it radiates away. At the same time it compacts the material in its core to greater and greater density and raises its temperature more and more. At long last a stage arrives when with continual compression the core becomes compressed either to its ultimate limit or blows itself to smithereens in a catastrophic explosion. Which of the two alternatives will come to pass depends on the mass of the star. If it happens to be less than about 1·2 times the solar mass, the so-called

'Chandrasekhar limit', the gravitational contraction is halted by rising internal pressure of degenerate electrons when its central density peaks to 10^6 gm./cm.^{3}. The star then passes into quiescence as a white dwarf. Nothing very spectacular can happen to it now. For it has practically no wherewithals of energy left – either nuclear or gravitational – because as a white dwarf it can no longer sustain any nuclear reactions within its core, and is too dense to contract any more being all degenerate electron gas. With all its energy engines shut it can do nothing but freewheel gradually into the oblivion of a dark invisible star.

But if, on the other hand, its mass happens to exceed the Chandrasekhar limit (1·2 solar masses), the star is likely to end its life in a catastrophe. The precise nature of the catastrophe is uncertain. But it has been predicted in broad outline by some very recent theoretical studies of John Archibald Wheeler, B. Kent Harrison, Kips S. Thorne and Masani Wakano based on our present understanding of atomic and nuclear theory. These studies show that the ultimate catastrophe the star suffers may be one of two kinds. *Either* it may undergo a gravitational collapse and be squeezed to extinction in the limbo of what is called its 'Schwarzschild singularity'; *or* it may shatter itself to bits as a supernova leaving perhaps a neutron star as its relic. We shall describe more fully its disappearance into the Schwarzschild singularity in Chapter 12, and shall confine ourselves for the present to its demise by supernova explosion.

Wheeler *et al.* believe that after a star has burned all its nuclear fuel, having undergone supernova explosions or other violent vicissitudes of its life and after its relic has cooled to absolute zero temperature, it settles into a final state wherein its material content is in that peculiar condition which they call 'cold matter catalysed to the end point of thermonuclear evolution'. Naturally the behaviour of the star in these last throes of its evolution depends on the equation of state of such cold, catalysed matter, that is, the relation subsisting between its pressure and density. Wheeler and Harrison have managed to derive such an equation by dint of sheer ingenuity.

The Harrison-Wheeler equation of state is believed to represent reliably the behaviour of 'cold, catalysed matter' at densities ranging from 7·86 gm./cm.^{3} to about 10^{13} gm./cm.^{3}. Although it does tie up pressure and density for values of density exceeding 10^{13} gm./cm.^{3} also, the reliability of the extrapolation is not very high. The reason is that at densities above 10^{13} gm./cm.^{3} the effects of certain complications like nucleon-nucleon interactions and of hyperon* formation, which are not now well understood, do not permit a more reliable calculation. Differing assumptions about the behaviour of nuclear matter in this realm of density yield widely differing equations of state. Nevertheless, these differences are not very material for our present purpose. They all seem to suggest that the stellar core is squeezed to such high densities during the last stages of its gravitational contraction that catastrophic nuclear processes occur before rising internal pressure can halt the contraction. These processes cause the star to explode with so great a violence that it becomes a supernova, that is, a nova which shines for a brief while with 100 million times the solar luminosity.

We have observational evidence of such happenings. For instance, the Crab Nebula (see Plate 6) in the constellation of the Bull is the debris of such a catastrophic explosion of a star in our own galaxy in the year 1054 when the 'guest-star', as the Chinese who recorded its observation called it, flared into such a phenomenal brilliance that it outshone every other star in the sky and for three weeks on end was visible even during daylight. In our own day similar supernovae explosions in other galaxies occur regularly but require special techniques for their observation because of their vast distances from us.

We still do not know for certain what precise physical processes trigger a supernova explosion. But there is little doubt that it is the outcome of some extremely catastrophic nuclear activity in the interior of the star. One surmise attributes it to what may be called the neutrino catastrophe. It is now known that all processes of nuclear energy generation which power the stars produce neutrinos – elementary particles without mass or charge possessing incredible ability to penetrate astronomical thickness of matter. These particles act as an irreversible energy sink. That is

* A species of elementary particles more massive than protons.

to say, they carry off energy without leaving a trace.* In normal stars, during the major part of their history, only a negligible fraction of the energy is carried off by neutrinos. But it has recently been realized that in the late stages of stellar evolution the energy drain by neutrinos may for a short while become a major fraction. Such a situation, however, can arise only at the end point of nuclear burning with the conversion of all the hydrogen in its interior to iron 56. At this stage, when the temperature in the core of the star reaches five billion degrees, pair creation, that is, production of electron and its antiparticle, positron, by the collision of high energy photons, occurs in such abundance that they begin to collide thereby annihilating themselves to form neutrino-antineutrino pairs. The new factors that now come into play make the energy drain via the neutrino sink intolerable. According to H. Y. Chiu and Philip Morrison, a star can pump out all its energy in this way in little more than a day. To compensate for such an ultra-rapid escape of energy through neutrino flux, the star must fall back on its gravitational energy, as in the earlier stages of nuclear burning. It begins to sink rapidly thereby raising its core temperature still more. The rise in temperature in turn accelerates still more energy dissipation through the neutrino sink causing the star to contract even more rapidly. Inevitably the gravitational collapse of the star becomes a runaway implosion of its core and explosion of its outer envelope which takes barely a second. As a result the exploding supernova shines with light equal to 100 million suns for two or three weeks. But whether such an explosion-cum-implosion leaves a relic of the parent star or completely obliterates itself is still in doubt.

One speculation is that the imploded core becomes a neutron star, that is, a superdense pack of neutrons of almost one solar mass enclosed within a sphere barely ten miles across! Since supernova explosion in our galaxy is observed to occur at the rate of one to one fifth as few per century, during the ten billion

* In fact, this is why W. Pauli had literally to invent them in 1930 in order to make the process of nuclear β-decay obey conservation laws of energy and momentum. In the study of β-decay energy was disappearing without a trace and it was shown that energy books could be balanced if neutrinos were supposed to exist. See also Chapter 15.

years of its existence there should now be $\dfrac{1}{5}\dfrac{10 \times 10^9}{10^2} = 20$ to 100 million neutron stars. Observation of such superdense neutron stars would therefore be a test of the theory.

Unfortunately neutron stars do not emit any of the usual radiation – radio or optical – by which we observe ordinary stars and galaxies. But they could emit very short X-rays which are completely absorbed by our atmosphere. The only way to detect them is therefore to scan the sky for possible sources of X-ray radiation by means of rocket-borne instruments. Such scanning has indeed been attempted recently. It has so far revealed two powerful sources – one in the Crab Nebula and the other in the constellation of Scorpius – which may also be the site of a supernova explosion. But by lunar occultation of the Crab Nebula source, that is, by observing the X-ray source when it is behind the moon, H. Friedmann has been able to show that the source is a region of about one light year in diameter within the nebula. This rules out the possibility of it being a neutron star. It has therefore been suggested that the observed X-ray emission is due to synchroton radiation from high-velocity electrons spiralling round lines of magnetic fields with the speed of light – a mechanism that will be explained more fully in Chapter 5. Although such high-energy electrons might well result from a supernova explosion, electron energies would decay below the necessary threshold in a much shorter time than the age of the Crab Nebula. It is therefore necessary to provide an energy source for a continual acceleration of electrons to relativistic energies in the Crab Nebula.

Till recently there seemed no plausible way of explaining how a neutron star, the putative debris of the exploded supernova, could power the Crab radiation. But it is now surmised that a rapidly rotating neutron star in an intense magnetic field might just about do it. The surmise derives its cogency from the observation in the Crab Nebula as well as elsewhere of an entirely new species of celestial objects first reported only in February 1968. These new arrivals in the astronomer's firmament are the rapidly pulsating radio sources, or pulsars for short, which emit unusual radio signals of very high frequency with an extremely regular but short rhythm. Although so far only 42 such objects have been observed, it is believed that there may well be over a million within the confines of our galaxy. Those further away from us are much more difficult to observe so that their apparent scarcity is only a selection effect. However, even the few that have been hitherto observed have raised more problems than can be solved at present. The reason is that their behaviour is a puzzle on many counts of which we will enumerate here only three. First, while they emit their unusual high-frequency radio signals with an ultra-precise periodicity, the amplitude of the signals received shows wide fluctuations. One is then hard put to understand how such a precise timing of the pulses can be combined with the great fluctuations in their amplitude. Secondly, the pulsation period is extremely short. Most of the pulsars so far observed have oscillation frequencies of the order of one second or less. They range from 0·033 to 1·962 seconds, although a pulsar with the relatively high frequency of 3·746 seconds has also been observed. Thirdly, the radio pulse occupies a very small fraction of the cycle – only about 0·04 seconds in a period of 1·3 seconds in one case that is fairly typical of them all.

While it is not easy to devise a model that can fill the pulsar bill, it is reasonably certain that they are very compact dense bodies not much larger than our own earth situated at distances 100 to 4,000 light years away. The reason is that a radiating body cannot emit a pulse of duration shorter than the travel-time of radiation across it. Since the pulse width of most pulsars is around ten to twenty milliseconds and a ten-millisecond pulse must originate in a body of radius less than the distance light can travel in ten milliseconds, namely $186,000 \times \left(\dfrac{10}{1,000} \right) = 1,860$ miles, pulsars cannot obviously be much larger than small planets comparable to the earth. But it seems that their actual size is much smaller. For the extreme regularity and brevity of the pulsed emissions suggest some astronomical clockwork mechanism like radial pulsations, orbital motion, or axial rotation of a compact massive body. Now, in any of these cases, the shortest possible period is controlled by gravity and has an order of magnitude

$(\gamma\rho)^{-\frac{1}{2}}$ where γ is the constant of gravitation and ρ the density of the body. It therefore follows that a pulsar such as NP 0532 found in the Crab Nebula, having an oscillating frequency of 0·033 seconds, must have a density of the order of $\left(\dfrac{1,000}{33}\right)^2 \dfrac{1}{\gamma} \approx 10^{10}$ gm./cm.³, nearly 100 times greater than that of white dwarfs. But matter at such density appears to be too unstable to stay put for any length of time. As we have seen already, white-dwarf material is degenerate matter in the *penultimate* stage of condensation wherein electrons and positively charged nuclei of simple elements like hydrogen, helium, carbon, etc., lie cheek by jowl. If degenerate matter is further compressed to densities around 10^9 gm./cm.³, the electrons begin to glue themselves to protons to form neutrons. With the drastic reduction in the electron population, the kinetic pressure they exert disappears. No force then exists to restrain further gravitational compression and the density increases until neutron kinetic pressure is adequate to combat gravity. This occurs at a density near 10^{14} gm./cm.³ and a body of stellar mass would then be only ten to 100 miles across. But such a high density requires a period of rotation as brief as a millisecond instead of the thirty-three milliseconds of the Crab Nebula pulsar.

Paradoxical as it may seem, even such short periods as thirty-three milliseconds are too high for pulsars to be neutron stars as the expected periods are of the order of only one millisecond. The difficulty, however, appears to have been resolved by the observations of F. Drake at Arecibo who has found that the rapid Crab pulsar is slowing at such a rate that the period would double in about 2,400 years. On the other hand, we find that for slower pulsars the increase of period with time corresponds to a doubling in 10^6–10^8 years. On the basis of these results, it seems that pulsars may, indeed, possess periods around one millisecond in the early stages of their formation. But this phase is so short-lived that the chance of finding a very rapid pulsar at the moment of its very birth is quite remote. By the time the pulsar is detected, its period has slowed down considerably with loss of energy from the star.

If the present consensus that pulsars are fast-spinning neutron stars, whose spin is slowing almost imperceptibly, as first suggested by T. Gold, is confirmed, it is not difficult to postulate a mechanism for the release of radio energy over a wide frequency range which causes the star to flash at the rotation period. Several suggestions have been made. But the most plausible of them seems to rely on the existence of intense magnetic fields of strength as high as 10^{12} gauss at the surface of the neutron star. Such intense fields have interesting effects, one of which is that any ionized gas which escapes from the surface of the star is able to move only along the lines of magnetic field. Thus it is whirled around at the angular velocity of the star and, at a distance of about 1,000 miles from the Crab Nebula pulsar, the tangential velocity approaches the speed of light. At this speed Gold speculates that the plasma will radiate by synchrotron mechanism if sufficient charge-bunching is present. Under such conditions the neutron star appears to a distant observer rather like a lighthouse flashing its signals with a precise regularity. Gold further suggests that since the plasma particles on attaining the speed of light must break away from the constraining field and stream outwards from the star, they continually supply the Crab Nebula with enough energy to emit the diffuse light which is a characteristic visual feature of the nebula.

While Gold's pulsar model has no doubt received a wide measure of assent because of its many attractive features, it remains to be proved that his 'slingshot' mechanism can accelerate particles to sufficient energies before they break loose from the star. Consequently theoretical astronomers are still busy spewing out a wide assortment of ingenious and imaginative new ideas. But most of them are not yet thoroughly developed. If none of the new ideas now in the field works, we shall face serious difficulty in explaining even the optical radiation of the Crab Nebula – let alone its synchrotron radiation – unless we revert to an earlier surmise of Ramsey for its purely optical part.

Some years ago Ramsey suggested the decay of radioactive elements produced in the original explosion as the source of its optical radiation because of the absence of any central star in the nebula. For the nebula is comparatively transparent and ordinary stars both within and beyond it are easily observed. All but one of them can be excluded as a likely source of its illumination because their proper motions differ significantly from that of the nebula. Accordingly many competent observers like Baade and Minkowski were inclined to believe that this central star *is* the remnant of the Crab supernova explosion. Although their belief has now received some support from the very recent observation of visible flashes from this central star having the *same* period as the Crab Nebula pulsar NP 0532 at two observatories in the U.S.A., Ramsey, unaware of subsequent developments, rejected it for the reason that the central star is 600 times less luminous than the nebula. This and other difficulties led Ramsey to deny the existence of any central star in the nebula. He concluded that the nebula shines by the radioactive decay of unstable nuclear isotopes or varieties of elements formed during the explosion of the parent supernova. While most of the unstable isotopes will be short-lived, some of the more slowly decaying isotopes may nevertheless remain in a supernova relic such as the Crab Nebula.

Of those radioactive isotopes whose lives are longer than thirty years and whose atomic mass less than seventy, there is an isotope of carbon, carbon 14, whose mean life is neither too short nor too long. It is 7,300 years. Ramsey showed that the mass of carbon 14 need only be 0·05 per cent of the nebula for its gradual decay to maintain the latter's radiation. If this explanation is correct, its luminosity is likely to fade by about one magnitude per 7,000 years, or 0·01 magnitude per seventy years. Such a gradual extinction of the nebula is unobservable in a human lifetime. But it may prove possible to check observationally some other consequences of the theory in a shorter time. Since the Crab Nebula could be detected with present telescopes even if it had been twelve magnitudes fainter, clearly it is likely to fade out of view in 12 × 7,000 or 84,000 years.

The observed rate of supernova explosions in our and other galaxies is not yet certain. It may be anywhere between one supernova to one fifth as few per century. Consequently we should expect to find between 200 to 800 relics of superova explosions of the past 84,000 years in our galaxy. Thus, as Oort has suggested, the great Network Nebula in Cygnus shown in Plate 2 may well be such a remnant of a supernova explosion about 30,000 years ago. We also have records of supernova explosions observed by Tycho Brahe and Kepler. But they have left no region of turbulent gas like that of the Crab Nebula, although small spots of nebulosity have been noticed close to each site. There are in all about 400 planetary nebulae which may perhaps be relics of supernova explosions of the past 84,000 years. Although their number is well within the expected range of 200 to 800 supernova relics, some authorities like the Soviet astronomer Shklovsky regard them as vestiges of former novae. If so, planetary nebulae would represent a genetic link between novae and white dwarfs at least in the case of some types of stars. The similarity of the spectra of novae at a certain phase of their evolution, as also of former novae to those of planetary nebulae, would seem to support this view. But some serious difficulties have to be overcome before it can be regarded as fully established.

First, even in cases where all luminosity is known to emanate from its central star, the total amount of visible light radiated by the nebula is forty to fifty times greater than that emitted by the star. This discrepancy may be explained on the basis of Zanstra's theory. Zanstra assumes that the central star is so hot that most of its energy is radiated as invisible ultraviolet light, which is completely absorbed in the interior of the nebula and re-emitted as visible light by ionization of nebular hydrogen and its subsequent recombination with released electrons. This enables Zanstra to calculate the temperature of the central stars. The calculated temperatures which are found to be very high – over 20,000°C. – seem to resemble those of old novae. But the Zanstra mechanism of ionization and recombination cannot explain the existence of 'forbidden' lines in the spectra of planetary nebulae, though a supplementary assumption of electron impacts might.

Secondly, the envelopes of planetary nebulae are observed to be expanding at velocities ranging from twelve to fifteen miles per second, whereas those cast off by the novae expand much faster, their velocity range lying between 200 and 1,000 miles per second. The only way in which such huge velocities of nova shells could be retarded to some extent is Oort's suggestion of possible collision with near-by interstellar gas clouds. But the theory of such cosmic collisions lacks any agreed basis. It is therefore difficult to confront the conclusions of the suggestion with observation. Oort's own rough and tentative calculations show that the reduction of a shell's velocity to a tenth of its initial velocity would take a few centuries in the case of an ordinary nova shell but some twenty to thirty millennia in the case of a supernova.

According to Oort's view the Cygnus Nebula is a remnant of a supernova explosion of about 20,000 to 30,000 years ago. But even if the original shell expanded with an initial velocity of 1,000 miles per second (it is unlikely to be much less), the nebula should still be expanding at some 100 miles per second, whereas it is now observed to be expanding very slowly if at all. One could no doubt avoid the discrepancy by sufficiently predating the original explosion, but this would aggravate the difficulty about the source of its present illumination. It is doubtful if the supposed heating caused by the collision could last that much longer, especially as the later phases of retardation would occupy the bulk of the time. Finally, a study of five planetary nebulae by Aller has shown that their central stars are less hot and dense than former novae. It therefore seems possible that the formation of a planetary nebula is yet another way in which a star manifests its instability which, though analogous to that of a nova or supernova outburst, may have no direct connexion with either.

CHAPTER 4

Stars in the Making

IN this chapter we shall describe how stars are born. Any theory of stellar births must also explain why, like birds of a feather, stars flock together in such huge herds as the galaxies. It is therefore linked with the theory of the origin and evolution of galaxies. Just as in our search for clues concerning stellar evolution we classified stars according to their various attributes such as mass, luminosity, spectral class, and the like, so we must select some large-scale attributes applicable to a galaxy as a whole. But here we face a difficulty – the incredible distances of the galaxies from us. They are so far off that any trace of the separate identity of their constituent stars is completely obliterated and they appear as mere misty nebulosities among the clean, pointlike stars of our own Milky Way. Even when seen with the aid of our most powerful telescopes, only a tiny proportion of the billions of galaxies scattered all over the sky is close enough to show any details of their structure. For this reason, the astronomy of galaxies – the systematic survey of their distribution, size, luminosity, structure and so forth – has had to await the construction of more powerful telescopes able to plumb the depths of our space to distances of not a few hundred or thousand light years, but a million times as great.

One of our most powerful telescopes for galactic explorations – the 100-inch Mount Wilson telescope – came into operation barely fifty years ago. A few years after its construction, E. P. Hubble used it to complete his first systematic survey of the nearer galaxies in the Northern Hemisphere. Since then several systematic surveys both in the Northern and Southern Hemispheres have been undertaken, and some more are still under way. They show a strong, almost universal tendency for galaxies to occur in clusters. The old picture of a more or less uniform 'general field', that is, a random distribution of isolated galaxies interspersed here and there by an occasional cluster or group, is now seen to be the result of observational selection owing to scanty sampling with small-field Newtonian reflectors. The new picture yielded by large-scale surveys conducted with wide-field photographic refractors and Schmidt cameras shows galaxies belonging to a large number of supersystems, regular clusters or irregular clouds ranging from dense groups to huge agglomerations of galaxies of several million light years in width.

Our own Milky Way is no exception to this clustering rule. Some years ago Gérard de Vaucouleurs assembled impressive data regarding the distribution of bright galaxies in our vicinity to show that the Milky Way is part of a big cluster known as the Local Group. During the last thirty years the membership of this group has grown from an original half-dozen to nearly twenty-four. The concentration of nearly all the twenty-four galaxies within a broad band running roughly along a great circle of the celestial sphere in galactic longitudes 90° to 270° shows that they all form part of a single local supergalaxy. The further conclusion that this Local Group is but a minor structural detail in a still larger collection is the outcome of an investigation into the distribution of the bright, or nearer, galaxies in the Harvard thirteenth-magnitude survey (see Figure 7). This conclusion has now been confirmed by recent radio-astronomical studies of cosmic radio noise by means of radio telescopes.

Besides the general clustering of galaxies, recent surveys have also revealed a rather wide diversity in their other morphological features such as size, luminosity, mass, structure and so forth, although the diversity of galactic conditions in these respects is not quite so wide as in the case of stars. Thus, while our own Milky Way with its enormous spread of some 100,000 light years is a giant among galaxies, there are also midgets like the faint

Figure 7. Distribution of Shapley-Ames galaxies in galactic coordinates. Each dot represents a galaxy, or rather its projection on the plane of the galactic circle; the upper circle is for those in the Southern Hemisphere and the lower one for the Northern. The figure also shows the zones of total (black) and partial (grey) obscuration by the Milky Way. The small white spot at $l = 15°$, $b = +5°$ indicates the north pole of the local supercluster. The figure also shows the supergalactic equator and the circles of latitude $\pm 30°$. Three external clouds of superclusters can be seen in Hydra, Pavo-Indus and Fornax. (Drawing by Gérard de Vaucouleurs.)

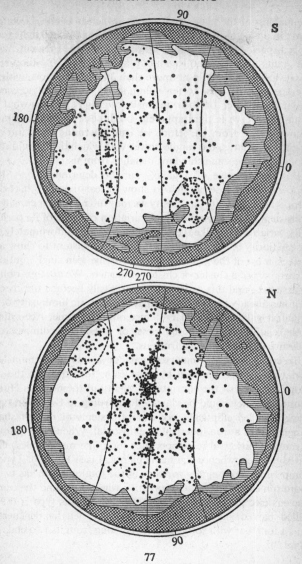

companion of the giant Andromeda, which is barely 4,000 light years in diameter, the average size being about 12,000 light years. This disparity in size naturally reflects itself in their masses. While the giants are estimated to contain as many as 200 billion suns, the dwarfs may contain no more than 300 million suns, with the average running close to ten billion suns. We are not so certain about the scatter of their intrinsic brightness. Since what we actually observe is their apparent brightness, we need to know their distance from us before we can determine their intrinsic brightness, that is, the candle power with which they would shine if placed at the same standard distance from us.

But the distance of a galaxy can be determined only if it happens to be close enough to show at least some of its bright stars, whose absolute luminosity we may infer by other considerations such as the period of light variation of some of its periodically variable stars, like the Cepheid variables. Unfortunately, the vast majority of the galaxies we see are too distant to show any. In fact, most of the galaxies cannot even be seen singly, but only as members of a cluster or cluster of clusters. We are thus obliged to assume – and this assumption may easily become treacherous as it has already proved once – that the intrinsic luminosity of the brightest galaxies in all large clusters is the same. Accordingly their relative apparent brightness, or rather their dimness, is a measure of their distance from us.

To conclude this summary of galactic surveys: we find a large variety of forms exhibited by galaxies as seen through the telescope. In spite of the enormous diversity of these forms Hubble found that they could be classified according to two main types. Type I, called ellipticals, consists of systems of regular shapes ranging from a true spherical to ellipsoidal. When photographed through a powerful telescope, they look like more or less elongated ellipses, whence the name elliptical (see Plate 7). Type II, comprising about eighty per cent of the total number has two or more somewhat irregular spiral arms winding around the central nucleus (see Plate 8). A peculiar subclass of this type is the so-called 'barred' galaxy in which the central nucleus has degenerated into a long bar with spiral arms emanating from its two ends (see Plate 9).

On the basis of Hubble's classification of the shapes of galaxies, Jeans constructed a theory of galactic evolution. He claimed that the observed shapes of galaxies formed an ordered sequence which might be represented by a Y-shaped diagram like the one shown in Figure 8. He then proceeded to show by mathematical reasoning that such a sequence could be produced by the continuous variation of a single fundamental parameter, namely the

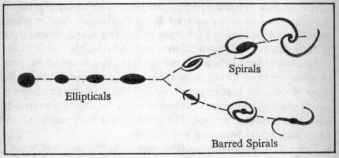

Spirals

Ellipticals

Barred Spirals

Figure 8.

rotation of the galaxy. The main idea underlying Jeans's cosmogony was the one Kant had used to explain the origin of our own solar system.

The idea is that in the beginning the sun and all the planets formed one gaseous sphere with a radius that stretched to the outermost planet. This mass was in rotation. As a result there was a flattening at the poles very much like the polar flattening of our own earth because of its diurnal rotation. The main mass of the gaseous sphere contracted, but as it did so a ring of gases around the equator was sloughed off. This process of sloughing off gaseous rings was repeated as often as necessary to account for the required number of planets. Although the great mathematician Laplace himself had a hand in formulating this theory, it never really worked.*

Nevertheless, Jeans took up the theory and suggested that Laplace's process on a vastly greater scale might help explain the origin of galaxies. According to Jeans, the zero state of a galaxy

* For fuller explanation see Chapter 16.

was a rotating sphere of gas with no stars. Mathematical theory shows that if such a rotating sphere is of the size of a galaxy, it is unstable. In other words, it cannot stay put. Local irregularities that might be present initially would continually exaggerate and the whole mass would break up into a large number of separate irregular clouds. Each cloud would in turn break up into subclouds, and for the same reason: gravitational instability of the parent clouds. We can imagine this process to be repeated until condensations as dense as stars begin to emerge. At this stage, further condensation would cease as the increasing temperature at each successive condensation would now be high enough for the proton-proton fusion process to occur, thus keeping the mass in what we have called radiative or convective equilibrium. In other words, the energy generated in its interior by this process would exactly compensate the loss of energy resulting from radiation from its outer surface. The mass would then cease to condense and would become a normal star.

This condensation process is an ever-recurring theme in Jeans's cosmogony. He actually used it to push a stage further the zero state from which we started. If we consider the whole of space as being filled with a tenuous intergalactic gas, gravitational instability requires that galaxies should condense out of this material in exactly the same manner as stars condense out of galactic gas. Jeans showed by mathematical reasoning that a diffuse gaseous mass spread uniformly through infinite space would break up into separate masses of the size of galaxies, which themselves would evolve through a sequence beginning with a globular gaseous mass and ending with a great star cluster.

Since Jeans's attempted rationalization of Hubble's classification of nebular forms, two new and far more powerful instruments of galactic exploration – the 200-inch Mount Palomar telescope and a series of new radio telescopes in different parts of the world – have come into operation. As a result, a vast amount of new observational material concerning the galaxies has already been gathered. The accumulation of new data, some of which we will describe more fully in the next chapter, was in any case expected to complicate the original, rather simple cosmogony of Jeans, but it has subverted it altogether.

The reason is that one of the strongest pieces of observational evidence in favour of Jeans's Y-shaped evolutionary tree (Figure 8) has now been abandoned.* At the time Jeans formulated his theory it was possible to resolve the arms of spiral galaxies into their constituent stars, but none of the elliptical galaxies, nor the central core of any of the spirals, could be so resolved – not even by the then most powerful 100-inch Mount Wilson telescope. Owing to this inability to resolve elliptical galaxies into their component stars, it was possible to identify them plausibly with the gaseous clouds of Jeans's theory in a stage prior to the condensation of stars. Recent observations, however, have shown that elliptical galaxies as well as the central cores of the spiral galaxies are no mere gaseous clouds without stars, but as chock-full of them as the spiral arms.

The earlier difficulty of resolving galaxies into their constituent stars arose from a fundamental difference in the character of the stellar populations of the two types of galaxies. While the spiral arms can be resolved into their constituent stars by using blue-sensitive photographic plates, the ellipticals as well as the central core of spirals require for this purpose the use of red-sensitive plates. The reason for this is that the former are a class whose brightest stars are blue supergiants, while the latter form a sequence whose brightest members are red giants. This difference in colour of the stars making up the two contrasting systems is no superficiality. It is far more organic, extending to almost every other attribute such as age, origin, distribution in the galaxy, and so forth. Thus the red giants in the elliptical galaxies are known to be about 7,000 million years old as against the paltry fifty to 100 million years of the spiral-arm blue giants. Because of the difference in the origin of the stars in the centre of our galaxy (and also in elliptical galaxies) and the origin of the stars in the arms of our galaxy, the former are known as stars of population II and the latter as stars of population I. The stars in the centre of our galaxy were formed from the primordial hydrogen and are

* Another reason is that Jeans's theory requires 100,000-fold longer time for the evolution of galaxies than that allowed by present-day observations. As we shall see later, our Milky Way appears to be about 10^{10} million years old. Jeans's calculations put it at 10^{14} years, or about 10,000 times longer.

generally poor in metals, whereas the stars in the arms were formed from the dust ejected from the population II stars near the end of their lives. These population I stars are therefore richer in metals. Our sun is an example of a population I star.

The aforementioned difference in the character of the stellar populations of the two types of stars is now believed to be linked with the existence of clouds of interstellar dust and gas in galaxies which the earlier cosmologists like Jeans and Eddington had ignored. We mentioned before that our own galaxy (which, by the way, is a spiral type) contains vast gaseous clouds. Almost 2 per cent of the total mass of our galaxy, equal to two billion suns, is contributed by them. This is also true of all other spiral galaxies though not of all ellipticals which are free from dust and gas.

It is in these interstellar gas clouds of spiral galaxies that new stars keep arising phoenix-like from the ashes of burnt-out ones, that is, from the dust and gas in the spiral arms continually replenished from nova and supernova outbursts. This is confirmed by recent investigations of stellar associations of our own galaxy – of what are now called O-associations.

Consider, for example, the nearest such association, Zeta Persei in the constellation Perseus. If the observed movements of stars in this association are projected backward in time, we find that they must have started from a common centre barely 1·5 million years ago. This is no mere isolated case. A similar extrapolation of the observed movements of stars in the Orion association shows that they too started racing from a common centre about 2·8 million years ago. In all probability these stars were all born together at the time of their coincidence at their common centre.

There are in our galaxy probably thousands of such associations, though only a few have been surveyed so far. Few though the surveys are, they have shed a new light on the character of the stellar population of galaxies as well as on the cosmic processes leading to the formation of stars. As mentioned before, young population I stars are formed from dense clouds of gas and dust which condense by gravitational attraction in a manner to be described presently. As the cloud shrinks, it becomes what would look like a roughly spherical blob. One may actually see these blobs of dark matter – known as Bok's globules – in the sky, especially if they happen to be projected against the luminous background of bright nebulosities, as for example the Rosette Nebula (see Plate 10).

Dr Hoffleit has counted no fewer than 150 such globules and at least ten conspicuous large 'holes' in rich star fields indicating the presence of many units of dark matter with diameters running from a few light days to three light years or more. But whether these globules are really embryo stars in condensation is still in doubt. We do not yet know with absolute certainty whether they are mere concentrations of dust grains, or of dust grain *and* interstellar gas. Without the latter the globules would not have enough material to form a star.

Recently, however, a much more conspicuous example of stars in embryo has been found in a class of very peculiar stars which are sometimes embedded in small luminous nebulae known as Herbig-Haro objects, as for example the T Tauri star shown in Plate 11. Briefly, they are semi-stellar nuclei with very peculiar spectra observed in the dense dark cloud not far from the Orion Nebula. That these objects are stars in formation has received a good deal of added confirmation by the recent emergence of two new apparently stellar objects in one of the brighter Herbig-Haro objects sometime between January 1947 and December 1954.

Plate 12 is a photograph of these objects at three epochs – 1947, 1954, and 1959. The 1947 photograph of the object shows a complex structure consisting of a pair of stars of seventeenth magnitude 8″ apart, and probably three still fainter ones all enveloped in several masses of nebulosity in an area of about 20″ across. The 1954 plate shows *two additional* stars each lying 3″ to 4″ from a component of the original pair of 8″ separation. These new apparitions have brightened up by about three magnitudes, that is, at least fifteenfold between 1947 and 1954. There has been no further change in these objects since 1954 except some further brightening in one of the 'new' nuclei in Herbig No. 2. The 1959 photograph is not directly comparable with the other two, being taken with a different telescope and in a different colour.

It is a fair surmise that in their emergence we have witnessed the birth of a prototype of hot young population I stars from the gas and dust of the cloud in which they have appeared. There is reason to believe that before evolving into normal population I stars these prototypes – the quasi-stellar Herbig-Haro objects – become T Tauri stars, the typical stars of what are known as *T*-associations which are found in the nuclei of all the nearest *O*-associations. These stars belong to a class of faint variable stars which exhibit irregular variation in their light and spectra. That they are young stars freshly condensed out of interstellar gas is all but certain now. For they have been found only in association with nebular material, both bright and dark.

For several reasons it is unlikely that these stars are merely pre-existing objects that have accidentally strayed into dust and gas clouds. First, there are far too many of them to be accounted for by random encounters of star fields with the clouds. The spatial density of T Tauri stars in such clouds exceeds that of ordinary stars of similar luminosity in the vicinity of the sun by a factor of 5 to 15. Secondly, these stars show no evidence of the infall of material that might be expected to occur if they were outside interlopers wandering into the dust and gas cloud. On the other hand, spectroscopic evidence is in favour of ejection of material from the stars into the cloud rather than the other way about. Thirdly, there is fairly strong observational evidence to show that they emerge at the sites of Herbig-Haro objects, because the latter's spectra resemble in many respects those of T Tauri stars. According to G. H. Herbig, if the source of the spectrum in the brightest of these objects were to brighten about four or five magnitudes, the result might be a replica of a T Tauri spectrum. Finally, recent studies by Merle F. Walker of extremely young clusters of stars in NGC 2264 containing T Tauri stars support this conclusion (see Plate 13). The scatter of stars in this cluster on their luminosity spectral-class diagram drawn by Walker indicates that the cluster possesses normal main-sequence stars extending from Class *O* to *A*, below which the stars fall above the main sequence (see Figure 9). The deviation of these stars from the main sequence is due to the fact that they are still 'baby' stars in the process of contracting gravitationally from the pre-stellar medium, and have

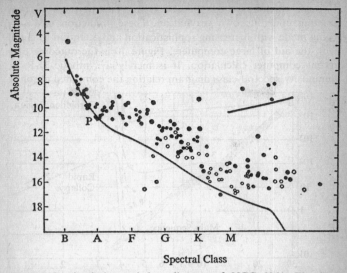

Figure 9. Luminosity/spectral-class diagram of NGC 2264. The lines represent the standard main sequence and giant branch. (From Merle F. Walker, 'Studies of Extremely Young Clusters. I. NGC 2264', *Astrophysical Journal*, Supplement Series, Supplement No. 23, Vol. II, October 1956.)

not yet reached the main sequence. In fact, the diagram agrees approximately with that predicted theoretically for young clusters containing stars still in the making. The point *P*, where the cluster begins to depart from the main sequence, enables the age of the cluster to be calculated. It is found to be about three million years.

These conclusions have been further corroborated by a still later study by Harold L. Johnson of faint young stars in the *O*-association I Orionis. The occurrence of T Tauri stars in them and their location in the luminosity/spectral-class diagram indicates that stars of this type are still in the gravitational contraction stage. We may safely rely on the age estimates of Walker and Johnson as they have recently been reinforced by astrophysical considerations. For they do enable us to predict many of the salient structural changes and successive equilibrium states through which a bloated cloud of gas and dust passes while contracting under its own gravitation. These predictions are now being made with increasing sophistication and attention to detail with the aid of large computers. Figure 10 is the outcome of a recent computer calculation. It is merely a synthetic H-R or luminosity/spectral-class diagram relating the computed absolute

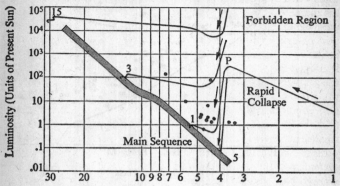

Figure 10. Black dots represent T Tauri stars on their way to the main sequence (thick band). The diagram shows three paths (black curves) followed by stars of masses 15, 3 and 1 times the solar mass to arrive at the main sequence. The first part of track 1 up to point P shows the rapid-collapse phase of a star of one solar mass. (From G. H. Herbig, 'The Youngest Stars', *Scientific American*, August 1967. Copyright © 1967 by Scientific American, Inc. All rights reserved.)

luminosities of stars to their spectral class or surface temperature. But the really valuable bonus of the calculation which is not shown in the diagram is that it makes it possible to determine the ages of the stars from their observed position in the diagram. It turns out that the T Tauri variables fall in the domain that marks them out as very young stars – so young, indeed, that they must still be contracting and have not even begun to produce any significant amount of their energy by burning hydrogen. In other words, they are still in a state of transition between gravitational contraction and the settled steady state of a mature hydrogen-burning star like our sun.

Consider, for example, a gas cloud of one solar mass contracting under its own gravitation. Although, as we saw in the last chapter, it will take some fifty million years before it ceases to contract and becomes a steady main-sequence star, more refined calculation shows that this contraction process is very uneven. Thus when the cloud has over a rather long time shrunk to about the diameter of our solar system, it is still a comparatively cool, dark cloud. But as gravitational attraction shrinks it further, a new phenomenon emerges. Some of the energy released by the contraction begins to go *not* into heating the gas *but* in breaking hydrogen molecules and ionizing the resultant atoms. As a result the gas pressure supporting the outer layers falls abruptly so that the cloud collapses rapidly. Calculation shows that under these new conditions it will shrink from the size of the solar system (10^4 times the solar radius) to a ball whose radius is 100 times smaller in barely half a year (see Figure 11). But in the meanwhile pressure within the cloud builds up again thereby restoring the system to structural equilibrium. This is the stage at which the embryo star begins to be visible for the first time, perhaps like Fu Orionis that suddenly appeared in 1936. The reason is that the point P on the theoretical evolutionary track (1) of the pre-stellar cloud shown in Figure 10 marks the end of its rapid collapse and its first emergence as a luminous object.

A glance at Figure 10 shows that once arrived at P its surface temperature will be about 4,000°C. and absolute luminosity 100 times that of the present sun. The conjunction of both these attributes in Fu Orionis shows that it may well be due to such a collapse. As will be observed from Figure 11, the whole process from the onset of the rapid collapse of the cloud to its first emergence as a self-luminous body takes just about 20 years. The young star gradually contracts and for a time diminishes in luminosity. But as the thermal energy in its interior is transported to the surface mainly by the convective movement of rising hot gas, the surface temperature remains nearly constant, though the interior grows hotter. This is the stage when the budding star turns into a T Tauri variable with its central temperature gradually rising because of the energy released by continual contraction. It is only when the central temperature rises to ten million degrees

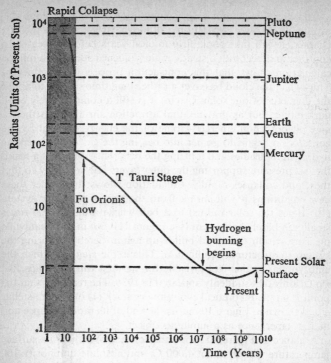

Figure 11. The radius of the sun is shown at various stages in its history. The initial stage of rapid collapse was completed in about twenty years. The subsequent slow contraction to a stable stage of hydrogen-burning took about fifty million years, the uncertainty arising from the fact that the conclusion of this stage is not well defined. Thereafter the radius has increased slightly as hydrogen has been consumed in the interior. The position shown for the star Fu Orionis (thirty years after the end of the collapse stage) is only schematic, since its present radius is only about twenty to twenty-five times the present radius of the sun, not sixty times as the curve indicates. Since the mass of Fu Orionis is unknown, it is not possible to say how significant this difference is. The general features of this diagram are probably about the same for other stars not too different from the sun in mass, except that the time scale is compressed for higher masses. (From G. H. Herbig, 'The Youngest Stars', *Scientific American*, August 1967. Copyright © 1967 by Scientific American, Inc. All rights reserved.)

K. that the star begins to obtain a significant part of its energy from nuclear fusion of hydrogen into helium. Thereafter the star contracts still more slowly till it reaches its final equilibrium state on the main sequence when contraction ceases and the star obtains its energy solely by nuclear burning of hydrogen.

Although there is no doubt that the foregoing theory of the contraction of very young stars from gas and dust clouds is fundamentally correct, its basic assumption that a normal star like our sun is an erstwhile T Tauri variable needs to be substantiated more fully in detail. For there is nothing in it to account for many features of T Tauri behaviour: their extremely active and luminous chromospheres, their massive ejections of surface material, their variability in brightness, their high lithium abundance, their excessive infra-red radiation. Despite these puzzles we should soon be able to put it to a crucial direct observational test. Thus the observed excessive infra-red radiation of T Tauri stars could be explained by assuming that the extra radiation emanates from large clouds of dust with a surface temperature of around 700°C. lying near the stars. If so, there is every likelihood of observing at least one such dark dust cloud on the threshold of starhood. For E. E. Becklin and C. Neugebauer have recently discovered in an infra-red survey of the Orion Nebula an object that could represent a cool infra-red radiating cloud in exactly the pre-collapse stage of formation envisaged by the theory. It is a cool, dark, body about 1,500 times the size of our sun and is emitting infra-red radiation similar to that from T Tauri stars but much greater in amount. As this corresponds to the predicted size of a pre-stellar condensation of the sun's mass just prior to collapse, it should complete its transition from a large cool body to a visible star in about twenty years. We should thus be able to see within the next decade or two a new star like Fu Orionis at the site of the Becklin-Neugebauer object. Astronomers are naturally waiting with bated breath to observe it.

The predicted emergence of the Becklin-Neugebauer object as a new star within the next decade or two will, no doubt, be a great triumph for the blend of theory and observation outlined above. But whatever the outcome, it is all but certain that fresh stars do condense out of interstellar clouds. Thanks to many recent studies of spectral absorption and emission lines sponsored by interstellar material, we have a fairly good idea of the existing state of affairs as regards gas clouds, at least in the area from our immediate vicinity to a distance of 5,000 light years. The chemical composition of the interstellar gas is similar to that found for stellar atmospheres, with hydrogen by far the most abundant element, so that it is quite a suitable raw material for star making. The gas is largely concentrated in clouds whose average diameter is about thirty light years and average density ten hydrogen atoms per cubic centimetre. This is about ten times the overall average density of the Milky Way.

But such clouds are still far from being star buds. It is only when some of them shrink by further gravitational contraction to a tenth of their original size (so that they become a thousandfold denser like, for example, the cloud in the Orion Nebula) that they are ripe to blossom into a star cluster. Such a gas cloud may have a total mass of, say, 40,000 suns, of which about a quarter may condense in a shower of a thousand hot stars in the central nucleus of the cloud.

The newly formed hot stars heat up the inner regions of the cloud, to about 100 times their original temperatures. As a result, gas pressure in the inner part of the cloud increases a hundredfold. The pressure of the expanding gas as it pushes against the cool outer parts of the cloud creates a compressed zone of dense gas from which a new generation of stars might be born. Because of this expansion we may expect the new stars to start their life flying at a high speed away from the centre, such as we see in the case of Zeta Persei and other associations. But we do not have at present much quantitative data to confirm this pressure theory of the birth and flight of stars.

One thing, however, is certain. The theory has to be amended in an important detail to account for the observed motions of some young hot stars like AE Aurigae, MU Columbae, and 53 Arietis of the Orion association. It has been suggested that the much higher speeds of these stars might well be due to a rocket effect of the fusillade of hot gas shot at the dense cloud by the pressure of hot stars already formed therein. This fusillade might

accelerate the gas cloud (and the stars born in them) to velocities as high as thirty miles per second. Unfortunately, the speeds
with which the three stars of the Orion association mentioned
above are fleeing are considerably higher.

We thus see that a population I star makes its way from dust
and gas to fire and flame with the wheel of gravity rolling in
between. The chilled embers which the stars scatter as they march
to their white-dwarf doom or blow themselves apart in supernova
explosions are lured back to light and lustre if, perchance, some
of them band together in sufficient strength to rekindle the mass
once resplendent with celestial fire. However, for the celestial
fire (the star) to appear, much herding, and with it much nuclear
consolidation, is needed. This may take a long time to come to
pass. But once accomplished, the mounting tension of the cosmic
cloud of herded material casts into its depths whole showers of
stars like lightning frozen in a monsoon cloud. Star formation
therefore is a species of cosmic hailstorm, wherein drops of
lightning 'congeal' into a hail of stars.

CHAPTER 5

The Birth of Galaxies

IF the story of stellar evolution recounted in the preceding two chapters is astrophysical, its galactic counterpart – the theory of the birth and evolution of galaxies – is cosmological. For *all* the *distant* galaxies seem to conform to the first ever truly cosmological law that Hubble discovered some forty years ago. He found that spectral lines emanating from all the distant galaxies are shifted towards the red end of the spectrum. But as we saw earlier, a red-end shift of spectral lines of a celestial object means that it is receding from us at a speed proportional to the magnitude of the shift. The red shift of an object is usually measured by the ratio $z = \dfrac{d\lambda}{\lambda}$, where $d\lambda$ is the absolute magnitude by which any spectral line of wavelength λ is shifted – the ratio z being the same for *all* wavelengths. The recessional velocity v of the object is then given by the relation $v = \dfrac{cz}{\sqrt{1 + z^2}}$, where c as usual is the velocity of light. For *small* values of z the approximation $v = cz$ usually suffices. But since c is large, a red shift even as small as 0·1 will mean a recessional velocity of 18,600 miles per second. Hubble's observation, therefore, implies that all the distant galaxies are receding from us at prodigious speeds. For example, the Virgo cluster at a distance of 40 million light years is receding at 750 miles per second, the Corona Borealis at a distance of 700 million light years, at 13,400 miles per second, and the Hydra cluster at 2,000 million light years, at 38,000 miles per second (see Plate 14). This increasing red shift and therefore recessional velocity with increasing distance seems to hold even for those more distant extragalactic objects – the radio galaxies and quasars – that the newly installed telescopes both radio and optical have recently revealed. The fastest recessional speed so far measured for an optical galaxy is about 85,000 miles per second corresponding to

$z = 0.46$. It pertains to a faint galaxy called 3C 295 which is estimated to be about 4,000 million light years away.

If we plot the recessional velocities of galaxies against their respective distances, we find that the points fall on a straight line as shown in Figure 12. This means that for every increase of a million light years in distance the recessional velocity leaps by about nineteen miles per second. Thus a galaxy at ten million

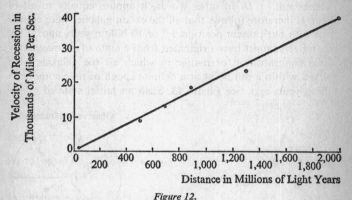

Figure 12.

light years recedes at 190 miles per second; a galaxy at 100 million light years recedes at 1,900 miles per second. In general, the recessional velocity of a galaxy at a distance D is given by Hubble's law

$$v = HD \qquad (1)$$

where H is a constant of proportionality called Hubble's constant. We may easily determine its numerical value from the empirical fact already noted that a galaxy at a distance d million light years recedes with a velocity of $19d$ miles per second. Substituting in equation (1)

$$D = 10^6 d \text{ light years,}$$
$$= 10^6 \times 365 \times 24 \times 3,600 \times 186,000d \text{ miles,}$$

and $v = 19d$ miles per second
we find that

$$\frac{1}{H} = \frac{D}{v} = \frac{10^6 \times 365 \times 24 \times 3,600 \times 186,000d}{19 \times d} \text{ seconds}$$

$$= 9{\cdot}7 \times 10^9 \text{ years.}$$

We note that the period $\left(\dfrac{1}{H}\right)$, that is, the time taken by the galaxy to travel its present distance D at its recessional velocity v, is independent of D. In other words, it applies equally to all of them. It therefore follows that all the extant galaxies were coincident with our present position 9·7 or 10 billion years ago. If so, our universe must have originated from a state of unimaginably dense concentration of matter in which all the galaxies were packed within a pin point at a definite epoch of time some ten* billion years ago. See Figure 13. Such an initial state of quasi-

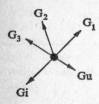

Initial State of Galaxies
Ten Billion Years Ago

Present State of Galaxies

Figure 13. G_1, G_2, G_3 ... Gi in the diagram on the right are galaxies at their present distances d_1, d_2, d_3 ... di from our own galaxy at O. Since their recessional velocities are v_1, v_2, v_3, ... vi they would be at our present location 0 at d_1/v_1, d_2/v_2, d_3/v_3 ... di/vi years ago. But Hubble's Law requires $d_1/v_1 = d_2/v_2 = d_3/v_3 \ldots = di/vi = 1/H$. It follows therefore that $1/H = 10^{10}$ years ago all the galaxies were coincident at O as shown in the diagram on the left. The observer is at O.

infinite concentration looks at first sight too queer to be credible. However, its credibility gap was greatly reduced when the application of Einstein's relativistic equations to the universe as a whole

---

*This estimate may perhaps stretch to thirteen billion years, if we allow for the inevitable uncertainty of observational data on which the graph of Figure 12 has been based.

predicted that the zero state of our universe was precisely such a singular condition of hyperdense concentration as Hubble's observation of receding galaxies seemed to suggest. As we shall see more fully in Chapter 7, only for the first million years after the epoch of coincidence of all the galaxies at a single location, or 'creation', ten billion years ago, the universe remained a veritable flood of light and lustre with matter playing the second fiddle. But thereafter matter got released from the sway of radiation and came into its own so that the gravitational influence of matter began to dominate over that of radiation. After the expiry of the reign of radiation of one million years' duration, the expanding universe got fragmented into distinct but extremely *tenuous* blobs of primeval hydrogen gas. Tenuous because continual expansion diluted greatly the original high density at the epoch of 'creation'. The central problem of cosmology is to determine the physical processes, parameters and conditions that led to the evolution of such blobs of hydrogen gas into galaxies with the observed distribution of mass, size, luminosity and other features including their tendency to occur in clusters. Considering the immense complications of the problem it is no wonder that it is still largely unsolved. Indeed, it is a riddle likely to remain perhaps forever beyond human comprehension. Nevertheless, the attempt to divine it is an adventure in ideas which many spirits find too fascinating to resist.

Earlier we considered one such attempt by Jeans. Although subsequent observations have not confirmed the evolutionary sequence he sought to deduce, the central concept of his cosmology remains valid in spite of some expressed doubts regarding its applicability to rotating cosmological gas clouds, which may be expected to be ionized and permeated by magnetic fields. For these doubts have now been shown to be largely unfounded by recent investigations of Chandrasekhar and others. However, it has to be emphasized that earlier attempts to explain the sequence of galactic forms on the basis of gravitation alone are oversimplifications of a vastly involved situation where magneto-hydrodynamics and turbulence play an equally important role.

Consider, to begin with, magneto-hydrodynamics. The first intimation of the importance of magneto-hydrodynamical effects in cosmic processes came with the revelation of Hall and Hiltner that light from certain stars that happens to pass through considerable clouds of dust is strongly polarized.* For such polarization could only result from extended interstellar magnetic fields in our Milky Way. Since these effects are most pronounced for stars behind extensive clouds, it is reasonable to assume that this polarization is caused by the orientation of ferromagnetic grains in interstellar dust clouds caused by magnetic fields, even though this theory raises many unsolved problems. We accept it, nevertheless, because no other explanation is even remotely conceivable at present.

How such extended magnetic fields arise in interstellar clouds is not yet certain. One view – that of Batchelor and Heisenberg – is that any small magnetic field that may be created in the cloud by temperature differences producing electric currents will be greatly amplified by the turbulent motions of the clouds. The cause of this amplification is the fact that these motions tend to stretch out the lines of magnetic force and bend and entangle them, thus increasing their concentration, which means that the average magnetic field increases. But the amplification is limited by the fact that increasing magnetic strength tends to make the turbulent motion more and more ordered, so that further entangling of magnetic lines by turbulent motion is halted. It happens that fields of strength up to 10^{-6} gauss† can arise in this way, but not more. This is indeed an incredibly feeble field. By comparison the strength of the earth's magnetic field which directs the mariner's

* It is possible to endow ordinary or natural light with a new quality which, unlike direction, intensity, and colour, the eye cannot detect. Thus a glass or water surface may refuse to reflect natural light at a certain angle of incidence if it has been passed beforehand through a certain kind of crystal. Such light is said to be polarized. As explained in the Appendix, light is the outcome of periodic alterations of the intensity of electric and magnetic fields with regular frequency. In ordinary or natural light the directions of the oscillating electromagnetic forces change too rapidly to be detected by any means at our disposal. But if it were to pass through material in which the molecules are all oriented, as for example by a magnetic field, the vibrations of the electromagnetic forces become of fixed type and orientation, and the light becomes polarized.

† Gauss is the unit in which the intensity of a magnetic field is measured. A unit north pole is attracted by a force of one dyne in a field of one gauss.

compass is 0·2 gauss, that is, more than a hundred thousand times that of the cosmic clouds.

Nevertheless, even such feeble fields do influence significantly the behaviour of interstellar gas clouds. This is because the electromagnetic phenomena in gases and liquids differ from those in solids such as conduction cables in that mechanical forces deriving from electric currents produce hydrodynamical motions in the conducting fluids themselves. But the latter in turn give rise to electromagnetic effects such as the self-amplification of pre-existing magnetic fields in the manner already described. In other words, there is a coupling between the ordinary electromagnetic and hydrodynamical phenomena.

Because of this coupling, the fluid motions in cosmic environments are not governed by the classical hydrodynamical laws but by more complex magneto-hydrodynamical laws. Unfortunately, the latter are not yet understood sufficiently well. Therefore it has not been possible precisely to delimit the role magneto-hydrodynamics plays in the formation, structure and evolution of galaxies from the primeval gas cloud, although a number of piecemeal results have been derived. For example, Chandrasekhar and Fermi have shown that the observed properties of the spiral arms of our own galaxy require a magnetic field of 10^{-5} gauss along each spiral arm. Likewise, a field of similar strength permeating the interstellar medium will be important in determining the characteristics of stars formed by gravitational contraction of the interstellar gas clouds, for the field will modify the nature of fluid motions which otherwise would take place under gravitation alone. While further research will no doubt amend the present theory in important details, it seems that the existence of magnetic fields in the extended original zero-state intergalactic gaseous cloud does not directly affect Jeans's idea of gravitational instability. It is, therefore, permissible to treat the evolution of galaxies from such a primeval gas cloud without regard to magneto-hydrodynamical phenomena, or at any rate to obtain a first approximation.

Our knowledge of the second of the three possible motive forces of galactic evolution – turbulence – is a shade better than that of magneto-hydrodynamics, thanks to recent studies of aerodynamics stimulated by the needs of aviation and meteorology. But here there is a vital difference of opinion between leading cosmologists as to the role of turbulence in directing the course of galactic evolution. One school led by Weizsäcker assigns to turbulence as fundamental a role as to gravitation. We shall revert to his viewpoint in the next chapter after we have expounded the simpler theory of galactic evolution wherein gravitation alone is its main architect.

As we have already seen and will see even more fully later, there is strong evidence now to show that the zero-state of our universe was the explosion of a giant primeval atom within which all the matter of the extant galaxies was condensed some ten billion years ago. Except for the brief tenure of the radiation regime during the first million years after 'creation', the fragmented primeval hydrogen gas clouds have been under the sway of gravitation. But before gravitation could assert its sway the fragments of the explosion became tenuous blobs of gas as continual expansion diluted greatly their initial high density. The central question then is to examine how a tenuous and extended cloud of gas behaves under the influence of its own gravitation. The question is not new. Newton himself asked it in attempting to explain the origin of the sun and the fixed stars when he suggested that 'matter evenly distributed through infinite space . . . could never convene into one mass [but] into an infinite number of great masses scattered at great distances from one another'. Why should an extended homogeneous gaseous medium splinter off into a series of distinct accretions around separate nuclei instead of coalescing into one huge central mass? To understand the reason for this, imagine a slight fluctuation in density at any point P of the medium such as may arise by a small local condensation of the gas at P at the expense of an equivalent rarefaction at Q. If the gravitational attraction of the gas particles on one another is negligible, as will be the case if the medium is closely confined, what happens is merely the passage of a sound wave through the gas, like that of an engine whistle through air, until it is dissipated by viscosity. The upshot of the passage of the disturbance is an increase in the internal energy of the medium.

In the case, however, of a gaseous medium extended over the vast distance of, say, a million or more light years, the mutual gravitation of the gas particles cannot be neglected. If we take this into account, an increase in the internal energy of the medium as above is accompanied by a decrease in its gravitational energy. Now if P is sufficiently distant from Q, the decrease is numerically greater than the increase, so that the total energy of the medium is decreased. There is now no excess energy to travel in the medium until it is dissipated by viscosity. Instead, the medium is able to acquire a greater and greater amount of kinetic energy, which is another way of saying that it has become unstable. This instability expresses itself by the occurrence of pairs of condensation and rarefactions at sufficiently distant points. In other words, a vast extended homogeneous medium cannot stay put. It begins to condense into a series of distinct accretions around separate nuclei. These separate accretions, the denser first-generation subclouds of the original parent cloud, in turn fragment into a series of second-generation subclouds, and so on. Thus each successive generation of subclouds of gas shrinks *as a whole* until it is dense enough to cease further shrinkage. It is then ripe to begin fragmenting into a number of next-generation subclouds. One may imagine this process of condensation and fragmentation as continuing indefinitely. But a stage ultimately arrives when two new factors emerge to limit the cycle.

First, a system formed in the way described would inevitably start rotating as a result of the intermingling of currents in the original gas, just as whirlwinds arise when currents of wind collide. But any rotation that the system might acquire initially would tend to increase continually with further shrinkage, as that is how the system can conserve its angular momentum. It is this conservation law that is used when a skater whirling about with extended arms manages to spin faster by pulling in his arms. The increase in rotation would increase the centrifugal force until it became large enough to balance the gravitational attraction.

Broadly speaking, two courses of evolution are then possible. If the system happened to start with low initial rotation, the condensation process would last longer, for it would take longer for shrinkage to increase rotation sufficiently to provide enough centrifugal force in the plane of rotation to balance gravitational attraction. As a result, it would contract far more in the plane of rotation, so that it would very likely become extremely dense and condense almost completely into stars forming a closely compacted system such as an elliptical galaxy. A faster initial rotation, on the other hand, would severely limit its contraction in the plane of rotation because centrifugal force sufficient to balance the gravitational attraction would now be built up sooner. Consequently, the contraction in this plane would cease much earlier than contraction in the perpendicular plane. Such a system would therefore evolve into a loose, disc-shaped galaxy with a great deal of uncondensed gas especially in its outer parts, which is what we call a spiral galaxy.

The second cut-off process limiting the condensation of the subcloud arises from the fact mentioned earlier that condensation of any material by its self-gravitation releases energy in much the same way as a falling cascade of water generates hydroelectric power. A part of the gravitational energy released by the shrinkage of the subclouds is lost by radiation as heat. But how much of it is lost in this way depends on the radiation capacity of atomic hydrogen, which varies widely with its temperature. At temperatures below 10,000°C. it is negligible, but it mounts sharply when the temperature rises beyond this level, and particularly when it exceeds the one million mark.

Now calculation of the supplies of gravitational energy likely to be available as a result of the condensation of subclouds of the dimensions under review shows that the gas temperature inside them is unlikely to be below 10,000° or more than 1,000,000°. Between these two rather wide limits all ranges are not equally probable. There are reasons to believe that the most likely temperature ranges are (1) between 10,000° and 25,000° and (2) between 150,000° and 1,000,000°.

At the lower temperature range practically none of the gravitational energy released by possible shrinkage is radiated as heat if a cloud contains a mass significantly in excess of ten billion suns. Consequently, all of it remains stored in the subcloud. This means that the cloud cannot shrink further as a whole because even if it did shrink a bit, the stored gravitational energy would merely re-expand it to its initial dimensions, though perhaps in a different direction. Therefore it fragments into a number of small subclouds. When the subcloud at last attains a mass of the order of ten billion suns, which is the average galactic mass, about half the gravitational energy released by its shrinkage is radiated as heat. This continues until the cloud shrinks to about one-third of its original size, which on the average is roughly 100,000 light years across. It is now too dense to shrink further as a whole and thus begins to yield its own progeny of subclouds. The further fragmentation of the subclouds proceeds at an ever-accelerated pace until they become so dense and opaque that they cannot lose any more heat by radiation. As a result, the fragmented subclouds can break up no further and become embryo population II stars. When this happens, the primeval galactic cloud begins to burst into myriad stars:

> Huge fragments vaulted like rebounding hail
> Or chaffy grain beneath the thresher's flail.

It is thus that gravity conjures the star-spangled heavens out of the Cimmerian darkness of the deep. Hoyle finds a corroboration of this theory in the mass range of population II stars which is observed to lie between 0·3 and 1·5 times that of the sun. For according to his calculation, when gravity has herded material to this extent, it becomes so opaque as to stop losing any heat by radiation and thus ceases to fragment further.

Once the galactic mass has broken up into a loose system of stars in the manner described, mere passage of time suffices to compact it into a condensed system similar to what we find in the elliptical as well as in the central cores of spiral galaxies. Although most of the stars of the system cluster by mutual encounters into a central amorphous mass, some do remain at the outer periphery of the parent subcloud where they were born. These remaining stars are the very extensive haloes surrounding the galaxies. We observed earlier that a portion of a cloud with an average galactic mass of ten billion suns would condense to about one third its size before it begins breaking up into subsystems. This is in agreement with the occurrence of halo stars extending to distances thrice the average diameter of the central nucleus.

We have now to consider the higher temperature range mentioned above. It has been estimated that for a gas cloud to increase its internal temperature to the million mark, its total mass must exceed 100,000 billion suns, that is, about 10,000 times the average mass of a galaxy spread within a sphere of about six to seven million light years in diameter. A cloud of these dimensions begins by contracting as a whole until it develops a temperature of one million degrees. At this temperature heat losses by radiation leap up suddenly to a prodigious level, leading to a sudden fall of temperature. The outcome of this cataclysm – a temperature surge of about a million degrees – is a shower of thousands of dwarf protogalaxies with an average mass of only 300 million suns. Each protogalaxy is then well on its way to becoming a galaxy by further fragmentation into stars in the manner already described.

However, a close cluster of 30,000 dwarf galaxies cannot remain in the state of loose anarchy in which it is born. A war of consolidation ensues. Some of the closer dwarfs coalesce and become powerful enough to devour others. As always, the big become bigger until a few giants emerge by the consolidation of some two to three hundred dwarfs. It is in some such way that the giants like Andromeda and our own Milky Way in the Local Group were built up.

In spite of some satisfactory features the foregoing account of galactic evolution is, no doubt, a gross oversimplification of a very complicated state of affairs wherein magnetic fields and turbulence play a role that cannot be ignored. It is a consequence of this oversimplification that it cannot even explain the spiral arms of spiral galaxies. Moreover, the recent discovery of radio galaxies and quasars seems to show that besides magnetohydrodynamics and turbulence there may well be other factors moulding the course of galactic evolution. Merely as a preamble to the more sophisticated theories of the origin of galaxies, which incidentally have yet to be formulated, we shall take up the study of turbulence and other preliminaries in the next chapter.

CHAPTER 6

The Birth of Galaxies:
A Turbulent Alternative

When we consider the motion of any *large* fluid mass, it is intuitively clear that the mass must inevitably be more or less turbulent. For all particles of the main mass in motion are unlikely to have the same regular motion, such as is the case with the laminar flow of water in a pipe where all its particles move uniformly in the same way along parallel stream lines like the orderly route march of a regiment. Even such an orderly march may easily turn into the hurly-burly of a rowdy mob under certain conditions. These conditions were empirically formulated by the British physicist Osborne Reynolds. He showed that laminar flow of any liquid in a pipe becomes turbulent if the Reynolds number of the flow, that is, the ratio ud/v, exceeds the critical value 2,000, where u is the velocity of the liquid flow in the pipe, d its diameter, and v the viscosity* of the liquid.

The reason laminar motion under these conditions turns turbulent is that this type of motion, where the movement of the whole determines that of each one of its parts, is a macrostate of low statistical probability in the sense that macrostates of higher degree of organization are statistically less probable.† If, therefore, laminar motion changes somehow, no matter in how small a detail, it soon turns into some form of turbulence exactly as a state of higher internal order (or low entropy) yields to one of lower order (or of higher entropy).‡ This is why orderly trade winds are apt to become, at slight provocation, headlong hurricanes whipping up mountain waves with tempestuous fury, and placid streams swollen by summer rains become turbid

---

* Viscosity is the physical magnitude which measures resistance of the fluid to deformation. Thus a fluid like honey, which puts up a stiff resistance to deformation by, say, gravity, is more viscous than water.

† For a fuller explanation of this point of view see Chapter 14.

‡ For an explanation of these terms see Chapter 14.

storm-flushed torrents heaping wreckage on forest and plain. It is thus that the laminar motion of wind and water changes into turbulence, and turbulence into heat. This explains Weizsäcker's maxim that turbulence is the gateway through which large fluid masses in ordered motion march to their heat-death doom.* But the mathematics of the march is very complicated. We have hardly begun to understand it on the cosmic scale.

Nevertheless, a number of statistical studies of cosmic turbulence have been made by Weizsäcker, Chandrasekhar, von Kármán, Hoerner, and others. They show that, hydrodynamically speaking, if turbulence spreads uniformly in all directions, the energy of turbulent motion tends to be transported from large eddies to produce smaller ones. The quantitative formulation of this tendency is given by the so-called spectral law of turbulence which defines the kinetic energy associated with each eddy as a function of its wave number, exactly like the spectral law of light radiation, which associates the radiant energy of each wave with its appropriate wavelength as in Planck's radiation† law.

The spectral theory of turbulence has its empirical counterpart. For the spectrum of turbulence in a turbulent stream may actually be measured by special techniques, such as studying an ersatz turbulence generated in a wind tunnel by installing a grid of cylindrical rods upstream of the testing section. Such empirical observations confirm the theoretical derivation of the spectral law. But for cosmological applications, the spectral law of turbulent flow is better expressed in another form which is completely equivalent to it. This is the famous Kolmogorov law of turbulence, which states that the relative velocity v between two points of a turbulent stream at any distance D apart is, on the average, proportional to $D^{1/3}$.

Recently the theory has been verified empirically even on the cosmic scale. Hoerner has shown that the pattern of the observed radial velocities of the cosmic cloud in the Orion Nebula, measured at eighty-five different points by the Doppler shifts of the emission lines in the blue part of its spectrum, conforms fairly closely to Kolmogorov's law. We may therefore apply the theory

---

* For a fuller explanation of this point of view see Chapter 14.
† See Chapter 2, page 37.

of turbulence to the evolution of galaxies with some confidence in the manner suggested by Weizsäcker, especially as the high value of the Reynolds number for cosmic clouds does indicate that their motion is turbulent.

Weizsäcker considers that the zero state of our universe was a certain state of condensation of all the galaxies, to which Hubble's recession law (described earlier) leads. The condensed galaxies were in the form of a gaseous cloud in motion ever since the initial explosion which set the galaxies fleeing from one another. This primeval motion of the cloud wore the garb of cosmic turbulence probably right at the outset, but certainly very shortly thereafter, because of the tendency of all large-scale motion to turn turbulent. As a result, the original gaseous cloud would begin to splinter off in a number of smaller eddies. This effect of turbulence would be superimposed on that of the initial explosion that put the galaxies to flight with their existing uniform recessional speeds in accordance with Hubble's law.

It therefore follows that the relative velocities of galactic matter at any two points in space a distance D apart would be composed of two components arising from two different effects, namely those of turbulence and recession. Now as we have seen, the recessional velocity is proportional to distance D in accordance with Hubble's law. But the turbulent component of velocity would be proportional to $D^{1/3}$ provided Kolmogorov's law held in those early days. Even if it did not, however, it is unlikely to have been proportional to a power of D as high as unity. If the law held, then for large distances the recessional velocity of expansion would greatly exceed the velocity of turbulence and for small distances it would be the reverse.

There is thus a limiting distance D_0 such that for distances exceeding it, expansion would dominate, whereas for smaller distances the recessional effect would be masked by that of turbulence, which, as we saw before, tends to concentrate the main mass of the cloud in a number of smaller eddies. For this reason, the material of the primeval cloud that managed to condense into smaller eddies of linear dimensions less than D_0 would remain united by the action of turbulence while larger eddies would be dispersed by recession.

We have empirical evidence to show that this is indeed the case. The observed irregular motions of the galaxies in our immediate neighbourhood are of the same order as the recessional velocities which should be ascribed to them by Hubble's law so that they mask the latter completely. On the other hand, in the case of galaxies ten or more times as remote we find that their recessional velocities, which because of Hubble effect would now be ten or more fold greater, completely overshadow the irregular velocities due to turbulence.

It therefore seems that as the primeval gaseous cloud exploded it produced both turbulence and recession. The turbulence broke up the main cloud into a number of irregular eddies – the galaxies. Those cosmic eddies which happened to be close together initially have stayed together, while the distant ones have receded and will continue to do so as they are doing at present. But each eddy or galaxy that is thus formed will not have its mass distributed uniformly all over within the limits of its own confines but rather spread in a large number of separate subclouds, which we still observe in interstellar matter.

The reason for this is that fluctuations of velocity arising from continual churning of material by turbulence produce fluctuations of density as soon as turbulent velocities exceed that of sound. For so long as the velocity remains below that of sound the pressure and therefore density changes in the gaseous medium that accompany its motion will be exceedingly small. In fact, strictly speaking, the velocity of sound is merely the speed with which *infinitesimal* pressure or density changes, whether audible or not, are propagated in the gaseous medium. This means that for subsonic motions density changes in the medium are negligible. But supersonic motions are a different matter altogether. They cause violent and abrupt changes in the pressure and density of the medium. We know this to be the case from our experience with high-speed aircraft, rockets and missiles which give rise to shock waves, that is, violent and abrupt fluctuations of pressure in those regions of atmosphere through which they travel.

What happens in these regions is similar to what tidal bores do at the mouths of large rivers flowing into the sea, such as the

Hoogly in India or the Severn in England. That is why Lord Rayleigh called these shock waves of supersonic flight 'aerial bores'. On the basis of our experience with shock waves of supersonic flight we can no doubt make a model of the conditions prevailing in cosmic clouds in turbulent motion. But the actual state of affairs here is infinitely more complicated. We no longer have to do with a single shock wave as in the case of the supersonic flight of a projectile. Instead, what we have is a statistical mixture of several shock waves in the same region of space, caused by the mad carmagnole and tarantella of many clouds in supersonic relative motion. As a result, cosmic eddies of all sizes arise and intermingle. All of them, however, remain embedded in a single vaster eddy – the galaxy – into which the primeval gaseous cloud broke up just before the recessional velocities of other similar large eddies (galaxies) separated them from it. This internal mixing of smaller cosmic eddies *within* the larger eddy, the galaxy, throws it into non-uniform rotation exactly as the lateral component of the motion of two wind currents meeting even a little aslant of each other makes them spin around each other like a whirlwind.

Weizsäcker has made extensive calculations to solve the hydrodynamical equations of such a non-uniformly rotating gaseous mass under the influence of its own gravity and turbulent friction. These calculations show that an irregular gaseous cloud passes through a sequence of rotatory forms of various kinds the final outcome of which is a sphere via a flattened rotatory disc such as the spiral galaxies like our own Milky Way have. The reason why a rotating irregular gas cloud develops into a rotating disc with spiral arms is that a cloud, once it is set spinning by turbulence, cannot lose its spin because of what is called 'rotational obstinacy', ordained by the mechanical law of conservation of angular momentum or spin.

Angular spin acquired by the cloud as a result of internal turbulent motions will, therefore, persist. In fact, it will increase progressively, thus flattening the cloud into a rotating disc, because there is continual coupling between turbulence and rotation. The turbulence generates rotation which in turn stimulates turbulence anew, forming gaseous clouds again and again.

The clouds in turn are distorted by non-uniform rotation into a spiral arm like milk stirred in a cup of coffee. If the theory is true, there is scope for an infinite variety of forms which the spiral shapes may assume. This is indeed what we find in the heavens. Just as they have seen the man in the moon and canals on Mars, astronomers have detected an unending kaleidoscope of patterns in the spiral galaxies, from the uncurling fern to the nautilus shell and the hurricane.

Nevertheless, the theory does run into one difficulty. It requires the arms to be wound around the core many times whereas actually we never find them coiling more than once or twice. However, we may surmount this difficulty by appealing to turbulence once again. For turbulence not only creates the clouds which rotation spools around the core but also destroys them. It is therefore unlikely that spiral arms wound around the centre several times could persist for long. But this St Vitus's dance of cosmic gas clouds cannot go on forever. Gradually but inexorably, the energy of turbulence is degraded into more and more disordered and random motion of the mass. Weizsäcker's calculations show that this takes place by the changing of the central part of the main mass into a uniformly rotating core by turbulent friction and by the escaping of its outer peripheral parts into the cosmic space from where they came. The spiral galaxy then evolves into an elliptical one.

The time scale of such an evolution is roughly estimated to be the diameter of the system divided by its rotational velocity and multiplied by a factor which may lie between ten and 100. For a galaxy of diameter d completing one revolution in T years, the rotational velocity is $\pi d/T$ per annum so that the period of evolution is ten to $100 \times d/\pi d/T$ or T/π years. Our own Milky Way, which rotates in a period of about 230 million years, may be expected to become an elliptical galaxy in $230/\pi \times 100$ million, or about seven billion years. Since its present age is reckoned to be about ten billion years, it should have evolved into an elliptical galaxy already. As this is by no means the case, Weizsäcker explains that actually the period will be much longer because his calculations apply only to purely gaseous systems without any stars. Star systems where conditions are quite different can go on rotating non-uniformly for a very long time as is the case with our own planetary system.

Where in this hierarchy of galactic forms – irregular clouds, spirals, and ellipticals – is the place of barred spirals? It can be shown that elongated bodies can be figures of equilibrium under the common influence of gravity and rotation if their rotation is strictly uniform. This is the basis of Weizsäcker's suggestion that barred spirals may well be galaxies whose rotation became uniform before they evolved into an elliptical form by the loss of their outer parts. He finds a corroboration of this suggestion in the fact that barred spirals in general show the bar structure in the inner region while the bar is changed into a spiral arm in the outer region. The reason is that in the immediate neighbourhood of a maximum gravitational potential, rotation is always uniform.

In sum, the two main forces directing the evolution of galaxies – turbulence and gravitation – are moulding irregular clouds into spinning forms. Since spinning forms cannot easily lose their angular momentum in view of the mechanical law of its conservation, they evolve into flattened non-uniformly rotating discs. As far as the formation of stars in such a system goes, they evolve along paths all their own but the interstellar matter of the system will keep forming and dissolving spiral arms till it streams completely out of the system into cosmic space. The spiral then becomes an elliptical galaxy. In other words, irregular galaxies evolve into elliptical galaxies via the spirals.

It will be seen that the foregoing account of galactic evolution actually reverses the order of evolution derived by Jeans and depicted pictorially in Figure 8 on page 79. But it is already obvious that even this revised evolutionary sequence of galactic forms is not likely to survive the accumulation of yet newer data yielded by deeper explorations of intergalactic space by means of recently installed radio telescopes in conjunction with giant optical ones. These explorations have shown that the original Hubble classification of galaxies is now outmoded as it is solely based on the shapes of galaxies as observed in large optical telescopes. If we observe the galaxies by means of radio telescopes, we find that some of them are extraordinarily strong sources of radio waves. A case in point is the radio source known as Cygnus A which has been identified with an optically visible galaxy some 700 million light years away (see Plate 15).

Cygnus A is by no means the only one of the new species of objects called 'radio' galaxies, that is, visible galaxies that emit radio waves besides light waves. So far about 100 such discrete radio sources have been identified with visible galaxies. They have loosely been grouped into two categories: the 'normal' and 'peculiar', according to their radio and optical characteristics. The 'normal' galaxies are those spiral and irregular galaxies which are weak radio emitters. The total energy emitted in the radio region of the spectrum by a normal galaxy is of the order of 10^{38} ergs per second. This is about 1,000 times the radio output of the most powerful radio source within our own galaxy such as the Crab Nebula, the debris of an exploded supernova. But it is still only about a millionth of the energy that galaxies emit at optical wavelengths. However, more interesting for our present purpose are the 'peculiar' radio galaxies which, though of standard appearance, emit about 100 times more radio power than the normal galaxies. The peculiarity of their radio emission is its concentration in a source of small diameter at the centre of the galaxy. If there is any emission from an extensive halo, it is too weak to be detected by the present instruments. The peculiarity of their optical spectrum is the existence of intense, broad emission lines that suggest a high degree of energetic chaotic activity in the nucleus of the galaxy. A case in point is the Seyfert galaxy NGC 1068, a late spiral whose nuclear region seems to be the site of a titanic explosion (see Plate 16). Perhaps the nuclear explosion is responsible for the enhanced radio emission actually observed. But several other spiral galaxies whose optical spectra exhibit the same features do not show enhanced radio emission.

NGC 1068 is one of the weaker of the galaxies that have come to be known as 'peculiar' radio galaxies. There are others which are even more powerful radio emitters. The radio emission of the well-known elliptical galaxy M 87 is almost 100 times more intense than that from NGC 1068. The emission appears to come from two concentric sources, approximately centred on the visible galaxy: a small and intense core, and a large and less intense halo.

On long-exposure photographs M 87 appears as a normal giant elliptical galaxy (see Plate 17). Short-exposure photographs, however, reveal the presence of a bright jet extending from the galactic centre as shown in Plate 18. The jet is suggestive of material being ejected with high energy from the nuclear region of the galaxy. Observations made with the 200-inch telescopes have shown that the light radiated from the jet is strongly polarized, indicating the existence of magnetic fields in the galaxy.

While a number of other intense radio sources have been identified with elliptical galaxies that have normal appearance, some comparatively close giant elliptical galaxies have no detectable radio emission. It now appears that only a small fraction of normal looking elliptical galaxies emit enough radio energy to be called 'peculiar'. Whether or not the peculiar ones also have jets similar to that seen in M 87 is not known. The reason is that the jet of M 87, which is only some thirty-six million light years away, would not be easily detectable at the much greater distances of other elliptical galaxies that are strong radio sources. It seems likely, however, that the phenomenon responsible for the jet in M 87 is also the cause of its intense radio emission.

Many radio sources have been identified with galaxies that cannot easily be classified by the usual scheme based solely on form. Thus Centaurus A (NGC 5128) is a galaxy whose visual features cannot be readily interpreted (see Plate 19). It looks like an elliptical galaxy with a huge lane of dust running through its centre. Among several other galaxies whose appearance is similar to NGC 5128 the most conspicuous is Cygnus A which appears visually as two nuclei in contact, surrounded by a faint common envelope. This was formerly interpreted as two galaxies in collision. But it is now realized that even such a cosmic cataclysm as a galactic collision will not suffice to power the extremely strong radio emission of the source. The right clue emerged from a study of the radio spectra of the radio sources which show that the intensity (I) of radio energy emitted by most galactic and extragalactic radio sources increases as some power (μ) of the associated wavelength (λ). That is, $I = \lambda^{\mu}$. The power index μ is called the spectral index of the radio source. It varies surprisingly little from source to source and has an average value of about

111

—0·8 in the case of peculiar radio galaxies, though normal galaxies have a rather wider spread.

Such a power law of energy distribution is called *non-thermal* because it is just the reverse of the energy distribution expected of a body of hot gas such as a star. It can be shown that in a hot gas the random thermal motions of its particles – electrons and protons – produce radiation both in optical and radio regions whose intensity remains essentially constant over a wide range of wavelengths. Only the radio sources associated with the emission nebulae within our own galaxy exhibit this type of thermal spectrum. All other discrete radio sources have a non-thermal radiation. The physical mechanism responsible for their radiation must therefore be something quite different from the random whirl of electrons and protons going on all the while in a hot gas.

The most plausible mechanism capable of producing such spectral distribution as the radio galaxies show was first guessed by the Soviet astrophysicist I. S. Shklovsky. Taking his cue from the known fact that ultra-fast electrons spiralling around lines of magnetic force in the magnetic field of synchrotron particle accelerators emit intense radiation somewhat akin to that of the Crab Nebula, he suggested that radio emission of the extragalactic radio sources is due to a similar process blown to cosmic dimensions. On the cosmic scale, radio emission occurs when relativistic electrons moving with nearly the speed of light spiral around interstellar magnetic field lines – the radiation leaving the electron at a tangent to its path as shown in Figure 14.

A characteristic of synchrotron emission is that the radiation is largely plane-polarized. When it was found that light from radio sources like M 87 and the Crab Nebula is indeed strongly polarized, little doubt remained about the synchrotron origin of the radiation. But, as is often the case in cosmology, one doubt removed is another revealed. For using the well-understood theory of synchrotron radiation we can calculate the minimum amount of energy which must be present in a radio galaxy in the form of magnetic fields and relativistic particles to produce the observed radio emission. Geoffrey Burbridge, who first made such a calculation, showed that whatever the assumptions made the energies involved were of the order of 10^{61} ergs. This output

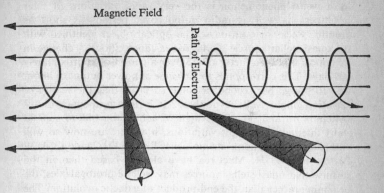

Figure 14. Synchrotron emission is a radiation in a narrow cone about the instantaneous direction of motion of an electron as it spirals around the line of a magnetic field. This radiation is linearly polarized in a plane perpendicular to the magnetic field irrespective of its direction of motion.

of energy is so enormous as to require the complete annihilation of an amount of mass equal to 5×10^5 solar masses. But no natural process can be expected to convert mass into energy completely. For example, the process of nuclear burning in stellar interiors yields only about one per cent of the mass energy. Consequently some fifty million stellar masses would appear to be involved in the process of energy generation. Since the mass of a galaxy is about 10^{11} suns, it is obvious that the transformation of a galaxy into a radio galaxy represents a major disturbance. We still do not know how it occurs except that it probably involves a sudden and gigantic explosion in the nucleus of the galaxy as is shown by the jet gushing out of the nucleus of the Seyfert galaxy M 87 or the violent eruptions of filaments from that of the irregular galaxy M 82.

The problem of energy is even more of an enigma in the case of quasi-stellar sources or 'quasars' which are yet more powerful radio emitters. Thus the radio emission from the quasi-stellar source 3C 47 is nearly 1,000 times stronger than that of Centaurus A NGC 5128 even though it is believed to be 4,000 million light years away as against only fifteen million for the latter. But the most weird phenomenon is the very rapid variations of their brightness as well as radio output over periods as short as months, weeks or even days. The optical object identified with the quasi-stellar source 3C 48, for instance, shows a change in its optical flux by a factor of $1 \cdot 5$ over a period of approximately 600 days. This corresponds to a *change* of power output of about 4×10^{44} ergs per second, a value so large that the reality of variation might well be doubted. It is like near-synchronous switching on and off of the light of a hundred billion suns at short intervals. The light variations, however, are now so well authenticated that some cosmologists like R. D. Cannon, M. V. Peston and W. H. MacCrea have already based thereon the surmise that quasi-stellar sources may well be protogalaxies, the beginning rather than the end-product of galactic evolution. The underlying rationale of the suggestion is that at this stage the galaxy would consist of a number of fully formed stars as well as a large number of thick opaque clouds of protostars that have yet to emerge as luminous stars. The light variation then is explained as occultation effect, that is, periodic obscuration of some stars by protostars in the course of their rotations round the protogalactic centre. The nature of such processes as the metamorphosis of a 'quasar' into a radio galaxy and a radio galaxy into a normal galaxy (or is it the other way about?) will not be understood unless we resolve the 'quasar' puzzle.

As we have already noted, they are indeed a strange species of celestial objects. Though looking like stars on photographs taken with giant optical telescopes, they exhibit large red shifts, even larger than the most distant luminous galaxies. Consequently if Hubble's law applies to them, they would seem to be at the very limit of our observable universe. A case in point is the quasar 3C 9 with red shift ratio $z = 2$. Its recessional velocity is therefore

$$\frac{2}{\sqrt{1 + 2^2}} = \frac{2}{\sqrt{5}}$$ times the velocity of light. Hubble's law then

requires that it be at a distance of $\dfrac{2}{\sqrt{5}} \times \dfrac{186,000}{19} = 8,732$ million

light years from us. The distance derived is so large that cosmologists are still debating the validity of deriving quasar distances by recourse to Hubble's law. Obviously the issue cannot be decided unless we know what quasars are and what physical processes power their enormous energy output. A lot of novel theories have been mooted in recent years to account for them. But as we shall see more fully in Chapter 12 none of them is really satisfactory. We still do not know what quasars are. But whatever they may turn out to be, they do seem to be at the farther end of the scale, if we arrange in ascending order of their radio output all the aforementioned celestial objects ranging from normal galaxies through radio galaxies to quasars. Figure 15 shows a linear ordering of some fifty such objects. It will be observed that we may classify them in seven major classes in supersession of Hubble's earlier threefold classification based solely on optical form. They are:

(i) Spirals including irregulars. Cases in point are galaxies like the irregular galaxy M 82 or the spiral M 31. They are weak radio sources. The Seyfert galaxy NGC 1068 is the strongest of the weak sources.

(ii) Ellipticals (E) as, for example, the galaxies M 84 and M 87. They are spheroidal with peculiar appearance which is not easy to recognize.

(iii) D galaxies. These galaxies have an elliptical nucleus surrounded by an extensive envelope. They are of great interest in connexion with the classification of radio sources. Although some of the optical forms of these sources might be classified so on the Hubble system, they form a distinct class by themselves.

(iv) Galaxies having form types intermediate between E and D. They are abbreviated DE or ED with the preceding letter indicating the dominating characteristic.

(v) The 'dumb-bells' (db). Allied to D galaxies is the class labelled dumb-bells in which two nuclei are observed in a common envelope. They may well be related to the galaxies which have one or more fainter companions in their envelopes, the dumb-bells being the extreme cases of very close multiple galaxies when there are only two equal components.

(vi) The 'N' galaxies. These are galaxies having brilliant star-like nuclei containing most of the luminosity of the system with a faint nebulous envelope of small visible extent. They resemble quasi-stellar sources in spectral features and in having large red shifts. But they are not quite as star-like.

(vii) The quasi-stellar sources, that is, quasars or QSRs.

Figure 15 also shows that there is an apparent gap in the frequency of occurrence of sources with their radio luminosity in the range 10^{40} ergs per second to 2×10^{41} ergs per second. Although the sharp cut-off here is not real, being only a selection effect, it is taken as the critical point of division between 'strong' and 'weak' sources. The sources are classified 'strong' or 'weak' according as their radio luminosities are greater or less than 10^{40} ergs per second. Adopting this criterion we find that

(a) All spiral galaxies identified as radio sources with the exception of the Seyfert galaxy NGC 1068 are in the 'weak' group.

(b) The great majority of the strong sources – the 'radio' galaxies – with powers in the range 10^{40} to 10^{45} ergs per second are spheroidal. No highly flattened galaxy has been found among the radio galaxies. A large fraction are in clusters – 50 per cent in rich clusters with more than fifty members within a range of two magnitudes below the brightest galaxy in the cluster. This suggests that the strong sources are intrinsically different from the weak ones. In any case there seems to be a clear distinction between the strong and weak sources in their optical appearance and in their radio structure.

(c) All identified radio galaxies of the types D, db, N and QSR are strong sources.

(d) The range in radio luminosity of D-type radio galaxies is more widespread, being of the order of $10^4:1$.

(e) The quasi-stellar sources are the most powerful sources. The four sources exhibited in Figure 15 have radio luminosities ranging from 2×10^{44} to 2×10^{45} ergs per second. They therefore form a much more compact group in luminosity than D-type radio galaxies. But there may be different selection effects in discovery for the members of the two groups.

(f) The N radio galaxies, which most closely resemble the quasi-stellar sources in their optical appearance on direct plates, are of considerably lower radio luminosity than the latter.

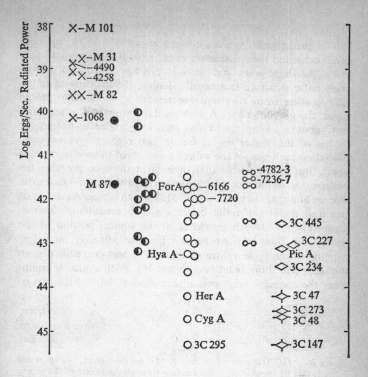

X — Spirals and Irregulars

● — Ellipticals (Class E)

O — Bright Nuclei and Extended Envelopes (Class D)

◑ — Intermediate between Classes D and E

◇ — Brilliant, Star-like Nuclei and Less Extensive Envelopes (Class N)

◇ — Quasi-stellar Sources

∞ — Dumb-bells (Related to D Systems)

Figure 15. The relationship between optical-form class and radio luminosity. The horizontal location of the plotted points has no significance; the form classes have been separated horizontally for clarity. (From Thomas A. Matthews, William W. Morgan and Maarten Schmidt, 'A Discussion of Galaxies Identified with Radio Sources' in I. Robinson, ed., *Quasi-stellar Sources and Gravitational Collapse*, University of Chicago Press, 1964.)

The strong radio sources cover a range of 2×10^5 in radio luminosity and their diameters range from a few thousand to a few million light years – a range of 10^3. In contrast to weak sources their radio structure is generally double, the majority of radiation coming from two separate regions symmetrically placed about the parent galaxy. A striking illustration of this tendency is Cygnus A, where the radio emission originates not in the region of the visible object but in two regions symmetrically placed on each side of the galaxy about three million light years apart. But it is possible that their great distance prevents the observation of the more detailed pattern of their radio emission. For we find that the galaxy NGC 5128 which is much closer to us has four regions of radio brightness, two resembling those in Cygnus A and two lying closer to the central portion of the galaxy. All four emissive regions lie approximately on a line running through the centre of the galaxy and perpendicular to the dark absorbing band (see Figure 16). With available equipment the complex brightness distribution of NGC 5128 can be

Figure 16. NGC 5128 shows four regions of radio brightness, two small and intense, that lie roughly on a line running through the centre of the galaxy and perpendicular to the dust band.

observed only in near-by sources. It seems that the formation of a double source is going on in M 82, an irregular galaxy, in which gas streams out from the ends of the minor axis with a velocity of about 1,000 km./sec. relative to the centre of the galaxy.

In short, we have during the past ten years acquired lots and lots of facts about galaxies and their possible precursors, the quasars, largely because of the tremendous progress in observational techniques of both radio and optical astronomy. But unfortunately our theoretical understanding of the cosmic processes that the new astronomical techniques have revealed has not yet advanced in step. We still do not know what energy sources can power the 'quasars', the strong radio galaxies and the medium strong sources like M 87 and Seyfert galaxies. Nor do we know what physical processes distinguish the weak from the strong radio sources and even which of the two is an earlier stage in the evolution of a galaxy. Before we can formulate an acceptable theory of galactic evolution we must first decide what quasars are and what is involved in observing them. We shall revert to the problem in Chapter 12.

PART II

Cosmological Theories

CHAPTER 7

Space, Time and Gravitation or Relativistic Cosmology

SINCE modern cosmology as a new scientific discipline is an offshoot of Einstein's reformulation of Newton's theory of space, time and gravitation, it will be appropriate to begin our survey of cosmological theories with an account of the latter and the reasons for its revision by Einstein.

As is well known, Newton discovered the mathematical form of action at a distance, the famous inverse square law of discrete gravitating masses,* which grafted on to his own laws of motion enabled him to explain the motion of planets in the sky as well as apples in orchards. Although both the contributions of Newton to mechanics – his inverse square law of gravitation and the laws of motion – were essential to the grand synthesis of planetary and terrestrial motions he achieved, the latter were a more revolutionary innovation. Before Galileo and Newton, bodies were supposed to move under the influence of some force – 'gravity' if moving downwards, 'levity' if like smoke rising upwards. But force was related to *velocity*, not *acceleration*. The reason why Newton had to replace velocity by acceleration as the sole consequence of force producing motion is clearly seen when we consider the motion of the earth (see Figure 17). If the force responsible for this motion is related to its velocity, as was the belief in pre-Newtonian days, then one should be able to locate the source of this force by looking in the direction of the earth's velocity. But looking in the direction of the earth's velocity yields no clue. We see sometimes one insignificant looking star, then another, and at other times virtually nothing. But if, on the other hand, we look in the direction of the earth's *acceleration*,

---

* The gravitational force (F) exerted by one body of mass (m) on another of mass m' is $\gamma\,\dfrac{mm'}{r^2}$ where r is the distance between them and γ a universal constant, the constant of gravitation.

123

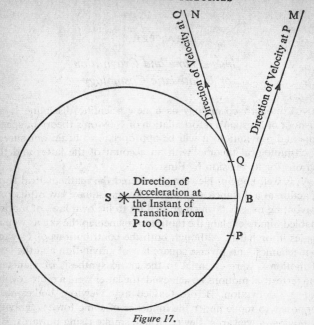

Figure 17.

we see the sun all the time. It is therefore but natural to regard this important body as being in some way the cause of the earth's motion in its orbit. It was by linking this kinematical observable – the acceleration of the earth – with something that could be optically observed – the direction of the sun – that Newton arrived at his concept of force embodied in his second law of motion, namely the force producing motion in a body is equal to mass times the acceleration.

However, the pre-eminence of acceleration over velocity in Newtonian dynamics had one awkward consequence – an awkwardness that gradually grew to be intolerable by the end of the nineteenth century. For acceleration of a body is rate of change of velocity and velocity is change of its position. But when we consider change of its position, we cannot escape the question – change of position relative to what? It is well known that by choosing an appropriate frame of reference the state of motion of a body may be changed at will. A body at rest in a coach of a uniformly moving train is at rest relative to the train but is moving with a uniform velocity with respect to the earth and in a much more complicated manner relative to the sun and still more so to the system of 'fixed' stars. Since a body may be made to acquire almost any velocity arbitrarily by an appropriate choice of the frame of reference relative to which velocity is measured, it can also be made to acquire arbitrary acceleration. After all, acceleration is merely rate of change of velocity. But acceleration and its sole determinant, force, in Newtonian dynamics are something essentially real and therefore cannot be arbitrary. It therefore follows that all arbitrary frames of reference to which motion is related are not equally valid. Only those systems of reference that give the correct value of acceleration – that is, wherein 'force-free' particles are *not* accelerated with respect to the system itself – are admissible for our observations. Any such frame is called inertial.

Unfortunately there is no unique inertial frame that we may adopt. For it can easily be shown that if there is one such admissible frame of reference S, there will be an infinity of them. Thus if acceleration in S yields the true value, then acceleration relative to any other frame S' in unaccelerated motion relative to S, that is, moving with any constant velocity in a fixed direction, will also lead to the same value of acceleration. The reason is that although velocities measured in the second system S' will differ from those measured in the first system S, yet the difference between the two being constant the accelerations measured in both are the same. This is because a shift in the frame of reference affecting all velocities *equally* will *not* affect their rates of change or accelerations at all. The second system S' is therefore equally valid as a standard frame of reference. This is merely a way of saying that all motion is relative and that any inertial system of reference is as good as another because in any such inertial frame acceleration is simply connected with force in accordance with Newton's second law of motion.

Since the measure of acceleration in any inertial frame is the same while that of velocity is not, it results that any observer in an inertial frame will agree with another as regards the magnitude of force acting on a body but *not* as regards that of its velocity. This led in course of time to a serious conflict with observation. For it was found that velocity of light as measured by all inertial observers in relative unaccelerated motion is the same. It is in fact a fundamental constant of nature. Such constancy is clearly contrary to what our common sense leads us to expect. For normally if you chased a car moving, say, at thirty miles per hour on a motor cycle at ten miles per hour, the car's velocity relative to you would be only $30 - 10 = 20$ miles per hour. But sophisticated observations show that this simple rule will *not* work if you chase a ray of light. Its velocity relative to you would still remain the same, namely 186,000 miles per second, which you would have found if you had stood still on earth. In fact, no matter whether you moved on a motor cycle at twenty miles per hour or in a jet rocket at 20,000 miles per hour in pursuit of a ray of light, it would still continue to escape you at the same even pace – 186,000 miles in a second. The attempt to explain the puzzling peculiarity of the velocity of light – its refusal to mix with other speeds in the usual high-school manner – led Einstein to question the tacit assumption of a universal absolute time that Newton and following him all classical physicists had made.

According to the classical or common-sense view of time every one of us can arrange the events that we perceive in an orderly sequence. That is to say, we can tell which of two events perceived occurred earlier and which later. By means of physical appliances such as a watch we can even say how much earlier or later. Suppose I observe the occurrence of two events – *A*, the emergence, say, of a nova in the constellation Hercules, and *B*, a lunar eclipse. Seeing them I could tell whether *A* occurred before, after, or simultaneously with *B*. I could even measure the time interval between their occurrence by means of a clock. Similarly another observer, say, Voltaire's Micromégas, looking from his native star Sirius, could also observe the same two events. He too would place them in a time order. Now Newton adopted the common-sense view that the temporal order in which I would place the events *A* and *B* would be precisely the same as that of Micromégas, the Sirian observer. Furthermore, my reckoning of the time interval between the occurrence of the two events by means of my watch would be identical with the time reckoning of the Sirian measured, of course, with a Sirian watch.

Unfortunately, this assumption would be valid only if the time at which an event is observed were simultaneous with its occurrence. That would be the case if light travelled, as our ancestors believed, with infinite velocity so that news of any event (the birth of a nova, for example) could be flashed instantaneously everywhere. There would thus be no time lag between its occurrence at one place and its observance from another, however far. We know, however, that this is not the case. Instead of instantaneous propagation of light, we find that light takes millions and billions of years to come to us from some places. When we allow for this time lag between the occurrence and observance of an event caused by the finite velocity of light, we find that my reckoning of the time interval between the events A and B is not the same as that of our imaginary Sirian. It is possible that I may find A and B simultaneous while the Sirian would find that A occurred before B. It is not that the Sirian watchmakers are in any way inferior to their terrestrial counterparts. It is because the velocity of light is found to be constant with respect to any moving observer. Einstein therefore held that there was no basis for the hypothesis that all observers must ascribe the *same* time to any given event like the emergence of a nova even if they reduced the *subjective* time of their perception to the *objective* time of its occurrence. Indeed, he gave very cogent reasons based on the observed constancy of velocity of light by all inertial observers why this hypothesis must be rejected.

If the Newtonian idea of an absolute even-flowing time underlying his mechanics became by the end of the nineteenth century untenable, the parallel Newtonian notion that space too has an absolute structure, which is over and above that of the matter it contains, began to run into trouble. Newton had taken over the pre-scientific concept of space characterized by the sentence: 'We can think away things but not the space which they occupy.' In other words, he pictured space as a self-abiding featureless container of infinite dimensions, where matter may be put or withdrawn without in any way affecting the inside of the container exactly as we do not apparently affect the hollow of our room by rearranging or even emptying it of its furniture. This container concept implies two assumptions which the classical physicists made explicit. First, the size of the container is infinite, that is to say, there is nowhere where we can locate its walls, top and bottom. Secondly, its character is uniform throughout, which merely means that any chunk of its hollow is as good as any other. This common-sense view of space was mathematized by Euclid and taken over *in toto* by the classical physicists mainly because of the influence of Newton.

How does it square with the facts of our actual universe? Our answer depends on what we believe would be the outcome of the following imaginary experiment. Suppose we imagine ourselves moving outwards in a straight line past the planets, stars, galaxies, and beyond them. There are only three possibilities. First, we may ultimately find ourselves in an infinite empty space with no stars and galaxies. This means that our stellar universe may be no more than a small oasis in an infinite desert of empty space. Secondly, we may go on for ever and ever meeting new stars and galaxies. In other words, we may find an infinite number of stars and galaxies in an infinite space. Thirdly, we may be able to return ultimately to our starting point, having, as it were, circumnavigated the universe like a tourist returning home after a round-the-world trip.

If we reject the last possibility as utterly fantastic, we have to do some pretty tough thinking to get over the difficulties which the other two possibilities raise. A finite stellar universe in an infinite space would either condense into a single large mass because of gravitation or gradually dissipate – dissipate because, if gravitation failed to produce condensation, the stellar radiation as well as the stars themselves would pass out into infinite empty space without ever again coming into interaction with other objects of nature. For as Lucretius, innocent of gravitation, divined long ago, finite matter in infinite space cannot endure for one brief hour: 'The supply of matter would be shaked loose from combination and swept through the vastness of the void in isolated particles or rather it would never have coalesced to form anything since the scattered particles could never have been driven into union.'

We find the second possibility, that of an infinite number of galaxies in infinite space, no more satisfactory. It leads, as Seeliger showed, to the rather ridiculous conclusion that gravitational force is everywhere infinite. Every body would find itself attracted by an infinite pull – infinite not in the metaphorical sense of something very large but the mathematical sense of being larger than any figure we care to name. We could hardly survive such a terrific pull. It is true that we could avoid this difficulty by an *ad hoc* amendment of Newton's law, but there is a strong feeling against tinkering with well-established physical laws every time we meet an objection.

Besides the amendment would be of no avail because there is another objection which could still be raised. For if space were really infinite with an infinite number of stars and galaxies therein, then the night sky would be a blaze of light with no dark patches between the stars. To understand the reason for this, suppose you were in a big, sparsely sown forest. No matter how sparsely sown it might be, provided only it was big enough the horizon would be completely blotted out of view by the trunks of the trees. In the same way, however sparse the nebular distribution in space, provided space is infinite, the entire sky would appear as Olbers pointed out over a century ago, to be studded with stars and galaxies leaving not a single point uncovered.

We could, of course, explain the darkness of the night sky by assuming that the light of some of the distant galaxies is absorbed en route or that the galactic density becomes progressively less as we recede from our present position or that the vacant spaces of the sky are occupied by dark stars or galaxies that we cannot see. But all these assumptions would be quite arbitrary because our present experience is to the contrary.

What then is the way out?

A deeper probe into the reasoning underlying the deduction of these paradoxes reveals that an assumption regarding the nature of our space has tacitly been made. It is that space on the large scale is exactly what it looks to us on the small scale – the container-type space of common sense. In more technical language, it has been assumed to have the properties Euclid embodied in his *Elements* by a process of logical deduction. Can it be that our actual space when taken as a whole is not what Euclid made it to be?

Until as recently as about the time Olbers deduced his paradox (1826), there was never any doubt that there could be any kind of space other than Euclidean. Why? Because Euclid's process of deduction is such a marvel of perfection that for centuries geometry has been considered the most certain of all the sciences. It was for this reason that Plato considered God to be a geometer and Hobbes regarded geometry as the 'onely science that it hath pleased God hitherto to bestow on mankind'. No wonder Spinoza was driven to essay geometry to provide a secure foundation for his *Ethics*, or that his example has been widely imitated by others down to our own day, from Newton in his *Principia* to Woodger in his *Axiomatics of Biology*. Nevertheless, even though philosophers were convinced that Euclid's deductive method had crossed the threshold of perfection, some mathematicians thought that its rigour could be improved. In what manner did they hope to do so? To see this we have to examine the grand strategy adopted by Euclid in proving his theorems.

In his *Elements*, Euclid had shown in detail how all the geometrical theorems could be logically deduced from a dozen-odd axioms or postulates. If the truth of his postulates were granted, that of his theorems would automatically follow. There was no difficulty in accepting the truth of almost all his postulates because they appeared 'self-evident'. For instance, one of these postulates was that if equal magnitudes are deducted from equals, the remainders are equal among themselves. No one can seriously deny the truth of this proposition, and other postulates of Euclid were equally plausible and self-evident. But there was one exception which ultimately upset the entire Euclidean applecart. This was Euclid's fifth postulate about the behaviour of parallel lines, which signally failed to appear self-evident to his successors.

For over 2,000 years after Euclid, mathematicians tried in vain to prove the truth of the fifth postulate till about 120 years ago the Russian Lobachevski and the Hungarian Bolyai showed indisputably that such a demonstration could not be given. They pointed out that no *a priori* grounds exist justifying belief in the disputed postulate and that an equally consistent system of geometry could be constructed by replacing it by its contrary.

Now what is Euclid's parallel postulate which has been discussed so much by mathematicians for over two millennia? It may most simply be explained by means of a diagram (Figure 18). Let *AB* be a straight line and *P* a point outside it. Naturally we believe that the straight line *AB*, if produced, would go on forever

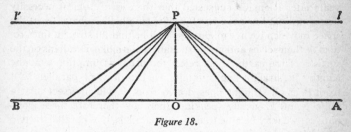

Figure 18.

without coming to an end. In other words, it is infinite. If we draw straight lines joining *P* to various points of *AB*, we get a pencil of lines radiating from *P* to points more and more remote from *O*. In the limit, when one of the lines, say *l*, meets the line *OA* at infinity towards the right, we say *l* is parallel to *OA*. Similarly, if we consider the points of the line towards the left, there will be another line *l'* through *P* which will meet the line *OB* at infinity towards the left.

Now Euclid assumed that both the lines *l* and *l'* would in effect be in one and the same straight line. In other words he assumed that the angle between these two lines *l*, *l'* through *P* would be 180°. But Bolyai and Lobachevski argued that it was not *logically* necessary that the limiting positions of the right and left parallels *l*, *l'* must be in the same straight line. The only reason for assuming it was that in ordinary diagrams such as we draw on paper the two parallels do appear to lie in one straight line.

But suppose we draw (in imagination, of course) a diagram on a really cosmic scale by taking *P* as the centre of Sirius and *AB* as the line joining say, the sun and Polaris. Will the left and right parallels in this case be in the same straight line, that is, enclose an angle of 180°? They may or may not do so. *A priori* there is no reason why they should. The argument of Bolyai's opponents, based on the behaviour of parallels in accurately drawn paper diagrams, is no more applicable than the argument that the surface of a continent must be flat merely because a football field appears so.

Both Bolyai and Lobachevski believed that a straight line in space extends to infinity in both directions. About twenty-five years later Riemann suggested that there is no logical necessity that a straight line need be of infinite length. The character of the space in which we live might very well be that all straight lines return to themselves and are of the same length like meridians on the surface of the earth. If we reject the usual assumption that the length of a straight line is infinite, we can draw no parallels to it from *P*, as all straight lines drawn from *P* will intersect it at a finite point. Logically speaking, there are therefore three alternatives: we may be able to draw no parallels, one parallel, or two parallels to any given line from a point outside it.

Euclid assumed the second. Bolyai and Lobachevski suggested the third, and Riemann the first alternative. New geometries, logically as impeccable as Euclid's, were constructed by assuming the first and third alternatives. For instance, in the geometry of Bolyai and Lobachevski, also known as hyperbolic geometry, it can be proved that the sum of the three angles of a triangle is less than two right angles, the deficiency being proportional to the area of the triangle.

Gauss, the celebrated German mathematician, used this theorem to test the geometry of the actual world. He measured the angles of a triangle formed by three distant mountain peaks and found that the deviation of the sum of its angles from 180° was well within the limits of experimental error. The experiment, however, was inconclusive, as the size of Gauss's triangle, though large compared with triangles drawn on paper, was small compared with the dimensions of the universe. However, we may easily remedy this defect by taking a larger triangle whose vertices will be formed by three distant stars, or better still, galaxies. But here we introduce a fresh complication. For the measurement of angles would depend on the observation of light rays and consequently on the physical laws governing the propagation of light through space. The experiment would therefore tell us more about the behaviour of light rays during their voyage through interstellar or intergalactic space than about the nature of physical space itself, and we could interpret the result in several ways. How then can we determine the geometry of our actual space?

Mathematicians have evolved what may be called a coral-reef technique which, if it could actually be carried out, would reveal the true character of our space. Its leitmotif is the principle of gaining knowledge of the external world by consolidation of that obtained by an examination of the behaviour of its infinitesimal parts, very much as the marine polyps build up whole islands by the accumulation of coral. To understand its underlying idea, we may simplify our problem a bit by considering only two-dimensional 'spaces' even though our actual space is three-dimensional.

Let us therefore imagine a two-dimensional space such as the surface of a sheet of paper, a sphere, a horse saddle or any other

Figure 19. A network of rectangles formed by lines of longitude and latitude on a spherical globe.

surface whatever. We may suppose it covered with a network of a large number of very fine meshes somewhat like the network of squares on a graph paper or tiny rectangles on a geographical globe formed by its lines of longitude and latitude (see Figure 19). A simple way of doing this is to cut small bits out of thin veil and paste them all over the surface leaving no portion of it uncovered.

It can then be shown that the distance between any two points within any single mesh can be computed from a knowledge of only three ratios which may be obtained by actual measurements in the mesh by means of a very small measuring rod.

These three ratios do not change so long as the two points chosen remain within the same mesh though they may vary from mesh to mesh. In other words, each mesh has its own set of three ratios which enables calculation of distances between any two points enclosed within it. But what about two points lying in different meshes? In this case the line joining the points crosses several meshes and is thus divided into as many bits, each bit lying in some single mesh. The total distance then is merely the sum of all these bits, each of which can be computed from the three ratios of the mesh to which it belongs. This means that all that we need for a complete determination of the distance (or metrical) relations of our surface is a network of meshes and the set of trios of ratios pertaining to each mesh. Such a set is called the *metrical ground form* of the surface because it is the open-sesame of its entire geometry. In particular, a knowledge of the ratio numbers of the ground form enables us to calculate its curvature.

The case of three-dimensional spaces is analogous – a lattice of three-dimensional space cells in which honeycombs in a beehive or brickwork in a wall take the place of a network of meshes (see Figure 20). The only difference is that the introduction of an additional dimension in our reckoning makes the computation of distances within each cell more complicated. We now require six ratios for each cell instead of three for a two-dimensional mesh. Each such sextuple of ratios has as before to be derived by actual measurements in the cell concerned by means of a small measuring rod. A knowledge of the set of such sextuple of ratios pertaining to each cell then suffices to determine the character of our three-dimensional space. But once again all regions of our actual space are not accessible to us and we cannot always explore all of them with our measuring rods in order to discover the value of ratios in these far-flung cells. To obtain them we therefore actually proceed in a different manner.

One way would be to *assume* any arbitrary set of functions to

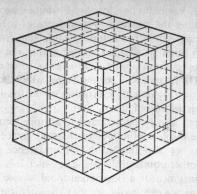

Figure 20. A lattice of spatial cells in three-dimensional space.

define the values of the ratios of the ground form and then work out the geometric proporties of our space such as its curvature at various points. We thus obtain a rich abundance of geometries and we may choose out of them the one that best fits our actual space. Another way is to look around and make some *plausible* assumptions regarding the geometric properties of our actual space and then see to what values of ratios they lead us. Surprising as it may seem, just one simple assumption that appears eminently reasonable regarding the character of our actual space ensures that the cell ratios can have only a very restricted set of values and no others. All the rich abundance of geometries that can be invented by giving the cell ratios arbitrary values thus disappears at one blow and we are left with only a few types of geometries that are applicable to our actual space.

Now we know that a rigid body may be transferred from one place to any other and put in any arbitrary direction without alteration of its form and content or metrical conditions. This means that space is everywhere homogeneous in all directions, that is, one chunk of space is as good as any other. But if space is homogeneous, its curvature must be the same at all points. If we make this assumption, the ratio numbers comprising the metrical ground form of our space can take only certain restricted values. In fact, Riemann developed a formula whereby the ratio numbers defining the metrical ground form of a homogeneous space of constant curvature could be calculated from its curvature alone. In other words, a knowledge that the curvature of space is a given constant everywhere enables us to derive all the ratio numbers of its metrical ground form, which, as we saw before, suffice for a complete description of all the intrinsic geometrical properties of our space.

This fact may also be appreciated in a more intuitive manner by considering the case of a one-dimensional 'space' of constant curvature. The set of points on the circumference of a circle as distinct from other points of the plane in which it is drawn can be taken as an instance of a one-dimensional 'space' of constant curvature. Here we can easily see that the curvature is constant because the circle bends uniformly everywhere in the two-dimensional plane in which it is embedded. A knowledge of its curvature (or radius) tells us all there is to know about the internal geometry of this one-dimensional 'space' that is our circle. A sphere (i.e., the set of points constituting its surface as distinct from those lying within or without it in the three-dimensional space in which it is immersed) is the exact analogue of a two-dimensional 'space' of constant curvature. Here, too, we can see that the spherical surface bends uniformly everwhere in the three-dimensional space in which it is embedded. Again a knowledge of the curvature of this spherical surface tells us all that there is to know about its internal geometry.

When we come to a three-dimensional space of uniform curvature, we have to view it as embedded in a superspace of four dimensions before we can 'see' it bend uniformly all over. But that is what we three-dimensional beings cannot do. So we have to content ourselves with an extrapolation and conclude that what is true of one- and two-dimensional 'spaces' of uniform curvature is equally true of a three-dimensional space of uniform curvature where, too, a knowledge of its curvature suffices to give a complete knowledge of its intrinsic geometry.

Now if our actual space around us is homogeneous with constant curvature a, there are just three possibilities according to whether a is zero, negative, or positive. These correspond to the three possibilities we enumerated earlier concerning the behaviour of parallels drawn from a point P to any given line AOB (Figure 18), in which case there may be only one, two, or no parallels from P to AOB. If we assume with Euclid that there is only one

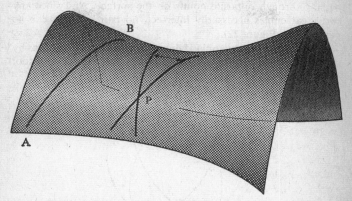

Figure 21. All lines through P within the limits of the two lines drawn through it are parallel to AB.

parallel through P, we have Euclidean geometry, corresponding to zero value of a, the space curvature. That is why Euclidean space is also known as 'flat' space. This is the common-sense space of our daily experience which Newton and the classical physicists following him adopted as the basis of their theories.

If we assume with Bolyai and Lobachevski that there are two parallels through P, we have hyperbolic geometry, corresponding to a negative value of a, the space curvature. A two-dimensional instance of a negatively curved 'space' is the saddle-shaped surface shown in Figure 21. On such a surface the analogue of the Euclidean straight line is the line of least distance between two points. As will be seen from the figure, many 'straight' lines can be drawn through a point outside any given 'straight' line without ever meeting it. These 'parallel' lines fall within the limits between the two intersecting lines shown on the right.

The final case, in which there is no parallel through P, gives rise to elliptical geometry and corresponds to a positive value of a, the space curvature. A two-dimensional analogue of a positively curved space is the surface of an ordinary sphere. On such a surface, as all mariners know, the shortest distance between two points follows a great circle. But a great circle is a closed line passing through opposite points on the surface. And since any two great circles necessarily intersect, no pair of them can be parallel (see Figure 22).

The collapse of Newtonian concepts of space and time was by no means the end. It was followed by that of his idea of mass too.

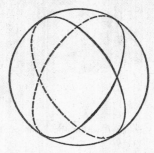

Figure 22. Any two great circles on a sphere always intersect. There are therefore no parallels on a spherical surface.

Newton had refined the earlier notion of mass as *quantitas materiae* – mass is the 'quantity of matter' whatever that may mean – by going beyond this medieval tautology even if some textbooks are still written as though Newton never lived. In equating the activating force to the product of mass and the resultant acceleration, Newton virtually defined the mass of a body as a measure of its inertia or resistance to being accelerated. In doing so he based himself on our daily experience. For obviously if we subject bodies of different masses to the same force, then the more massive the body the less will be its acceleration. This particular characteristic of mass, the resistance to being accelerated, is called the *inertial* mass of a body. It was a fundamental tenet of Newtonian mechanics that the inertial mass of a body remains constant no matter how it moves with respect to the observer who measures it. The new discoveries relating to the motion of elementary particles showed that their mass was very much dependent on the velocity with which they happened to move relative to the observer.

Besides the intellectual embarrassments due to the notions of absolute space, time and mass underlying Newtonian mechanics becoming increasingly untenable in face of new discoveries and experimental situations, there were others caused by the inverse square law itself. Chief among them was the difficulty due to the apparent necessity of admitting a mystical action-at-a-distance which propagated itself instantaneously in some unknown way across the gulf of space. Newton himself was quite puzzled by it. As he said at one time, 'That one body may act upon another at a distance through a vacuum without the mediation of anything else . . . is to me so great an absurdity that I believe no man, who has in philosophical matters a competent faculty for thinking, can ever fall into it.' But as it seemed to 'work', he overcame his philosophical qualms and refused to commit himself with his famous dictum: '*Hypotheses non fingo*' ('I make no hypotheses').

A way out of these difficulties of Newtonian mechanics was shown by Einstein. He suggested that physics is mainly concerned with events such as the arrival of a light ray at a particular point, the emergence of a nova, the quantum jump of an atom and the like. Now every such event occurs somewhere sometime. In more recondite language, it has two characteristics, a location in space and a date in time. Could these two aspects of physical pheno-mena – the spatial and temporal aspects – be somehow merged? This does not mean that Einstein contemplated the construction of some new equivalent of a philosopher's stone for turning a foot-rule into a clock. What he had in mind was quite different. Since both locations and dates of physical events can be denoted by numbers,* he envisaged the possibility of a *formal* procedure whereby these numbers could be combined to form a higher complex or 'unity' of which both the location and date numbers were mere separate aspects. Such a higher complex is the famous space-time continuum of Einstein whereby our eternally changing

---

* For instance, the great Lisbon earthquake of the eighteenth century may be denoted by three numbers (4° West longitude, 45° North latitude, in A.D. 1755).

world of events is frozen, as it were, into a static manifold of what may be called 'point-events', that is, purely space locations having dates in time.

Einstein dealt with this abstract manifold of point-events, the space-time continuum, in two stages. In the first stage, he began by *denying* that there is only one universal even-flowing time available to all observers in relative motion for dating events observed by them. He suggested instead that every one of the several observers in relative motion has to have his own system of locating as well as dating them. Consequently the different systems of measuring the spatial and temporal relations of events would vary from one moving observer to another. But their variation is not wholly arbitrary. Given the space and time measurements (co-ordinates) of any event as determined by one observer, those of another observer in uniform relative motion with respect to the first can easily be computed. This rule of computation is the famous Lorentz transformation.

One consequence of the Lorentz-transformation rule is that a way of merging both the spatial and temporal co-ordinates of any two events can be devised to yield an entity called the 'interval', which remains the same for all moving observers even though the individual components (the co-ordinates) that go to make it do not. The situation is exactly similar in ordinary high-school plane geometry, where the merger of two co-ordinates of a point can be made to yield an invariant entity, namely their distance. The reason is well understood in geometry. The distance between two points has an intrinsic significance, while the projections of this distance upon the two co-ordinate axes depend on the co-ordinate system we choose and vary as it changes. Similarly in Einstein's special relativity the distance-analogue of two events, the 'interval', is something absolute. But its 'projections' on the space-time co-ordinates are arbitrary depending on the spatial-temporal frame of reference we adopt for locating and dating events. This is why the interval between two point-events plays exactly the part that distance between two points plays in ordinary plane geometry.

With the substitution of the space-time continuum in lieu of absolute even-flowing time and an all-pervading container type space, Einstein proceeded to the second stage of his theory by developing its geometry in accordance with two fundamental principles he had enunciated earlier. First was his principle of relativity embodied in the Lorentz-transformation rule already mentioned to which all the phenomena of light and electro-magnetism appeared to conform. Second was the principle of equivalence, which asserts the impossibility of distinguishing between a gravitational field and acceleration. Thanks to the progress of aviation and space travel, this impossibility is nothing very strange nowadays. It shows itself daily in such reports as that the 'pilot weighs half a ton as he pulls his plane out of a power dive', or that 'Gagarin remained in a state of weightless-ness as he hurtled around the earth for twenty-five hours'. Einstein's sophistication of this equivalence of gravity and acceleration is the heart of his generalized relativity theory, whereby he sought to abolish the notion of gravitational force and replace it by acceleration, that is, mere change in velocity and/or its direction. The acceleration of a gravitating body such as a falling apple or an orbiting satellite that shows itself as gravita-tion 'really' arises because of a particular feature called 'curva-ture' of the space-time in which it is embedded.

Einstein's geometry of space-time, based on the twin principles of relativity and equivalence of gravitation and acceleration, was really a switch-over to the Machian point of view whereby the local properties of both space and time were believed to be the direct consequence of the existence of surrounding matter, which properties in turn determined the motions of that matter. He was led to link the two because of a very simple consideration, just as earlier Newton was led to link force with acceleration by con-sidering the earth's motion. Einstein was deeply impressed with the fact that gravitation is an inalienable property of all matter. That is, there is no way whereby a body can be made immune from the influence of gravitation in the way that an object may readily be shielded from the influence of electric forces by inter-posing a sheet of copper or from that of magnetic influence by means of a plate of iron. In other words, materials like cavorite which enabled Cavor in H. G. Wells's *The First Men in the Moon* to reach the moon by simply screening his space ship from earth's gravity just cannot be made, because materials and gravitation are as inseparable as the twin poles of a magnet. Now the only entity with a similar ineradicable property that we can conceive of is space-time. For everything that is must be somewhere sometime. It will therefore be subject to the laws of space and time. What more natural then than to link the two by identifying the gravitational properties of materials with some appropriate geometrical construct of space-time in which they are embedded?

In his search for an appropriate geometrical property of space-time Einstein naturally proceeded to deal with the abstract manifold of point-events by the methods of geometry precisely as we do with points in ordinary space. Thus, for example, we can ascertain the ratios of its metrical ground form by a simple extension of the coral-reef technique described earlier. That is, we can discover them by point-to-point explorations of the space-time manifold by means of a small measuring tape and a clock, exactly as we mapped a two-dimensional surface within each mesh by a small measuring rod. These ratio numbers are a set of ten numbers associated with each point of the space-time continuum. In principle they are no different from the set of numbers mathematicians couple with each point in the interior of an elastic solid like a steel wire to describe the state of stress and strain of its material when it is twisted.

From these ratio numbers – a set of ten ratios for each point of the space-time framework – we can calculate any geometrical feature of our space-time continuum, in particular the ten components of its curvature at each point. It may be mentioned in passing that while a single-dimensional manifold such as a zigzag track has only one curvature at any point, a two-dimensional surface like the paraboloid reflector of a car headlight has two curvatures at each point simply because points lying therein have the freedom of movement in an additional dimension. It has therefore greater bending potentialities than a point of a uni-dimensional line. As the number of dimensions of our manifold increases, its points acquire more and more freedom to twist and turn exactly as an airplane has greater scope for somersaulting than an earth-bound automobile. This is why our four-dimensional space-time manifold at any point will have not just one curvature as an ordinary one-dimensional line has, but ten.

Consider now the material events that unfold in the space-time framework whose geometrical features we have outlined above. Taking the universe as a whole or even some large slice of it such as modern telescopes reveal to us, we find therein countless concentrations of matter, the galaxies. To compute the motions of these discrete masses separated by vast empty spaces under their mutual interactions is a problem bristling with extreme difficulties – difficulties which have still not been resolved. One way of mitigating them is somehow to water down the hard tenacity of these discrete galactic agglomerations of matter by dissolving* them altogether and distributing all their material content evenly throughout the intergalactic void as a tenuous gas. This is no doubt an extreme idealization, but it yields what can be regarded as a first approximation to the actual problem.

To describe fully the material features of such a smoothed-out universe as, for instance, the density, momentum, stress and energy of the thin medium pervading it, we require again a set of ten numbers, one such set being coupled with each point of the continuum. Thus one member of this set of ten numbers may specify the density of material there, another may specify momentum or stress in a particular direction, and so on. For this reason such a set of numbers is called the stress-energy-momentum tensor. Now Einstein obtained his field equations by equating each of the ten components of the space-time curvature to one of the ten components of the stress-energy-momentum tensor.

What is the justification for postulating the equivalence of the components of curvature and those of the stress-energy-momentum tensor? To understand the reason for this identification it is necessary to remark that any continuous medium such as a flowing stream or the material fog pervading our smoothed-out universe is in a state of perpetual flux. Beneath this ceaseless flux, moreover, lies an essential continuity, since any element of fluid stream conserves its mass no matter how it is carried away by the stream flow. That is why all fluid flows treated in hydrodynamics are subject to a condition of continuity which ensures this conservation or permanence of its material content. Its mathe-

---

* In imagination, of course.

matical formulation is the well-known equation of continuity in hydrodynamical theory.

Now if we consider our space-time manifold with its ever-changing metrical ground form, we can prove mathematically that an analogue of the hydrodynamical condition of continuity exists also for the components of space-time curvature. This is a way of saying that although the curvature components are in a state of flux, that is, vary from point to point of the space-time manifold, yet they do satisfy a condition which is the exact counterpart of the hydrodynamical condition of continuity or permanence of the *material* that flows. It therefore follows that in spite of all the shifts of our measuring nets whereby we discover the metrical ground form of our space-time, there is an element of *inherent permanence* in the geometrical construct we have created in the curvature components.

Now it happens that the components of the stress-energy-momentum tensor which, as we saw earlier, describe the material characteristics of our smoothed-out universe, also possess the same element of inherent permanence. That is, they too satisfy the counterpart of the hydrodynamical condition of continuity. Since components of space-time curvature behave in exactly the same way as those of the stress-energy-momentum tensor, it would seem plausible to identify the two. In fact, to some mathematical minds like Eddington's, the inherent permanence of space-time curvature is a sufficient argument in favour of its 'substantiality'. This enduring substantiality in turn is then taken to justify its identification with the corporeal features of the material content pervading the space-time framework. So complete is the belief in this that the adherents of this theory would not hesitate to shut up in lunatic asylums all those who dare deny it.

But unfortunately for the mathematical minds to whom the laws of nature are manifest by pure reason without appeal to experiment, the argument overreaches itself by the very ingenuity of its proponents. It happens that the derivation of such an invariant geometrical construct involves at one stage a mathematical process equivalent to that of integration. As a result there appears in the most general form of the invariant geometrical construct a constant (Λ), since named the cosmical constant, whose value is quite arbitrary. We thus acquire such a plethora of geometrical constructs possessing the same property of inherent permanence that one is rather hard put to it to recognize which one of them is to be identified with the material features of the universe embodied in the stress tensor. It is here that experiment and observation fill a fundamental lacuna. But as at the time Einstein was formulating his field equations there was no observational evidence to favour any particular value of the cosmical constant (Λ), he made the simplest assumption that it is zero.

Since Newton's grand synthesis of the dynamical and gravitational problems has been found so successful in correlating the observed motions of apples in orchards and planets in the sky, Einstein's field equations must preserve the hard core of Newton's system by proving themselves to be generalizations of the classical Newtonian equations if they are to provide a better substitute for them. This is indeed the case. To see it we first observe that in empty space all the components of the stress tensor are zero at each point. This is to be expected, for if there were no matter in any region there would be no density, momentum, stress or energy anywhere in that region. It can then be shown that the field equations in empty space are a simple extension of the classical Laplace equation which is merely a more recondite formulation of the Newtonian inverse square gravitational law in empty space.

There is also an analogue of Laplace's equation in space filled with matter. This is the well-known Poisson equation. If we assume that the fluid pervading the smoothed-out universe has a simple hydrostatic pressure (p) like the water pressure on a submerged submarine (which is the same in all directions instead of a variable stress), six of the band of ten numbers of the stress-energy-momentum tensor associated with each point of the space-time continuum vanish everywhere. We are thus left with only four out of the ten field equations. One of these four equations is an extension of the equation of continuity ensuring the conservation of the mass of any moving element of fluid. The remaining three are found to be generalizations of the classical hydrodynamical equations of fluid motion under pressure (p) in a gravitational field whose Newtonian equivalent can be computed.

Einstein's field equations thus provide the link between the geometrical properties of the space-time manifold and the corporeal characteristics of the material content of this space-time framework. In other words, they are an abstruse way of saying that the curvature of the space-time continuum in which material events occur is simply related to the distribution of matter therein. This is why it is possible in principle to discover a purely geometrical feature of our universe such as its spatial curvature by observing its material contents. For if (as the relativity theory assumes) the geometrical structure of the universe is conditioned by matter and if the distribution of matter on a sufficiently large scale is considered to be uniform so as to obtain a uniform geometrical structure, then our three-dimensional astronomical space must be homogeneous with the same spatial curvature everywhere.

A natural corollary of these assumptions is that the number N of galaxies within a sphere of radius r must be proportional to the volume V of the enclosing sphere. We have then only to examine the dependence of the number N as observed in a sufficiently powerful telescope on the distance r to determine the dependence of V on r. If N (and therefore V) varies directly as r^3, it means that the volume V of a sphere of radius r is proportional to r^3 as in ordinary high-school geometry. Our space is then Euclidean with zero curvature. But if N increases faster (or slower) than r^3, it implies that the volume V of the sphere of radius r is now no longer proportional to r^3. The measurement of the deviation of the increase in V from the Euclidean value, namely that of strict proportionality to r^3, enables us to compute the curvature. In the former case our space is spherical while in the latter it is hyperbolic. In other words, space curvature leaves its fingerprint on the pattern of galactic distribution in depth over the sky.

To identify this curvature we have merely to count the number N of galaxies brighter than a specified limit m of brightness or rather faintness. Since the limiting dimness m is also a measure of the distance r up to which the faintest galaxy included in the count has been seen, we have all the elements required for empirically determining the volume V of a sphere in our actual space as a function of its radius r. Now we can calculate by pure theory the volume V of a sphere in terms of its radius in each of the three kinds of uniform spaces described earlier. A comparison of the empirical dependence of V on r with its theoretical counterpart therefore enables us to find both the magnitude and sign of spatial curvature.

In actual practice, it is no simple affair to determine curvature in this way, as its effect on the distribution of galaxies gets entangled with a variety of other effects. But it would take us too far away from our present theme to go into the question of unravelling the tangle.* It will suffice for the present to remark that the coupling of the geometric features of the space-time continuum with the material distribution in the universe in the manner envisaged by relativity theory is a sophisticated way of combining both Laplace's and Poisson's equations with the conservation laws of classical dynamics into a single set of comprehensive equations.

They therefore naturally lead to results (as they must) in agreement with Newtonian theory except for a few deviations. These deviations are in fact the sole *raison d'être* of Einstein's relativity theory. For they enabled Einstein not only to explain a number of puzzling phenomena such as the planetary motion of Mercury which had hitherto defied Newton's theory but even to predict some new ones. This inspired confidence in the theory. So he proceeded to build a cosmological model of the universe as a whole on the basis of these ideas. But here Einstein encountered a difficulty. In dealing with planetary motions he could neglect other gravitating masses and confine his attention to a single dominating mass like the sun. But in an encounter with the universe as a whole he had to substitute for the problem of evaluating the space-time curvature of the actual universe the simpler one of computing that of the smoothed-out universe

---

* A review by Allan Sandage seems to show that there is no hope of deciding among the various possibilities on the basis of galaxy counts alone because the predicted differences among the various models are too small compared with the known fluctuations of the distribution even at the limit of the 200-inch Mount Palomar telescope.

postulated above. In addition he also assumed that the universe is static, that is, remains in one condition forever.

Unfortunately the latter assumption made things more difficult. Einstein found that his field equations would yield no solution that would permit a static cosmos. He was thus obliged to introduce the cosmological constant (Λ) that he had earlier ignored. Physically the insertion of a non-zero Λ in the field equations implied the existence of a new kind of force of cosmic repulsion at work among the galaxies. As we shall see later, it was a curious kind of force but its incorporation in the field equations enabled them to do the job they were required to do. The amended equations did yield two static homogeneous models to which the universe could conform – the Einstein model and the de Sitter model. But both models proved unsatisfactory and all attempts to fit them to the actual universe proved futile. The reason was that they were *limiting* solutions of the relativity equations rather than descriptions of the actual state of our universe. For in the former the universe contained as much matter as it possibly could without bursting the relativity equations, while in the latter it was completely empty and permitted the existence of neither matter nor radiation.

The difficulty of obtaining a feasible solution of relativity equations was shown by A. Friedmann to be due to the tacit assumption made by Einstein that the model universe must be *static*. But if we relax the restriction and make the more realistic assumption that the universe is *not* static, though uniform all over, we find that the hypothesis of cosmic repulsion introduced *ad hoc* by Einstein into his equations is no longer necessary for extracting solutions out of them. Indeed Friedmann deduced two non-static models from Einstein's original equations. Quickly recognizing the importance of Friedmann's work, Einstein admitted his earlier introduction of cosmic repulsion as the 'worst blunder of his life'. And no wonder, for cosmic repulsion is indeed a strange species of force. It is so cosmically egalitarian that it makes no distinction between an atom and a galaxy, being the same for them all – an atom, apple, asteroid, Antares, or Andromeda. In other words, cosmic repulsion experienced by a mass from any point O in the universe is quite indifferent to what happens to be at O. It is the same whether O is occupied by a big mass or a tiny particle or for that matter nothing whatever. Nor is this all: it increases directly as the distance between the inter-acting bodies as no other force in physics does. Nevertheless, many cosmologists like Eddington, Weyl, Schrodinger and Lemaître considered these vagaries of cosmic repulsion no bar to its eligibility as a respectable member of the comity of physical concepts like that of gravitation itself.

In his *The Expanding Universe* Eddington, for example, wrote:

Einstein's original reason for introducing cosmical repulsion was not very convincing, and for some years the cosmical term was looked on as a fancy addition rather than an integral part of the theory. Einstein has been as severe a critic of his own suggestion as anyone, and he has not invariably adhered to it. But the cosmical constant has now a secure position owing to a great advance made by Prof. Weyl, in whose theory it plays an essential part. Not only does it unify the gravitational and electromagnetic fields, but it renders the theory of gravitation and its relation to space-time measurement so much more illuminating, and indeed self-evident, that a return to the earlier view is unthinkable. I would as soon think of reverting to Newton's theory as of dropping the cosmical constant.

While the supposed unification of gravitational and electro-magnetic fields appeared in retrospect to be an illusion, cosmic repulsion is still retained by some cosmologists for reasons to be explained later.* We will, however, dispense with it for the pre-sent and describe the main features of relativistic cosmological models based on Einstein's original field equations, especially as there exist very good reasons – such as the peculiar behaviour of cosmic repulsion *vis-à-vis* other kinds of physical forces or, more subtly, an appeal to Mach's principle – for rejecting it completely.

The most interesting feature of such non-static or evolving models is, of course, their evolution in time or their version of the history of our cosmos as a whole. To be sure it is a heavily bull-dozed sort of history in which all details and nuances are ruth-lessly smothered. Nevertheless, it has still to render some account, no matter how bald and summary, of the past and future of at least a few of the very large-scale attributes of the universe such as the positions and motions of the clusters of galaxies, spatial density of matter, etc. The question then arises *whose* future and past the model is meant to describe, considering that according to Einstein every moving observer has his own system of time reckoning. Einstein's answer is that cosmic history is to be described with reference to a universal cosmic time, which, as it happens, is the same for all *galactic* observers. These are observers who are at rest relative to the average motion in their own galaxy because problems with time arise in cosmology only if the observer is moving very fast with respect to his own galaxy. Such galactic observers are not entirely fictitious. We ourselves qualify as one because the earth's motion happens to be not very far off the average for our galaxy. How will these galactic observers who keep the same cosmic time view the universe of expanding galaxies that Hubble discovered?

If we adopt the usual common-sense notion of absolute space that Einstein discarded, we may view the expansion of the material universe *into* outer empty space like the diffusion of a gas into a surrounding vacuum. But if we stick to the 'relational' theory of space adopted by Einstein there is nothing – not even empty space – outside the universe. Its expansion is simply a change in the scale relationship of the universe as a whole to the linear dimensions of its typical constituents, for example, the diameter of a typical atom or the radius of electron or proton. In other words, in the former case there is motion *in* space while in the latter there is motion *of* space. If we adopt the former view, as Milne did, our cosmological model would consist of a fundamental system of receding galaxies (assumed to be particles for mathematical convenience) in uniform relative motion from an initial state of superdense concentration at the epoch of 'creation' at cosmic time $t = 0$. Although Milne's* view attracted a good deal of attention about two decades ago, it is largely neglected now because of its inability to support an adequate theory of gravitation. We will therefore confine ourselves here to the latter point of view that the expansion of galaxies is really the expansion *of* space.

Consider two distant galaxies A and B and suppose that their

* For an exposition of Milne's theory see Chapter 8.

distance at a certain moment of cosmic time, say 'now', is one unit. Such a unit could be any large distance such as 100 million light years. Before this now-moment the distance AB will be less than one and thereafter it will be greater. In other words, the scale factor R for the distance between AB will be an increasing function of *only* cosmic time, t, with the proviso that R is unity when t is 'now'. But there are galaxies and galaxies and suppose we selected some other pair C, D, some different distance apart. No matter, if again we assumed CD to be unity at the present (cosmic) time, we would find that although the new distance CD is different, the *scale factor* R is the *same* function of cosmic time t in both cases. It is equal to, less than or greater than unity according as t is now, earlier or later than now. In other words, by assuming the distance between any given pair of galaxies as unity we can derive a scale function, R, of only cosmic time t, which is the same for all pairs of galaxies or clusters of galaxies in the universe. The scale function R will, no doubt, be different for *different* cosmological models. But for any given model it will be the same function of cosmic time t.

To appreciate the meaning of the scale function R better, consider the two-dimensional analogy of an expanding rubber balloon having a large number of dots marked all over its surface. In this illustration dots are the analogues of galaxies and the two dimensional spherical *surface* of the expanding balloon the counterpart of expanding space. It is not the dots that are moving but the geometrical space in which they are embedded. This is why each dot considers itself as the centre from which every other dot recedes (see Figure 23). In stage (b) which corresponds to cosmic 'now', the distance between any two dots (galaxies) A and B is clearly $r\theta$, where r is the present radius of the balloon and θ the angular separation between A and B, that is, the angle AOB. At any time t in the past or future, that is, the stage $t <$ or $>$ 'now' depicted in Figures 23(a) and (c), the distance AB will be $r'\theta$ so that the scale function or the ratio of the distance between A and B 'now' and at any time t is

$$\frac{r'\theta}{r\theta} = \frac{r'}{r}$$

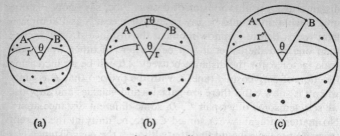

Figure 23. (b) is the state of the expanding balloon at the cosmic epoch 'now', whereas (c) and (a) are respectively its states at cosmic times greater and less than now.

In other words, the value of the scale function $R = \dfrac{r'}{r}$ is *independent* of which particular pair of dots (galaxies) we pick. It depends solely on the radius of the balloon at time t with the proviso that its value is unity at the epoch 'now'. Obviously the epoch 'now' may be located arbitrarily anywhere. But once located it will fix the scale function R uniquely. It is the same with our expanding three-dimensional space in which galaxies of the universe are embedded like the dots on the balloon except that the dots themselves are *not* supposed to share the expansion of the balloon. They retain the same size even as the balloon continues to expand.

Applying Einstein's original relativity equations to the universe as a whole, we can calculate the scale factor R as a function of cosmic time t. It happens that the precise functional relation between R and t depends on the nature of space. Since there are only three possibilities according as we believe our space to be Euclidean, open hyperbolic, or closed spherical, there will be three different scale functions. Figure 24 depicts three curves marked A, B and C which are the respective time-graphs of the scale function R according as space is Euclidean, hyperbolic or spherical. It will be observed that in all three cases the scale function R is zero at the initial epoch $t = 0$. Obviously when the scale function shrinks to zero, all galaxies separated at present by finite distances must come together. It therefore follows that in all three cases the universe started from an initial state of hyperdense quasi-infinite concentration in which all its material was squeezed within the eye of a needle. Now you may wonder whether such a state of origin is any more comprehensible than the 'darkness on the face of the deep' of which the Bible speaks.

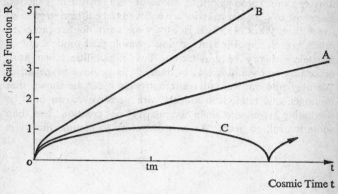

Figure 24.

That is why it is called 'singular', which merely means that it is peculiar or queer. While the existence of 'singularity' is an embarrassment for the theory, recent discovery of extremely weak microwave radiation seems to suggest that it may well have occurred. We will dwell on it later. Meanwhile we continue to explore the behaviour of our model universe.

Starting from such a singular state of zero or quasi-zero radius and infinite density it expands continually forever if space is Euclidean or hyperbolic (curves A and B). By a skilful blend of theory and observation it can be shown that the singular state must have occurred some seven and ten billion years ago accord-ing as space is Euclidean or hyperbolic. But if space is closed spherical, the scale factor oscillates between zero and a maximum as shown by the curve C. The universe therefore expands till R reaches a maximum. On reaching this maximum the expansion reverses and it starts contracting. The contraction thus initiated then continues until expansion begins again on reaching the initial state and so on *ad infinitum* with each cycle of expansion

153

following contraction taking some sixty billion years. If the universe is oscillating, the singularity or 'creation' must have occurred some six billion years ago.

In all cases the models have the handicap of making the universe much younger than some of its constituent galaxies. For some of the older galaxies are believed to be fifteen to twenty-five billion years old. This is why some cosmologists prefer to retain cosmic repulsion in carving cosmological models out of Einstein's theory of gravitation. Two possibilities then arise according as cosmological constant, Λ, is positive or negative. We may rule out the latter alternative as it can be shown that homogeneous relativistic models with negative Λ can only be oscillating types with their 'singularity' or 'creation' less than some ten billion years. We have therefore to assume that Λ is

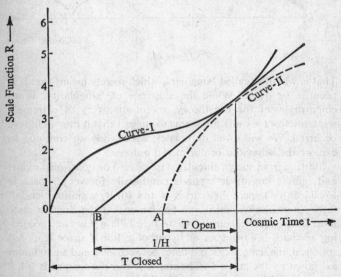

Idealized Expansion Curves for Open and Closed Models

Figure 25. Idealized expansion curves for open and closed models. (From Guy C. Omer and James P. Vanyo, 'Closed Models Universes', *Astrophysical Journal*, August 1966.)

positive. Homogeneous relativistic models with positive cosmological constants have a greater range of possible solutions. They include not only the oscillating models but also both types of exploding models. Figure 25 shows the graphs of the scale function R against cosmic time t for two cases according as our space is closed spherical, curve I, or open hyperbolic, curve II. It will be seen that for open models in hyperbolic space the model starts from a singular state with an infinite rate of expansion which continually decreases. But they all suffer from the handicap that their age cannot be more than Hubble's parameter $\left(\dfrac{1}{H}\right)$, that is, only some ten or at most thirteen billion years because their shape must conform to curve type II. Thus the origin of cosmic time must begin either at A or at most at B. The closed models are, however, possible with a shape similar to type I curve in Figure 25. Such models can easily have their origin well beyond B so that they may have ages considerably longer than $\dfrac{1}{H}$. In a recent study of such closed models, Guy C. Omer and James P. Vanyo have demonstrated the value of another degree of freedom provided by utilizing cosmological constant Λ. They show that closed cosmological models with putative ages from fifteen to twenty-five billion years are easily achieved by an appropriate choice of cosmological constant Λ conforming pretty well to the observed density of $\sim 10^{-30}$ gm./cm.^{3}.

However, ignoring the cosmological constant, Einstein's original field equations permit only two kinds of models: either oscillating or expanding. As we saw, an oscillating model can occur only in closed Riemannian or spherical space of positive curvature. But if, on the other hand, space is open, whether Euclidean or hyperbolic, the model must expand. If therefore we can decide by observation whether our actual universe is expanding or oscillating, we automatically come to know the character of our actual space. Now a decision between expansion and oscillation depends on whether or not the law of linear increase of recessional velocity with increasing distance continues to hold indefinitely, or at any rate to the very threshold of our barely visible horizon.

Figure 26 shows the recessional velocity of eighteen known clusters of galaxies plotted against their respective distances. If the velocity increase continues to follow Hubble's law, the velocities of these clusters should lie on the line C. If, on the other hand, the velocity line follows the curve between C and B, our

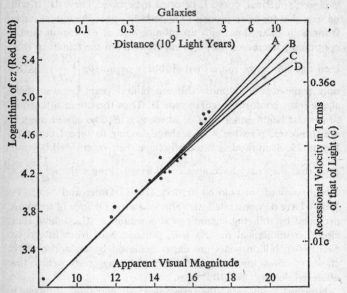

Figure 26. Velocity/distance chart of eighteen clusters of galaxies. (From Allan Sandage, 'The Red Shift', *Scientific American*, September 1956. Copyright © 1956 by Scientific American, Inc. All rights reserved.)

universe is expanding and space is open and infinite. Curve B would arise in the case of flat Euclidean space. If it lies to the left of B, it is oscillating and our space is closed and finite, the radius of curvature decreasing as we move to the left.

According to our present observations (still crude and inconclusive), our universe is oscillating with closed curved space, for line A is the trend followed by the six faintest clusters. Further, since the rate at which the expansion is slowing down (if it does) depends on the density of matter in the universe, it is possible to calculate the density from the observed deceleration of recession. Indeed, relativity theory provides a neat tie-up between the deceleration parameter (q) and the density of matter (ρ) in the universe in models with zero cosmological constant (Λ). It is given by the simple equation

$$q = 2 \cdot 7 \times 10^{28} \rho.$$

Since according to the most recent observation of Allan Sandage q is $1 \cdot 8 \pm 0 \cdot 7$, we find that ρ is $\dfrac{1 \cdot 8}{2 \cdot 7} 10^{-28}$ or $6 \cdot 7 \times 10^{-29}$ gm./cm.³.

Now the mean density of matter in the galaxies if diffused over all space comes to 7×10^{-31} gm./cm.³. This means that the total mass of the universe is 100 times the total mass in all the galaxies. If the observed deceleration of recessional velocities is confirmed, we should either have to scrap our existing estimates of galactic masses or conclude that that vacant immensity, the intergalactic void, is not so void after all and contains large quantities of undetected matter, possibly neutral hydrogen, which if present could have escaped detection but may in the future be detected by newer techniques of observation.

While our present observations seem to favour an oscillating model for our universe, another line of argument leads to the opposite conclusion, that our universe is an expanding universe with an infinite flat Euclidean or curved hyperbolic space. The reason is that the confirmed velocities of recession of the galaxies are about seven times the velocity of mutual escape. What it means is this: every gravitating system has a sort of Rubicon surrounding it. If the particle leaves with a velocity initially fast enough to carry it to this Rubicon in spite of the deceleration induced by the gravity of the system, it will never turn back but will escape from the system altogether.

For example, the escape velocity from the earth is seven miles per second, which means that any particle such as a bullet, if fired with an initial velocity exceeding this value, would never return. But if, on the other hand, it left with a velocity less than seven miles per second, it would fail to reach the bank of its Rubicon and fall back to earth after climbing a certain distance. In the same way it is possible from our present knowledge of the distribution of matter in the galaxies to calculate the velocity of mutual escape of galaxies from each other's gravitational field. Since the galaxies are moving with velocities seven times the critical velocity of escape, they are never likely to condense into coincidence again.

But this line of argument too has its weak point. For the calculation of the escape velocity is based on the assumption that most of the mass of the universe is concentrated in galaxies. If the intergalactic void contained matter whose total mass was more than seven times that in the galaxies, the corresponding velocities of mutual escape would increase sevenfold. Our provisional finding of the slowing down of recession suggests that the intergalactic void may contain 100 times as much material as in all the galaxies. If this is indeed so, we shall have to decide that the universe corresponds to the oscillating model. So far, however, there has been no evidence to show that there can be matter of this magnitude in intergalactic space (although it might evade detection if it were in the form of pure hydrogen unalloyed by other gases or dust).

If the universe should ultimately turn out to be an oscillating type, we should soon be with Sisyphus, rolling the stone up the rock. But should it be an expanding universe, we are doomed to an eternity of solitude. For if the linear increase in recessional velocity continues at the rate of 19 miles per second per million light years of distance, it is obvious that a galaxy at a distance of $186,000/19 \simeq 10,000$ million light years would be receding from us, or, in other words, we should be receding from it at 186,000 miles per second, the velocity of light. Any ray of light emitted by the galaxy would begin to chase us with the same velocity as that with which we are running away from it. It would thus be like a race between Achilles and the tortoise, in which the tortoise for once runs as fast as Achilles and therefore can never be overtaken.

If no ray of light, furthermore, emitted by it could ever reach us,* it would never be seen by us. Even though space might have

* You may find this puzzling in view of the statement previously made that the velocity of light remains constant with respect to any moving observer. This is true only in the special theory of relativity (which refers to a particularly simple system of space and time), but not in Einstein's general theory of relativity. The point is too technical to be elaborated further here.

an infinite number of galaxies in the infinite recesses of its depth, we could only see a few, namely, those within a radius of 10,000 million light years from us.

Now the mere fact of recession weakens the intensity of light which a galaxy radiates to us.* The total light we actually receive from it is less than what we would obtain were it to remain at rest at the same distance. The extent of this reduction can be calculated by pure theory. For instance, at a distance of 1,000 million light years the factor of reduction is about two-thirds, so that a receding galaxy at this distance appears only two-thirds as bright as it would if it were at rest at the same distance. As the distance and consequently the velocity of recession increases, this factor decreases, becoming zero at the critical distance of 10,000 million light years where, as we saw, the galaxies begin to recede with the velocity of light.

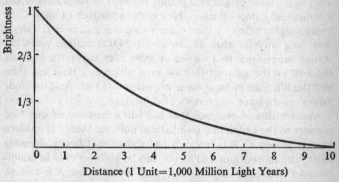

Figure 27. The curve shows the ratio of the brightness of a receding galaxy observed to that it would show if it were at rest at the same momentary distance. At the speed of light the ratio is zero so that a galaxy receding at this speed would be invisible.

We thus observe that every receding galaxy has its luminosity (as perceived by us) diminished more and more by the recession factor quite apart from that of distance until it eventually becomes invisible at the critical distance of 10,000 million light years (see Figure 27).

* See also Appendix.

The recession of galaxies therefore is the switch-off process which not only dims the galaxies all the way up to the afore-mentioned critical distance but also puts out all those lying beyond it. It thus deprives us of a magnificent celestial sight – an immense blaze of light spread all over the sky in the manner prophesied by Olbers if infinite space contained an infinite number of galaxies. But the receding galaxies do more than merely resolve Olbers's paradox for us. They will eventually rob us even of the sight of heavens that we still do enjoy. For since all the galaxies presently visible are also receding from us, one day they too would cross Olbers's threshold and thus pass out of our view.

It seems that not all their beauty or lustre can avert the inevi-table hour, for the paths of galaxies lead but to the cosmic grave – the Olbers's sink beyond some nine to ten billion light years from us. The universe of galaxies is thus doomed to be gradually but systematically impoverished. Nor is the approach of such Cim-merian nights when we shall cease to see any galaxy in the sky a very long way off after all. In about 20,000 million years the cosmic broomstick that we see at work may sweep the heavens clean of all the galaxies that we now observe, a time less than half the life span of an average population II star. And then our galaxy would have an eternity of solitude.

Another line of evidence that has lent a measure of qualified support to the expanding (oscillating) universe theory is modern nuclear research. This research has shown that heavy elements like uranium, thorium and so on, can be built up only in condi-tions of extremely high density and temperature. We saw in Chapter 3 how the nuclei of atoms of carbon, lithium, calcium, silicon, iron and the like, are synthesized in the hot interiors of the stars. We also saw how the process of nuclear consolidation without the use of neutrons could not proceed beyond the generation of the iron group of atoms for the reason that any such consolidation *absorbs* energy instead of releasing it and therefore can no longer serve as fuel to continue the chain of fusions. The build-up of heavier elements without calling upon neutrons therefore was at one time thought to require far more Draconian conditions than the stellar interiors provide.

Accordingly it was suggested that the point-source models of relativity theory offer a possibility of a place and time at which such extreme conditions could have occurred. If the universe did originate from the fragmentation of a primeval atom in which all the material of the cosmos was originally condensed, it must have been like the explosion of an infinitely magnified H-bomb. Lemaître has picturesquely described this cosmic conflagration in the following words: 'The evolution of the world can be compared to a display of fireworks that has just ended: some few red wisps, ashes and smoke. Standing on a well-chilled cinder, we see the slow fading of the suns, and we try to recall the vanished brilliance of the origin of the worlds.' Can we construct a more detailed picture of this 'vanished brilliance' by an examination of the cinder, ashes and smoke very much as archaeologists do by digging into the debris of lost civilizations?

It now seems that we can, thanks to the recent discovery of extremely weak microwave radiation that is believed to be the tell-tale residue of the cosmic conflagration of some ten billion years ago. For R. Dicke and his colleagues have shown that radiation emitted a few minutes after the initial explosion, when the temperature of the universe was about ten billion degrees, should still be detectable as a uniformly distributed cosmic background despite the lapse of ten billion years since its emission. Because of the continuing expansion the wavelength would have increased in proportion to the scale function and the effective temperature of the radiation correspondingly decreased. They predicted that at the present epoch it would appear as extremely weak microwave radiation with an intensity equal to that in a cavity kept at a temperature only a few degrees above the absolute zero. The prediction was verified by A. A. Penzias and R. W. Wilson, who using a highly sensitive radio receiver at a wavelength of 7 cm. for another purpose found quite incidentally some weak radiation they could not account for. Although it seemed possible that this was indeed the isotropic background radiation expected from the big bang of 'creation', further observations were necessary to see whether the intensity at other wavelengths is consistent with such an explanation. These have now been carried out at Princeton and at Cambridge. They do show that the spectrum is that of 'cavity' radiation with an effective temperature of three degrees absolute at 3·2 cm. Apparently we have been able to observe directly electromagnetic radiation generated in the initial cosmic conflagration. It is therefore no wonder that Gamow's cosmology which in trying to piece together the evolution of the universe predicted such radiation should have received an encouraging boost.

Taking his cue from Lemaître, Gamow had suggested that the zero state of our universe was a giant primeval atom which exploded some ten billion years ago. This explosion was perhaps in turn the result of a recoil from a state of prior cosmic collapse. For if the relativistic formulae embodying the present terrific expansion of the universe were continued beyond the commencement of the primeval explosion, it would appear that just prior to the state of maximum condensation the galaxies must have been collapsing as fast as they are now receding. In other words, in the infinite past long before its present 'beginning', or rather rebirth, the universe was in a state of cosmic collapse from a state of maximum rarefaction till it reached that of maximum condensation.

In this state of hyper-dense concentration at the first dawn of its creation or re-creation, the universe was a blaze of radiation, a veritable noontide of light and lustre that could only be equalled by the cumulative splendour of exploding galaxies of supernova stars, if that. But such a cosmic flood of fire soon spent itself. Taking its absolute temperature T as a measure of its fury, Gamow showed that T in degrees absolute would fall rapidly with time t reckoned in seconds according to the formula:

$$T = \frac{15 \times 10^9}{\sqrt{t}} \tag{1}$$

Consequently T tumbled from about 500 billion degrees a millisecond after the explosion to a mere billion degrees five minutes later.* Within a day it further dropped to forty million degrees

* The corresponding period according to Dicke's calculations is much briefer. He estimates that the temperature would have fallen by more than nine billion degrees absolute barely three minutes fifty seconds after the initial explosion. At that stage the universe would have had a radius of only forty light years compared to the present ten billion light years.

but took 300,000 years to fall to 6,000 degrees and ten million years to cool down to room temperature. It would thus appear that for the first few moments of its existence after the primeval explosion – barely thirty minutes according to Gamow's calculation – the universe was at a temperature sufficiently high to spark nuclear reactions. It was during these thirty minutes after creation that *light* elements were created as a result of nuclear reactions in the primeval *ylem*, that is, a mixture of protons, neutrons, electrons and above all high energy photons or light quanta. The reason is that at the very early stage of the universe and at the high ambient temperature matter must have been completely dissociated into elementary particles, being a mixture of free electrons, protons and neutrons tossed about by high energy light quanta. Under these conditions neutrons in the ylem would quickly begin to decay into protons, emitting electrons as they did so. Each proton promptly captured a neutron, the pair forming a deutron, a species of hydrogen or hydrogen 2. Some deutrons then captured another neutron becoming what is called hydrogen 3. This nucleus soon decays by emitting an electron and is thus transmuted into helium 3. In this way, by a rapid succession of neutron captures and decays, helium could be built up in the first burst of the universe's expansion. But the process could not be carried beyond to account for the synthesis of elements heavier than helium. Because of this failure Gamow's theory was derided as a 'wonderful way to build the elements all the way up to helium'! But with the accretion of fresh support from the discovery of microwave radiation its recently updated version neatly avoids the earlier difficulty in explaining the build-up of heavier elements.

The difficulty arises because of two gaps in the sequence of atomic weights at numbers 5 and 8. That is, there is no stable atom of mass 5 or of mass 8. We can produce an atom of mass 5, helium 5, in the laboratory by bombarding helium 4 with neutrons, but it immediately breaks down to helium 4 again. Likewise, we can produce momentarily an isotope of beryllium of mass 8, but it too instantly breaks down by fission into two helium 4 atoms. The question then is: how can the build-up of elements by neutron capture get past these gaps? The process could not go beyond helium 4, and even if it did somehow span the gap, it would be stopped again at mass 8. In short, if neutron capture were the only process by which elements could be built starting with hydrogen, the build-up would go no further than helium.

Gamow overcomes the difficulty by pointing out that recent progress in astronomical knowledge has shown that the earlier assumption of the chemical homogeneity of the universe is not quite correct. It was formerly believed that the composition of cosmic matter is more or less the same everywhere – on the surface of the earth, in stars, meteorites, interstellar space and the distant galaxies. This is not to say that there are no local deviations. For instance, hydrogen is a thousand times more abundant in the sun than here on earth. But such departures from the norm could be reasonably attributed to subsequent transformations of cosmic matter under the conditions of each region. Thus the earth's low gravity allowed all the hydrogen with which it began its life to escape, whereas the sun because of its much stronger gravity has been able to retain it all. If we ignore such local differences, the composition pattern of cosmic matter, it was believed, is remarkably uniform. Gamow points out that this belief was an extrapolation from 'parochial' observations made in our own neighbourhood to the universe as a whole. Apparently it has not turned out to be valid. Thus, as we saw in Chapter 3, while population I stars contain comparatively large amounts (up to one per cent) of the heavier elements, the amounts of these elements in the stars of population II are at least a hundred times smaller. It is therefore generally agreed that the stars of population II represent the original primeval material while those of population I have condensed out of latter-day materials impregnated with heavier elements synthesized in earlier stars. Consequently Gamow suggests that the observed abundances of elements are the result of two distinct processes – one cosmological and the other astrophysical. In the *cosmological* process which lasted some thirty minutes after creation, hydrogen and helium were produced in their presently observed abundances. The elements heavier than helium were produced later gradually for the most part by nucleosynthesis* in stars.

* For a fuller account see Chapter 15.

164

Recent more refined calculations based on better knowledge of nuclear reactions likely to arise at temperatures and matter densities prevailing during the first half hour of the universe's existence show that for a fairly wide range of densities of matter helium production remains more or less constant at about 27 to 30 per cent. Although the revised calculations have brought Gamow's theory into much better accord with observation, even this newly computed percentage of 27 to 30 is uncomfortably large since the observed helium abundance in the universe, around 26 per cent by mass, should virtually be an upper limit. For it is not possible for the initial helium abundance to be reduced, because our present knowledge of astrophysics and nucleogenesis suggests that very much more helium is likely to be produced astrophysically than destroyed. Luckily for the theory, the discomfort due to the predicted overabundance of helium is not irremediable. S. W. Hawking and R. J. Taylor have shown that helium production can be substantially reduced by a controlled departure from the presumed isotropy of the universe, which lasts briefly for only the first few years after its origin long before the appearance of galaxies that seem to be distributed isotropically now. Helium production can thus be brought into better conformity with actual observation while at the same time retaining the present large-scale isotropy of our universe.

While cosmic hydrogen and helium was thus forged within the first few minutes of the creation's birth, it remained a prisoner at the mercy of thermal radiation for a long while. As already noted, it took 300,000 years for the temperature of the universe to drop to 6,000 degrees, the present temperature at the surface of the sun. Although at this date the fury of the primeval flood of fire that gave primeval matter its birth had abated more than a hundred millionfold, the temperature of thermal radiation was still high enough to require attention to a feature that is of no consequence under ordinary conditions. This feature is the ponderability of the radiant energy of light quanta implicit in Einstein's mass-energy relation $E = mc^2$ quoted on page 44. It follows that radiant energy E possesses a ponderable or gravitating mass m numerically equal to E/c^2 where c as usual is the velocity of light.

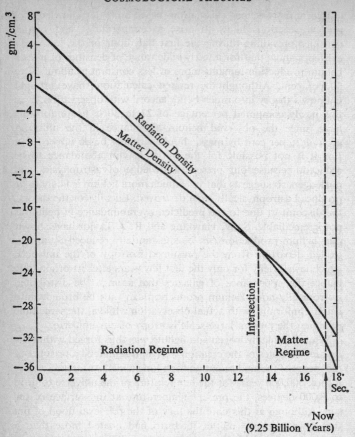

Figure 28. (From George Gamow, 'History of the Universe', *Science*, Vol. 158, 10 November 1967, pp. 766–9. Copyright © 1967 by the American Association for the Advancement of Science.)

Under ordinary conditions of our daily life, the 'weight' or pressure of light is negligible. Thus the total weight of light quanta traversing one cubic kilometre of atmospheric air at ordinary temperature such as may obtain on a bright sunny day is only one hundred millionth of a gramme. But as we mentioned on page 41 it increases rapidly with temperature (T), being proportional to its fourth power (T^4). Consequently at very high temperatures such as prevail in stellar interiors or in the case of an exploding atomic bomb, the weight of light quanta (which in this case are high frequency gamma rays) may be as high as one gramme per litre. Since the temperature T of the universe in the beginning was even higher in accordance with formula (1) quoted earlier, the density of radiation at that time must have been much greater than it is now. Gamow's calculation of the changing densities of radiation and matter with time is shown graphically in Figure 28. The ordinate represents densities of radiation and matter in gm./cm.^{3} while the abscissa represents the time in seconds since the primeval explosion, both on the logarithmic scale for the sake of convenience of representation. As will be observed from Figure 28, about one million years after the initial big bang the universe cooled off sufficiently to make the densities of radiation and matter equal. In other words, while radiation prevailed over matter during the first million years, about 0·01 per cent of the history of the cosmos, matter has prevailed ever since. It is only after the transition from the regime of radiation to that of matter during the remaining 99·99 per cent of cosmic history that the evolution of galaxies began to occur. For at this critical time one million years after creation the densities of radiation and matter became approximately equal to one another with the temperature of space falling to about 3,000°K. and the density of matter (or radiation) to $5 \cdot 10^{-22}$ gm./cm.^{3}, which is higher than the present density of matter in the universe ($7 \cdot 10^{-31}$ gm./cm.^{3}) by a factor of about 10^9. At this stage further expansion made matter gravitationally more important than radiant energy and gave rise to the first step in the differentiation of the originally homogeneous gas. In other words, the universe now became cool and dark enough to let matter show the innate property of gravitation that had lain waiting within it until the cosmic fires should have chilled. This transition from the reign of radiation to that of matter led to the formation of giant gaseous clouds from which the galaxies and their constituent stars evolved by gravitational break-up in the manner described in Chapter 5.

CHAPTER 8

A Flight of Rationalist Fancy

THE great danger that threatens the validity of the cosmological models of the last chapter is the possibility that the relativity theory on which they are based might not be applicable to the universe as a whole. To guard against this danger E. A. Milne has suggested recourse to philosophy in an endeavour to evolve a cast-iron system of physical laws that have the logical necessity of a geometrical theorem. Milne claims that if such a system of iron-clad laws is to be discovered in any science, it must become less and less empirical and more and more deductive. In other words, it must follow the trail geometry took when the Greek rationalists transformed it from a mere medley of empirical rules of Egyptian land surveyors into a deductive science of great beauty. Milne admits that so far it has not been possible to make any science other than geometry fully deductive but considers that the stage is now ripe for the transformation of the empirically derived dynamical laws like Newton's laws of motion and gravitation into theorems rationally deducible from a few fundamental world axioms. If the truth of these axioms were granted, that of the laws deduced from them would automatically follow. In this way Milne proposes to rid dynamical laws of their dependence on observational verification.

Now what are these world axioms which Milne assumes to be true *a priori* and adopts as a springboard for his deductive scheme? His first principle is what has since come to be known as the cosmological principle. It is in essence the Copernican principle extended to the scale of the cosmos in that it denies a central or privileged position to any observer in the universe. This means in effect that the large-scale aspect that the universe presents to any observer is the same whether he observes it from the earth, Andromeda, Magellanic Clouds, Virgo, Boötes, Hydra or anywhere else. Milne admits that it is very unlikely that the actual universe is such that its contents are described in the same way from every space location taken as an observing point. But his object is to construct a 'science of laws of nature in an ideal universe in which the various galactic nuclei or *fundamental particles* provide identical descriptions of its contents'. Once such a science has been created, it is relatively easy to 'proceed to a more realistic state of affairs, if we want to, by embroidering perturbations or variations on the ideal universe'. Taking his cue from Hubble's observation of galactic recession referred to earlier, Milne further postulates that this ideal universe consists of a swarm of particles (galaxies) which started receding from one another at some given time with uniform velocities. Milne calls this idealized system of mutually separating particles the *substratum*.

He next takes up the question of time. Leaping clear of the desperate conundrums with which philosophers from St Augustine and Kant to Robb and Dunne have become entangled, he simply assumes that our daily perception of events around us, not to speak of the physiology of our bodies, makes us aware of what we call the passage of time. This means that every one of us can arrange in an orderly sequence the events that we perceive. In other words, we can tell which of two events occurred earlier and which later. It is true that most events that we observe, such as a lunar eclipse or supernova explosion, have duration, so that the possibility that some of them will partly overlap cannot be excluded. But we can overcome this difficulty in a purely axiomatic theory by abstracting from this state of affairs the concept of point-event in the observer's consciousness.

Such a point-event is an event that takes place instantaneously, that is, at an instant of time. It is therefore an event of zero duration exactly as a geometrical point is a length of zero magnitude. We can then say of any two point-events that one of them must precede, follow or coincide with the other. But as we saw earlier, Einstein had already shown that this temporal order in which an observer may place events (or point-events) in his consciousness is not unique for every observer in the universe. It is possible for two events A and B to be simultaneous according to my reckoning, whereas one may precede the other when observed by Micromégas from his native star Sirius.

Milne accepts this result but goes on to add that there is no natural uniform scale of time measurement either. The idea of a uniform scale of measurement arose from the measurement of lengths. It is quite inapplicable to the measurement of time. The reason is this: If we want to measure a length, we have first to fix a standard such as a standard metre or a yard. The act of measurement then consists of superposing the standard metre or yard alongside the length to be measured and observing how many times the standard metre or yard 'goes' into the measured length. Thus the process of length measurement depends on the possibility of superposing one length over another so that the two ends of the superposed lengths coincide, or, in other words, on the possibility of producing equivalent lengths.

Now Milne has pointed out that there is no standard duration of time by which we can measure the lengths of various time intervals. To say that the period of time of the earth's diurnal rotation or of a pendulum swing is *uniform* and therefore can serve as standard for measuring durations is to beg the question. For obviously we cannot as it were, 'freeze' the period of a pendulum swing that is just finished and put it alongside another that it is about to execute and see if the two durations are really coincident. Nevertheless, in spite of the impossibility of placing one duration alongside another in order to establish their equivalence, Milne has shown that it is possible for different observers in different parts of the universe so to correlate their time reckonings as to make them in some sense equivalent. Although his method of correlating time measurements of different observers involves some rather recondite mathematics, the basic ideas are fairly simple. In fact, his theory is merely the arithmetization of the practice of an ordinary observer recording his own perceptions.

What it means is this: As already mentioned, any observer perceiving two point-events A and B can always tell whether B occurred after A, before A, or simultaneously with it. Between any two non-simultaneous point-events A and B we can interpolate an infinity of other point-events, and all of them will be after A but before B if B is later than A, just as we can interpolate an infinity of other points between two points A, B of a line such that any interpolated point is to the right of A but left of B if B is to the right of A (see Figure 29).

In other words, the flow of all the events in his consciousness is a linear continuum whose mathematical counterpart is a set of points on a straight line. We could therefore represent any event C as a point C on a straight line AB (see Figure 29).

O	A	C	B

Figure 29.

Choosing any point O as our origin, we can correlate all points to its right with positive real numbers and all points to its left with negative real numbers subject to the condition that the numbers t_1, t_2 correlated with the events A, B satisfy the relation $t_2 > t_1$ if B is later than A. Milne calls any such correlation of events in an observer's consciousness with real numbers a 'clock' 'arbitrarily graduated' and the real number t associated with any event C the 'epoch' of that event. Obviously there are an infinite number of ways in which the observer can thus graduate his clock. Any actual material clock such as a watch, clepsydra, sundial or calendar that he may use for the purpose of dating an event (i.e. associating some number with it) is only an example of such a clock graduation corresponding to some particular mode belonging to this infinite set.

Now it is possible to derive one mode of clock graduation from another by merely transforming the associated numbers or 'dates' of point-events to some other set of real numbers exactly as we may transform the numbers used to date a series of historical events from the Christian era to the Hegira or some other era such as that of the French or Russian Revolution. Such a transformation is called a regraduation.

So far we have considered a single observer. But there are an infinite number of other observers in the universe. Consider another such observer. He too can correlate events in *his* consciousness with the continuum of real numbers. That is, he too can set up another arbitrarily graduated clock in his own neighbourhood. Under what conditions may these two arbitrarily graduated clocks be said to keep the *same* time? To answer this question it is necessary for the two observers to be able to communicate with each other. One way of providing this intercommunication is to let the observers see each other's clocks, which we may assume to be arbitrarily graduated in any manner.

Now fundamentally an observer can record by *his own* clock the time of occurrence of any event taking place in his *own* neighbourhood, the time of occurrence and observation being the same. If the event occurs elsewhere the time at which it actually occurs and the time at which it is observed will be different. In this case the observer can only observe the time (by his own clock) at which the news of its occurrence reaches him by a flash of light or a radio signal or otherwise. If then he wants to observe another observer's time or clock, all he can do is to strike a light and watch for it to illuminate the second observer and his clock. If the second observer cooperates and reflects back to the first observer the light ray immediately after it has illuminated his own clock, the first observer can observe three times:

1. The instant of time t_1 (by his own clock) at which he flashes the light signal.

2. The *reading* t_2 of the second observer's clock at the moment it becomes visible to him.

3. The instant of time t_3 (also by his own clock) at which he perceives the second observer and his clock on return to him of the light ray after reflection from the second observer.

Since neither of the two observers (let us identify them as P and Q) is in any way privileged, things should remain the same if their roles were reversed. This means that the relation of t_2 to t_1 is logically the same as that of t_3 to t_2. For the epoch t_2 is the reading of Q's clock at the instant the signal which left P at t_1 by P's

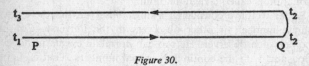

Figure 30.

clock reaches Q, and the epoch t_3 is the reading of P's clock at the instant the signal which left Q at t_2 by Q's clock arrives at P (see Figure 30).

Consequently t_2 stands in the same relation to t_1 *vis-à-vis* the observer P as t_3 does to t_2 *vis-à-vis* the observer Q. But since both the clocks are *arbitrarily* graduated, this *logical* equivalence between the numbers t_1, t_2 on the one hand and t_2, t_3 on the other will not in general be expressed by any *numerical* equality. To convert this logical equivalence into a numerical equivalence, we let P repeat this ideal experiment as follows:

On receipt of the light signal from Q at time t_3, P reflects it back to Q once again and thus obtains a second triplet of time readings t_3, t_4, t_5 exactly similar to the first, and so on indefinitely (see Figure 31). Now P can graph the readings t_2, t_4, t_6 ... of Q's clock against the corresponding readings t_1, t_3, t_5 ... of his own. In other words, he can plot the points (t_1, t_2), (t_3, t_4), (t_5, t_6), etc., on a graph and obtain a curve.

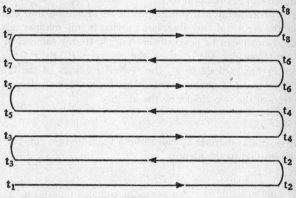

Figure 31.

But if he were to look at the state of affairs from the point of view of Q, he would find that the corresponding curve for Q would be obtained by graphing the readings t_3, t_5, t_7 ... against the readings t_2, t_4, t_6 ..., that is, by plotting the points (t_2, t_3), (t_4, t_5), (t_6, t_7) .... These two curves will not in general be the same.

If P, however, now tries various modes of regraduation of Q's clock to obtain new numbers T_2, T_4, T_6 ... corresponding to the earlier readings t_2, t_4, t_6 ... of Q's clock, he may stumble across one or more modes of clock graduation whereby the graph of T_2, T_4, T_6 ... against t_1, t_3, t_5 ... comes into coincidence with that of t_3, t_5, t_7 ... against T_2, T_4, T_6. ... When this has been achieved we have hammered the logical equivalence of the series of the readings typified by t_2, t_1 on the one hand and that typified by t_3, t_2 on the other into numerical equivalence as well. The proposed regraduation of Q's clock has rendered it equivalent to or congruent with P's. Observer P can now radio to Q the proposed mode of regraduation of Q's clock, asking him to substitute T_2, T_4, T_6 ... for the readings t_2, t_4, t_6. ... Such a substitution by Q will make his clock congruent with that of P. It is clear that this regraduation into equivalence of the two observers' clocks is only a mathematization of the logical symmetry of P's relation to Q.

The aforementioned procedure for regraduating into equivalence the clocks of two observers moving in any manner relative to each other can be generalized to apply to any number of observers. Consider, for instance, the observers P_1, P_2, P_3, P_4. ... After P_1 has radioed to P_2 the revised readings of the latter's clock to make it congruent with his own, we may denote the relation of the two clocks by the identity

$$P_1 \equiv P_2 \quad . \tag{1}$$

In a similar manner P_1 can radio a regraduation rule for P_3's clock leading to the identity

$$P_1 \equiv P_3 \quad . \tag{2}$$

Do the identities (1) and (2) imply the new identity $P_2 \equiv P_3$? In other words, if P_2 and P_3 regraduate their clocks to make them congruent with that of P_1 will they *ipso facto* become congruent with one another as well? The answer in general is no. For $P_2 \equiv P_3$, there must be some restriction placed on the motion of P_3.

Suppose P_3's motion satisfies this restriction. Then we may consider P_4 and after regraduating P_4 into congruence with P_1 we may consider the conditions which P_4's motion must satisfy so that $P_4 \equiv P_2$ and $P_4 \equiv P_3$. By taking the observers one by one we can determine the set of restrictions on their relative motions in order that all their clocks may be regraduated into equivalence taken two at a time.

This may seem to suggest that the set of restrictions on the relative motions of the observers piles up as more and more observers are admitted into the comity of congruent clock owners, so that a position may arise whereby no really large set of observers with congruent clocks fairly dispersed all over the universe may actually be possible. Milne, is, however, able to show that an infinite number of patterns of observers in relative motion permitting the construction of congruent clocks for the entire totality of the observers included therein can be constructed. Any such pattern of observers in relative motion which permits the regraduation of their clocks into congruence with one another taken in pairs is called an *equivalence*.

Two examples out of this infinity of equivalences are particularly interesting. First, consider an ideal universe consisting of a swarm of galaxies whose nuclei started receding from one another at some given time with uniform velocities from a common centre of coincidence. Milne calls such a model universe a substratum. It can be shown that if two observers located at any two of the nuclei of such an ideal universe or substratum graduate their clocks into congruence, then so can *any other* pair whatever of this universe. The possibility of equipping the whole army of observers located at each nucleus of the substratum with congruent clocks depends on their possessing in common two characteristics, namely coincidence at one point at the initial epoch $t = 0$ and uniformity of relative motion as observed by the clocks of any two of them.

Secondly, consider a set of observers, any two members of which remain at relative rest. A set of relatively stationary observers can also be equipped with congruent clocks. Milne shows that such instances of equivalences can be multiplied indefinitely. But the heart of his theory is the fact that at bottom all this infinity of equivalences reduces to a single kinematic entity, *the* equivalence. This means in effect that if there exists even one set of observers whose relative motions permit the graduation of their clocks into congruence, then an infinite number of other possible sets can be derived from it by a new regraduation of their clocks, the *same* regraduation rule applying to them all.

For example, the equivalence of stationary observers referred to above may be obtained from that of observers in uniform motion by transforming the time t of the latter set into time τ of the former by means of the formula

$$\tau = t_0 \log \frac{t}{t_0} + t_0,$$

where t_0 is a certain constant. In other words, if there is even one pattern of relative motion of a set of observers which permits them to set up congruent clocks, then this pattern of relative motions can be transformed into an infinite number of other patterns by arbitrary regraduations of their clocks.

Milne assumes that the relative motions of the galaxies, the main concentrations of matter in the universe, do conform to such a pattern. For if they did not and were purely random, there would be no possibility of constructing any rational self-consistent scheme of timekeeping in the universe, that is, of equipping the observers at their centres with congruent clocks. Assuming then that it is possible to associate with the matter of the universe a system of congruent clocks, the relative motions must by and large conform to a pattern which is represented by the equivalence. What mathematical form we use to denote such an equivalence is rather arbitrary, as all the forms can be deduced from one another.

Having defined equivalent or 'congruent' clocks, Milne proceeds to show that the idea of equivalent or standard measuring rods is neither valid nor necessary. Hitherto physicists had relied on two fundamental instruments for probing into the mysteries of the universe – a clock for measuring time and a rigid rod for measuring distance. Milne proposes to dispense with the rigid measuring rod and claims that distance can be measured solely by means of a clock. Thus, for instance, observer P can calculate the distance PQ from the time measurements t_1 and t_3 by which his own clock records the epochs of the departure of a light signal from P and its return to him after reflection at Q. The distance r of Q from P is then merely $\frac{1}{2}c(t_3 - t_1)$, where c is the velocity of light. Milne thus assumes that velocity of light signals is constant regardless of the motion of Q relative to P.

It may be mentioned that this is precisely the method now used by metrologists for measuring distance by means of radar technique, whereby the distance of a reflecting surface is measured by the time taken by a radio wave to return to the emitter. Thus 'equivalent' clocks give all basic measures of space as well as time.

Since the existence of observers equipped with equivalent or congruent clocks everywhere in the universe suffices to give all basic measurements of space and time, Milne assumes that the motion of the galaxies cannot be purely random as this would preclude any possibility of constructing a rational scheme of time-keeping in the universe. The relative motions of the galaxies must therefore conform to the pattern which enables every observer situated at each galactic nucleus to graduate his clock with that of every other observer. Such a pattern of relative motion is provided by Hubble's law of galactic recession.

Milne therefore starts with an ideal universe, the substratum, wherein the galactic nuclei recede from one another with uniform relative velocities from a common centre of coincidence at the initial epoch $t = 0$. In such a system, obviously the fastest-moving galaxies would recede furthermost and the slowest would be left behind closest to the original centre of coincidence, just as an observer at the starting-post looking at a horse race in progress would find the swiftest horses in front and the slowest ones in the rear closest to him. Further, such an observer would also find that the speed of every horse at any time was proportional to its distance from himself in a manner analogous to Hubble's law in respect of the receding galaxies, which postulates that every galaxy is receding from the original centre of coincidence with velocity V given by

$$V = rt,$$

where r is its distance from the observer at the centre at epoch t. Milne thus from the very beginning weaves Hubble's empirical law into the fabric of his deductive argument.

He next employs the cosmological principle according to

177

which the aspect of the heavens that each observer obtains in this system of receding galaxies is identical with every other. Suppose, for instance, an observer P counts the number n of galaxies moving with velocities lying in any small range v to $v + dv$. He may find that n is some particular function $f(v)$ of velocity. Now if another observer Q does the same, he will no doubt find that the galaxies included by P in his count no longer have their velocities relative to *himself* lying within the same range v to $v + dv$. But there will be other galaxies moving relative to Q with velocities lying in this range. Let this number be m. He may find that m is a particular function $\phi(v)$ of velocity. Likewise another observer R may count the number (l) of galaxies within the same velocity range and find it to be another function $\psi(v)$ of velocity, and so on.

Milne's cosmological principle requires that all these different functional laws $f(v)$, $\phi(v)$, $\psi(v)$ ... discovered by different observers from their own headquarters must be the same. In other words,

$$f \equiv \phi \equiv \psi. \ldots$$

The reason is that there is no particular privileged velocity frame and all of them are at par. Since the velocity with which any particle is moving in the system is proportional to its distance r from the observer, we can also derive from this law of velocity distribution that of distance or spatial distribution of galaxies. In other words, we can calculate the number of galaxies lying within any given distance range also.

Because of the identity of the laws formulated by all the observers in the system, this density distribution has a number of peculiar but interesting features. First, the observer in every galaxy considers his galaxy the centre of the entire system. Secondly, he sees every other galaxy as receding with a velocity proportional to its distance from himself, while regarding himself as in a permanent state of rest. Thirdly, the number of galaxies is actually infinite,* but the entire system appears to every observer, in his own private Euclidean space to occupy a finite

* For otherwise every galaxy would not be the centre of the system and some of them would have to be near the edge.

volume, namely that enclosed within a sphere of radius ct, where c as usual is the velocity of light and t the epoch of reckoning in t-measure. As a consequence, space appears to every observer infinitely overcrowded near the expanding edge of the universe, where galaxies move with the velocity of light, just as to an observer at sea all the oceanic waters seem to be concentrated at the edge of his horizon. But this is only an apparent effect, for to another local observer at or near the former's horizon everything seems normal, the concentration of oceanic waters now receding to his *own* horizon. Fourthly, this singularity or infinite density at the apparent horizon, though in principle observable by and certainly occurring within the experience of the observer remaining at the origin, is forever inaccessible to experience. For since the boundary or the observer's horizon is receding with the velocity of light, it can never be overtaken by an observer setting out from the origin with any speed not exceeding that of the light. Fifthly, the outward recession leads to a steady dilution of particles, for, as time passes, every observer is bound to disappear from the view of every other, leaving us the prospect of an eternal Cimmerian night when we shall cease to see any other galaxy save our own. Sixthly, at any fixed epoch the density increases outwards until it becomes infinite at the boundary of the system where the galaxies begin to recede with the velocity of light. Such is Milne's model universe, the substratum. How does it square with the facts of experience?

To show that this scheme does lead in many respects to the same results as classical physics, Milne proceeds to construct a new dynamics. Now dynamics begins with the motion of a free particle, that is, with the behaviour of a particle wrenched free from the action of all forces. But actually there are no free particles, as every particle is exposed to the gravitational pull of all matter in the universe. Consequently, the problem of the motion of the free particle boils down to the calculation of the total pull exerted on it by the rest of the universe. To do this Milne considers the model universe, the substratum, described above. In this smoothed-out cosmos a free particle is taken to mean a particle projected from any arbitrary position at an arbitrary time with an arbitrary velocity.

Consider now any observer located at some point O which we may take as our origin (see Figure 32).

Let the free particle be at any arbitrary position P at a distance r from O at any arbitrary epoch t. Let it move with any velocity v at time t. Our object is to discover the future pattern of its movement, or, in other words, the rate of change of v. Milne assumes here the very plausible principle of Galileo, namely that its acceleration is a unique function of the position, velocity and time, that is, of r, v and t. Let this function be $f(r, v, t)$.

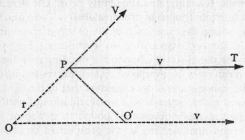

Figure 32.

But if we now consider any other fundamental observer O' headquartered at the nucleus of some other galaxy, his reckoning of the location, velocity and epoch of the *same* free particle P will in general be different. He may find that its location from him was r', velocity v' and time t'. Milne's cosmological principle again requires that the law $f(r', v', t')$ defining its acceleration be of the same form as before. This condition enables Milne to derive a function f defining the acceleration of the free particle. A simple geometric interpretation of the acceleration function f is as follows:

Let the free particle P move in observer O's experience with velocity v in the direction PT at time t. But the galactic nucleus at P^* which is a member of the substratum moves with a velocity V along OP. In general, the velocity of the free particle at P and that of the galactic nucleus of the substratum at P are different both

* Not to be confused with the free particle at P whose motion is being investigated.

in magnitude and direction. Since the galactic nucleus now at P was at O at the initial epoch of coincidence $t = 0$ we have $OP = Vt$. Now draw OO' parallel to PT and take $OO' = vt$. Then the free particle P (but not the galactic nucleus at P) will consider itself at rest in a frame whose apparent centre of the substratum is at O'.

For clearly the galactic nucleus at O' is moving in O's experience with velocity v along OO'. Consequently an observer at O' moving with the galactic nucleus at O' will regard the free particle P as at rest, since the velocities of both the free particle at P and the galactic nucleus at O' are identical in O's experience.

Milne's function for the acceleration of the free particle is simply the quotient of the distance $O'P$ and t^2, that is, $O'P/t^2$, provided that the free particle moves with speeds small compared with the velocity c of light and that the distance OP is small compared with ct. In other words, this formula gives the acceleration so long as the free particle moves with speeds small compared with light and so long as the particle is not too near the expanding edge or the receding horizon of the observer at O.

An important corollary of this result concerns the motion of a free test particle at any point P moving with the velocity of the galactic nucleus in its vicinity. In this case the magnitude v of the velocity of the free particle becomes equal to V, the velocity of the galactic nucleus at P, and its direction PT coincides with OP, the direction of V. It follows that the apparent centre O' of the substratum in which the free particle is at rest coincides with P itself so that the distance $O'P$ shrinks to zero. Consequently, if a free test particle has the velocity of the galactic nucleus in its vicinity its acceleration is zero. It therefore continues to accompany the same galactic nucleus forever. In other words, each galactic nucleus, too, behaves like a free particle. This means that the substratum, originally defined as a kinematic system with a *prescribed* motion, is also a dynamical system which will maintain itself in this motion of its own volition.

But it will be objected that this system of dynamics is very different from that of the everyday dynamics of Newton so amply confirmed by observation during the past three centuries. Milne, however, contends that his system is an absolute system from which that of Newton can be derived by an appropriate regraduation of clocks from the cosmic time t to the dynamical τ-time by means of the equation quoted on page 176. If we adopt the dynamical or τ-time, the rule for the derivation of the acceleration of a free particle given above transforms into Newton's first law of motion, according to which a free particle continues to remain at rest or in a state of uniform motion in a straight line.

Unfortunately, Milne's deduction of the law of motion of a free particle by the application of the cosmological principle to the substratum does not lead to a unique result. The acceleration formula it yields involves an arbitrary function which may be chosen in any manner whatever. To each choice of the form of this function there corresponds an acceleration law. There is therefore a whole infinity of acceleration laws which the free particle could obey. To rid himself of this *embarras du choix* Milne is obliged to consider systems more general than the substratum, his model universe of uniformly receding galaxies.

Now Milne's substratum in t-measure is a hydrodynamical system in the sense that there is a unique velocity of flow at each location, that is, the velocity with which the galaxy positioned there continues to recede from every other galaxy. In other words, in the experience of any observer O, every point P is the centre of a receding galaxy whose velocity of recession is OP/t, t being the the epoch of observation. But we could complicate this model universe, the substratum, a bit by imagining a system wherein each point P has in any observer O's experience *not* a single nucleus receding with velocity OP/t at any time t but wherein each point P is the centre of a whole *shower* of nuclei each moving with a different velocity. This means that each point or nucleus has not a single velocity of flow but a whole array of velocities, in fact, a velocity pattern or distribution.

Consider such a system, which Milne calls a statistical system, wherein at each point there is a whole shower of nuclei moving with all sorts of velocities. Can the structure of this system be such that it too satisfies the cosmological principle in that it also presents the same aspect to every observer? To see this, let us imagine an observer at O who makes a census of such a swarm or shower of nuclei located everywhere. Suppose he counts the number N of the nuclei in a small spherical volume w around any point P. Of these N nuclei he further sorts out those which have their velocities within a small range, say, from v to $v + dv$. Let the number of particles lying in the small sphere of volume w centred at P and having velocities in the range v to $v + dv$ be n. Then he will find that n is a certain function f of position P and velocity v. We can imagine another observer O' doing the same. He may arrive at another function ϕ and so on indefinitely.

Cosmological principle again requires that all the different functional forms like f, ϕ and so on, be the same. This has the consequence that the acceleration of a free particle can be interpreted as the sum of two terms. It can be further shown that of these two terms only one contains an arbitrary function depending on the population distribution of the shower imposed on the substratum, while the other is uniquely fixed. Milne therefore identifies the latter as being due to the effect of the substratum itself without consideration of the interposition of the statistical swarm of particles. It is in this way that Milne is able to determine the arbitrary function introduced by his earlier deduction of the acceleration law of a free particle in the presence of the substratum. In other words, he is able to transfer the arbitrariness encountered earlier into the second term.

The second term containing the arbitrary function of the variable defining the population distribution of the statistical system is the effect of the swarm of particles superimposed on the substratum. Milne therefore interprets it as a gravitational relation. His reason for doing so is that the second term, besides containing the arbitrary function of the variable defining the population distribution of the superimposed swarm, also contains an undetermined constant C which Milne chooses to regard as the 'irruption' into analysis of gravitational mass. If this is allowed, it is possible to argue with some show of plausibility that the second term is the gravitational effect according to the inverse-square law of a point condensation of mass CM'_0 at the apparent centre O', where the distance of the free particle is reckoned from O', the apparent centre of the substratum with respect to which P is at rest, and where the mass M'_0 is the mean apparent mass of the substratum per fundamental particle or galaxy.

What it means is this: if the observer at O fills the apparent volume enclosed within a sphere of radius ct with a density equal to the density near himself, he can obtain the apparent mass M_0 of the universe accessible to him by merely multiplying the apparent volume of the visible sphere, namely $(4\pi/3)(ct)^3$, by the local density ρ in his own vicinity. Why apparent? Because while in Milne's theory the actual mass of the universe is infinite, the part visible to any observer is finite, namely, that enclosed within a sphere of radius ct around himself. Further, this apparent mass M_0 is a constant, being independent of the epoch at which the observer evaluates the density ρ in his vicinity. The reason is that although at a later epoch the density according to his reckoning will diminish, yet the volume will increase in the same proportion so that what he loses on the swings he makes up on the roundabouts.

Now if the observer divides the apparent mass M_0 by the number of fundamental particles or galaxies within his ken, he will obtain M_0', whose product with the constant C gives the gravitational mass CM_0' which, if located at the apparent centre O', would give rise to the second term of the acceleration formula according to Newton's inverse-square law. But this results in the fact that Newton's constant of gravitation γ is no longer a constant and has to be equated to

$$\frac{c^3 t}{M_0},$$

so that fundamentally the so-called constant of gravitation is a variable in time, being proportional to the epoch t at which it is evaluated.

Milne concedes that the two models, the substratum and the statistical system, to which he applies the cosmological principle to derive the basic laws of dynamics and gravitation, are idealizations of two opposing aspects of our universe. The substratum represents a smoothed-out universe forming a continuous set, like the liquid masses in hydrodynamical theory. By the association of a unique velocity of recession with each point, the substratum idealizes the expanding universe of the galaxies. But this idealization destroys altogether the discrete character of the galaxies, the main aggregates of matter in the universe, which are separated by vast voids of intergalactic space. On the other hand, a statistical system 'over-represents' the discrete character of the galaxies by exhibiting the universe as an array of galactic nuclei separated by empty intervening spaces but with each nucleus itself being the seat of an infinite concentration of matter. Milne admits this to be an imperfection of the model but has not been able to construct a more balanced one embodying these two contrary aspects in the measure prevailing in the actual universe.

Since Milne's substratum is an ideal concept designed to provide *everywhere* a frame of reference in motion, that is, a standard of local rest against which any local departure of motion from that of a 'free' particle is to be measured, Milne is obliged to dissipate the discrete galaxies and distribute their matter evenly over all space so as to be able to associate with each point of space a 'natural' velocity in any observer's reckoning. However, this is an artifice which Milne's theory shares equally with relativistic cosmology, as it too replaces the discrete galaxies by a continuous distribution of matter though for other reasons. But the introduction of the statistical system, as we shall see more clearly later, does seem to be an interpolation designed to secure a coincidence with the already known law of gravitation. This inadequacy of Milne's reasoning to sustain a viable theory of gravitation has come in the way of its rearing a plausible cosmology. Consequently Milne's cosmology, though very much in vogue some thirty years ago, is now way off the main stream of cosmological thought.

The root cause of the inadequacy of Milne's theory of gravitation is his reliance almost exclusively on mere epistemology to deduce the basic laws of the universe virtually from the single assumption that a rational system of timekeeping is possible throughout the universe. This precludes the possibility that the main motions of the galaxies will be random. The only reason Milne assigns for the willingness of the galaxies to restrict their relative motions to a pattern compatible with the hypothetical observers' need to graduate their clocks into congruence and thus keep a rational cosmic time is his faith in the rationality of the 'Creator of the laws of the universe' who, 'limited by reason in the divine act of creation', could not do otherwise. Milne is obliged to make this apology at the outset as he does not wish to concede that he is basing his deductive theory on Hubble's empirical law of galactic recession. For Hubble's observation gives only the instantaneous distribution of velocities of galaxies and is therefore as compatible with Milne's expanding model as with oscillating models of the relativity theory. If Milne were to initiate his deduction from Hubble's empirical law, he would be faced at the very beginning with a bifurcation of possibilities which he wishes to avoid at all cost.

But why avoid the various alternative possibilities? It is because, in his words,

when we study so vast a problem and so deep a problem, as the *raison d'être* of the universe, we cannot afford to take over any results from currently accepted physical theory. Starting from first principles (like Descartes) we must pursue a single path towards the understanding of this unique entity the universe; and it will be a test of the correctness of our path that we should find at no point any *bifurcation of possibility*. Our path should nowhere provide any alternatives.

But the trouble is that whenever he does encounter an alternative logical possibility which he does not wish to consider he excludes it for no better reason than the 'uniqueness of the universe' or the 'rationality' of the Creator. Milne admits that this is a 'metaphysical argument transcending mere science' and therefore of the same genre as the now scientifically discredited ontological arguments of theologians like Anselm. Nevertheless, he essays it, since the 'why' of the universe is not a scientific question, only the 'how' of it is. Consequently, a deductive theory that seeks to probe into the 'why' of physical laws is in essence a metaphysical problem. This is why Milne's regress into theological paralogism seems to me no mere minor infringement of the rules of the intellectual game. It is, as H. W. Poole has remarked, a veritable murder of the umpire.

However, we need not stress unduly this initial weak point of the theory for even if we cannot logically exclude the alternative of an oscillating model, it is worthwhile examining how far Milne's deduction of dynamical laws from *his* interpretation of

Hubble's observation conforms to the accepted canons of scientific rigour. The most important of these laws is the law of motion of a free test particle in the presence of the substratum – Milne's model universe of mutually receding galaxies. But, as we have seen, the law that he deduces is not unique. His formula specifying its acceleration involves an arbitrary function. To each form of this function there corresponds a distinct acceleration formula, and Milne is hard put to it to make a choice among the plethora of rivals with which he is faced. The only way he is able to eliminate other pretenders to the dynamical throne is by a renewed application of the cosmological principle to a new model universe.

This new model is derived by superimposing on the substratum an interlacing set of particles in motion. In the substratum each fundamental particle, the galactic nucleus, has a unique velocity. In the new model – or as Milne calls it, the statistical system – there is associated with each galactic nucleus a shower of particles moving with all manner of velocities. In other words, each nucleus is the seat of an infinite number of particles, each moving with its own characteristic velocity. The cosmological principle is then used to ensure that the velocity pattern of this infinity of particles at each nucleus as observed by different observers from their different headquarters will be the same. But since any actual observer can only observe a single particle at the galactic nucleus, it is clear that the shower of particles is a hypothetical construct made up merely to provide a foothold for a renewed application of the cosmological principle.

It therefore follows that the 'sameness' of the velocity pattern from which the law of motion of a free particle as well as of gravitation is deduced is not something that the observers can actually see. It is, on the contrary, totally fictitious, conjured up by hypothetical observers out of the fictitious showers imagined by each one of them. It therefore seems that the ground frame of Milne's analysis is not merely that the *physical* universe *per se* presents the same aspect to all observers wherever they may be. It goes a step beyond. It also demands that if these observers were first allowed to *imagine fictitious* showers and afterwards required to describe the *phantom* velocity pattern of these *phantom* showers, then they would use identical mathematical forms or functions in their descriptions of it. This means that the deduction harbours a homunculus. That is, Milne not only invents a subject but even the object the subject is required to conjure up to make a headway with his deductive scheme. But lo! even in spite of the weird showers and their phantom velocity pattern, the law of gravitation does not emerge clear and undisguised from the analysis like Athena from Zeus's head. It has to be virtually teased out of it by a somewhat gratuitous identification of the constant C with gravitational mass.

Not that the little man, homunculus, with his conjuring tricks, appears at the end of an impressive logical chain to help smooth a minor difficulty in the argument. He makes his appearance at the very threshold of the argument. For he is needed to make possible communication (on which Milne's theory of space and time reckoning is based) between any two distant points of the universe. But communication, if it is to be intelligible, presupposes a certain identity of outlook or ideology between the beings making the communication. That is why the little men Milne creates and scatters so profusely all over the universe are infected with the habits of thought induced by the astronomical mechanics of our own planet – the earth. One rather awkward consequence of this procedure is that Milne's theory, instead of telling us what the universe is really like, leaves us to make of it almost what we like.

To see how it comes about, we may examine the ideal experiments which Milne conducts with light signals between observers headquartered at the nuclei of the galaxies. It is on these experiments that the most fundamental part of his theory, the theory of space and time reckoning, rests. At first sight the experiments seem to be ultra-scientific, for they appear to be an extension of the operational method which Einstein employed to show that absolute simultaneity was a meaningless concept. But the way in which Milne employs it is a caricature of the operational technique.

While Einstein's ideal experiments with light signals to prove the inadmissibility of universal simultaneity were valid (since they were only an abstraction from and simplification of the actual Michelson-Morley experiment), Milne's experiments are, as Born has remarked, weird fantasies. Moreover, Einstein's restricted relativity could be logically deduced from the observed constancy of the velocity of light at all places in uniform relative motion. It seems that Milne's ideal experiments take advantage of the concession granted in such cases of allowing a phenomenon which neither is actually observed nor can be observed but which could 'in principle' have been observed. It is clear that a limitation has to be placed on the *ad lib* extension of this 'in principle' postulate if we are to avoid certain very queer consequences.

In the case of Milne, his use of the operational method, combined with the assumption that the motions of the galaxies are not random, yields a *unique* kinematical entity, the substratum. As we have seen, this entity may be variously described. We mentioned in particular two of this infinity of descriptions, the t- and τ-times. In t-measure the universe is an expanding universe, and in the latter it is stationary. Both descriptions are equally valid, as are also many others which could be invented. This means therefore that there are no 'real laws' of the universe. What we call 'laws' are only alternative sets of descriptions or conventions and one can make one's choice of which particular set of conventions one is to adopt to rationalize his experience. No doubt this is in conformity with Poincaré's view that the question of discovering the 'real' laws of the universe is meaningless. It is only a question of what particular combination of conventions we choose to adopt – a choice essentially quite as arbitrary as that of the F.P.S. or the metric system for the measurement of quantities. It is true that this view commands widespread assent and has even a measure of truth. But it is possible to distort by exaggeration whatever germ of truth it may have. What I mean may be illustrated by an example.

We all know that the earth is a large sphere, and we have developed a geometry of the sphere, that is, a way of measuring distances on its surface. For certain purposes, we also represent the surface of this sphere on a piece of paper, as when we draw a map of the world, for example, by Mercator's projection. The Mercator map is quite a good substitute for the surface of the globe because it represents on a flat surface all essential terrestrial locations and distances. In other words, we could assume that the surface of the earth is 'flat' like that of a piece of paper on which we draw its map, *provided* we adopted a different system of geometrical laws.

For instance, on Mercator's chart the ordinary notion of distance does not apply. Thus it is impossible to specify a scale that will hold throughout the chart. A distance of one inch near the equator may represent 100 miles, while the same distance near the North Pole, as for example, in Greenland, may represent only, say, ten miles. In consequence, Greenland would appear on the map far larger than, say, Greece although a traveller may experience the same amount of fatigue in covering both the countries from one end to the other. Now Poincaré's thesis is that the global and Mercator's way of representing terrestrial distances and positional relations of continents, countries, and oceans are equally valid and that the facts of our experience can be fitted under both schemes, provided we adopt suitable geometrical laws appropriate to each way of representation. Hence the question of whether the earth's surface is 'really' round or flat is 'meaningless' – it is only a matter of convention.

Similarly, we could reckon time in t-measure, in which case the universe would be expanding, or we may assume the τ-scale of time, in which case the universe would remain static. It is again only a matter of convention. However, it will perhaps be agreed that it is stretching facts a bit too far to say that the question of whether the earth is round or flat or the universe expanding or static is 'meaningless' or a matter of mere convention, simply because mathematicians can find a functional form which can transform away the curvature of the earth or the recession of galaxies.

These idealist ramifications of Milne's theory are no mere accidents. They arise inevitably because the theory has been expressly designed to ensure that any ego, who observes what is out there in the cosmos and can see as well as plant over there his own contemplation of what ought to be there, would describe not only his *actual* view of the universe but also his own *contemplation* thereof in identical terms from all space locations. There may be some justification in adopting the cosmological principle with regard to an ego's *actual view* of the universe not as an *a priori* metaphysical truth 'transcending mere science' but purely as a working hypothesis to mitigate the enormous difficulties that would otherwise beset cosmology. But to adopt it also for an ego's mental constructs and contemplations that he himself projects into the universe seems to me without any justification. Milne is obliged to have recourse to it because he wishes to fire only on one cylinder. By shutting off the other cylinder – that of experiment and observation – he greatly emasculates the power of the deductive engine, which works smoothly only when tuned in unison with the empirical engine fed on the fuel of experience.

It is no doubt possible to build impressive deductive systems on a few axioms or what Whittaker calls principles of impotence, as for example in geometry or electrodynamics. But these axioms or principles express in a symbolic form certain features abstracted from human experience after a careful and profound analysis thereof. Consequently, axiomatization serves the purpose it is designed to achieve, namely that of liberating deduced theorems from the need of experimental verification, but only in those domains which have already been so well observed and thought out that universal agreement on the elements to be abstracted and expressed in the axioms can be obtained among mathematicians and physicists dealing with these concepts, as is the case, for instance, in geometry.

Where this condition no longer obtains, an involved deduction reared on doubtful principles does not necessarily confer greater confidence on the extrapolations to the scale of cosmos of empirical laws found to hold in our own tiny corner. This is not to deny the power of deductive theory. It is indeed an indispensable armour if we are to mount a successful assault on the vault of the heavens. But to rush into battle with it alone and without the protective shield of experiment and observation is to debase this invincible armour of victory – Mambrino's helmet – into a barber's basin.

CHAPTER 9

Cosmology and Continuous Creation

A s we saw in Chapter 8, Milne based his cosmology on the cosmological principle that, apart from minor small-scale irregularities, the universe by and large presents the same aspect to all observers from every space location. But this identity of aspect obtains only if the observers scan the universe at the *same* or *equivalent* time. Thus while all observers would no doubt have had the *same* view of the 'vanished brilliance' of the origin of the worlds had they been able to observe the universe at the epoch of its 'creation', that view has no more resemblance to the present view of contemporary observers than a flaming volcano has to the chilled crater it becomes after the eruption. Nor will the present view be anything like the Cimmerian nights that we shall see in the future when the last of the galaxies recedes from our view. In other words, the uniformity or homogeneity of the universe which the principle postulates pertains only to space but not to time.

But why restrict homogeneity to space locations only? Why not widen it to include homogeneity of times as well? Thus argue Bondi, Gold and Hoyle. They consider that of the two homogeneities, spatial and temporal, there is as much reason for postulating the one as the other. They therefore advocate the adoption of an extended form of cosmological principle. Such an extended form of the principle is known as the perfect cosmological principle in contradistinction to the *narrow* or *imperfect* principle adopted by Milne.

The perfect principle stipulates that 'minor small-scale irregularities apart, the universe by and large presents the same aspect to all observers from every space location *at any time*'. This means that the universe does *not evolve* in time but stays put in the same steady state forever. Not that there is no change. The galaxies are born and die like human beings but the aspect of the heavens that their totality presents to any observer at any time stays the same just as the aspect of street crowds remains unchanged from generation to generation in spite of the perpetual coming and

192

going of the individuals in the crowd. Like Tennyson's brook the universe may well boast that the galaxies may come and go but it goes on forever. In fact it actually overfulfils this boast, for the more the galaxies change, the more it remains the same.

This sameness is an attribute of the *flow* of its constituents rather than of the constituents themselves, as with the flow of a steady stream which remains the same even though its waters are in a continual state of replenishment and renewal. Since according to the perfect principle the universe is homogeneous both spatially and temporally (i.e., looks alike to all observers everywhere and at all times), it obviously takes for granted at the very outset the unlimited validity of our terrestrially discovered physical laws for all times. Thus instead of seeking to justify the extrapolation of physical laws to the scale of the cosmos by deducing them from some more fundamental plausible principles, the perfect cosmological principle in effect *assumes* the validity of this extrapolation by arbitrarily pegging the universe within a stationary framework. The warrant for thus strait-jacketing the universe in a stationary state is that otherwise we have to resort to even greater arbitrariness in choosing which of the physical laws vary and which remain invariable.

The reason is that if we do not extend the admissibility of the spatial uniformity of the cosmos to all epochs as well, we must take account of this lack of uniformity in some way or other. But to do so involves arbitrary assumptions as to which laws and constants of nature change on account of a general evolution of the universe and which do not. For without the specification of some invariable constants, no meaning can be assigned to such a change and evolution. The perfect principle proposes to end this arbitrariness by postulating that there is no change or evolution in the basic laws of the universe.

Now if the perfect principle is grafted on to the conception of an expanding universe as revealed by Hubble's observation, there blossoms in the cosmological garden a plant which in the opinion of some is an exquisite flower but in that of others a dangerous choking weed that has no business to be there. For as we saw earlier, an expanding universe is eventually doomed to disappear. Gradually but surely the galaxies would recede from each other's view until each galaxy would be entombed in its own void of Olbers's limit, a sphere of some 10,000 million light years. Accordingly the average density of matter would keep petering out unceasingly. But if the perfect principle is true, this cannot be. What then is the way out of the dilemma?

The only way to reconcile an expanding universe with the perfect principle is to postulate *continual creation of matter* to counteract the continual dilution of its material by the ceaseless drift to infinity of the receding galaxies. The situation is exactly parallel to that of a steady stream which has to have a continuous source of supply from its headwaters to compensate for the continuous drain of its contents into the sea. In the case of the river, the sea water is eventually transported to its source by the clouds. But in an expanding universe there can be no mechanism to establish such a cycle whereby departing galaxies can be brought back to our fold from what may be called Olbers's limbo. That is why Bondi and Gold are obliged to postulate that matter is being *continually created in the universe from nothing and appears everywhere from nowhere* to provide the basic materials from which to rear new galaxies to replace those which disappear.

But the conservation of matter-energy – its permanence and indestructibility – is one of the few basic laws of nature that has so far withstood all the changes and revolutions in physics. How is the hypothesis of continual creation reconciled with the absence of any experimental observation revealing its existence?*

* That the destruction of a small amount of matter yields vast quantities of energy we know already from atom-bomb explosions. But the reverse transformation of intangible energy into tangible matter on a large scale is something beyond our powers at present. The gamma rays emitted by magnetically accelerated particles in space may vanish and in their place matter in the form of pairs of electrons may appear, as has already been observed on an infinitesimally small scale in experiments with atom-smashing machines. Consequently the conservation law of matter is no longer true in the old sense of the term. But the newer revelations of nuclear physics have led to newer conservation laws – the laws of quantum additive numbers. These laws require that the algebraic sum of the values of the appropriate quantum numbers for a system of elementary particles and their antiparticles is a 'constant of motion'. The continuous creation hypothesis violates the conservation law of baryons, that is, elementary particles whose masses are equal to or greater than that of the proton.

If matter is being created continually everywhere, it should some-time reveal itself in some of the numerous experiments that have been conducted to test the law of conservation of matter ever since Lavoisier first demonstrated it. The answer is that we know roughly how much matter there is in a number of typical galaxies as well as the rate of their disappearance across the Olbers threshold. If we work out from these data the rate of creation of new matter required to compensate for the loss of material by their disappearance, we find it so infinitesimally small that it is forever beyond the power of detection of any experiment that we may devise in any terrestrial laboratory. Spread through the whole of space, it turns out to be on the average *one* atom of hydrogen per litre of volume per billion years.

But this is an overall average rate spread through all space. The actual rate of creation of new matter may conceivably vary from region to region depending on its material content. But if so, it has to have a strong 'anti-have' bias or at any rate it must *not* discriminate against the have-not or empty regions. For if the creation law acting on the injunction 'unto every one that hath shall be given' gives new matter to those regions that are already full, such as stellar interiors, it will confer on them an embarras-sing abundance – embarrassing because the bulk of new matter created within stellar interiors (and hence within the galaxies) will be carried along with their old matter into Olbers's limbo as the receding galaxies fade into it. In other words, the creation law would fail to do the very job it was intended to perform, namely help stop the gradual petering out of the material uni-verse on account of the dispersal of the galaxies. For this reason, the steady-state theory assumes that matter is being created continuously and at random everywhere in space independently of the local physical conditions, particularly the existing local concentration of matter prevailing there. In fact, but for this mode of creation, spatial uniformity of the universe assumed by the perfect principle could not possibly exist.

It is no doubt natural to inquire into the cause of this creation. The theory, however, at any rate for the present, does not treat it causally, preferring to regard it as spontaneous and not to be probed into further. The theory has been assailed for this failure to specify the ultimate cause of creation. But it seems to me that such an attack has as much (or as little) validity as that of Huygens and Leibniz on Newton's gravitational theory for the latter's failure to offer any explanation of the ultimate cause of gravitation. For the creation theory treats creation as a fundamental property of matter exactly as theories of gravitation and electrodynamics do. Consequently, its protagonists make the same answer as Newton did in his day: 'Hitherto I have not been able to discover the cause of those properties of gravity from phenomena and I frame no hypotheses.'

Taking therefore, creation as something given, the creation theory, at any rate according to its original* version, assumes the following further postulates:

1. Matter originates continuously and at random everywhere in space in the form of hydrogen atoms. The reason for this assumption is twofold. First, as we saw in Chapter 7, hydrogen is the simplest and most abundant element. Spontaneous creation, if it occurs at all, is likely to take place in this form rather than in any other. Secondly, the accretion process whereby according to these authors the young O and B stars are being formed continuously depends on the presence of vast clouds of hydrogen gas in interstellar space. If creation resulted in the emergence of some material other than hydrogen, there would be no suitable source material available to sustain the accretion process leading to the birth of such stars.

2. Newly created matter is statistically at rest relative to its cosmical neighbourhood. This means that in any observer's reckoning it possesses on emergence the same velocity as he would associate with a galaxy in accordance with Hubble's law if he were to find one in that position. In other words, any observer O will regard the newly created material at any position P as moving along OP with a velocity proportional to its distance OP from himself.

Granting continuous creation of new matter at the rate and in the manner described above, it is possible to show that the universe would conform to the de Sitter model, which we rejected in Chapter 7 as inapplicable to our actual universe for

------

* We shall describe the revised version later.

the reason that under general relativity such a model permits the existence of no matter at all. As we saw, it is an empty world, this one of de Sitter. But since in the steady-state theory the conservation of matter no longer obtains, the field equations of relativity implying such conservation do not apply. It can therefore serve as a model for a universe wherein the conservation of matter does not hold. The general picture of the universe then is as follows:

There is no beginning nor end. The universe remains in the same steady state it has always been in and will continue to remain so forever. Consequently the theory is not concerned with the formation of galaxies from an initially homogeneous medium spread through space as in evolutionary theories, but only with that of new galaxies in a universe already full of old ones. In other words, it does not have to account for the origin of Adam galaxies but only for that of their successors. It does so by identifying the birth process of new galaxies with the gravitational perturbation by the old pre-existing galaxies of the continuously created intergalactic hydrogen gas. The reason is that this perturbation of intergalactic hydrogen gas results in a systematic concentration of density in certain regions thereof in such a way as to lead to the formation of new galaxies both singly and in clusters.

It can be shown, as indeed D. W. Sciama has done, that the accretion process of Lyttleton and Hoyle, whereby massive stars *attract* into them the material of gas clouds in *interstellar* space into which they happen to run, leads in the case of galaxies to the birth of one or more child galaxies when a parent galaxy happens to stream past an intergalactic gas cloud. Because of the incomparably vaster dimensions of the intergalactic gas cloud compared with its analogue of interstellar space, the accretion process in the case of galaxies does *not* result in the perturbing galaxy's attracting into itself a significant proportion of the material of the gas cloud but rather results in its condensation into another galaxy. However, the galactic version of the accretion process, unlike its stellar counterpart, is a theme with many variations. For the conditions prevailing in the intergalactic gas cloud which a fully formed galaxy or cluster of galaxies leaves behind as it streams past, depend on several factors, such as the relative velocity with which the galaxy streams past the gas cloud, its mass, the density, temperature and dimensions of the gas cloud itself, and the like. The variations in these factors are fairly wide to allow a number of choices for the child galaxy left in its parent's wake to follow.

It may, for example, break loose from its parent's tutelage as it comes of age. Alternatively it may be too close to it to do so and thus remain tied to its parent's apron strings forever. If we assume with Sciama that chances of escape from and capture by the parent galaxy are fifty-fifty, then half the new galaxies formed would become a cluster of two. Now a double galaxy acts in turn as a parent in another birth process. But being gravitationally more powerful than a single one, it manages to retain one or both of the new child galaxies it breeds out of intergalactic gas. This further accretion, however, makes it gravitationally even more powerful, so that a cluster once grown rich must continue to grow richer and richer. On the other hand, any single galaxy that managed to break loose from its parent at the time of its birth would sooner or later be bound to one or more of its progenies.

Given the infinite time allowed by the steady-state theory, the accretion process on the galactic scale would result in the growth of only monolithic clusters with not even a single galaxy left in isolation. Consequently this mechanism for the formation of galaxies has to be amended in two ways to conform to observation. First, to prevent the indefinite growth of clusters beyond any limit, it is contended that by the time a cluster becomes large enough to acquire 10,000 galaxies an automatic shut-off process begins to intervene. For very massive clusters of a gargantuan size of this order induce in the surrounding gas such high temperatures that despite the cooling by radiation the gas cannot condense into new galaxies. It is too violently agitated to do so. Secondly, we need a source of single galaxies to replace the ones that are consumed by becoming tied up with their progeny. This source is found in the dynamical evaporation of single galaxies from clusters.

The justification for assuming such a phenomenon is the fact that from time to time some individual galaxies of a cluster can acquire sufficient energy from gravitational interactions to enable them to escape from the clusters. Breaking loose, they continue to plough their lonely furrow for a while until they, too, are dissipated by becoming linked with one or more of their progeny. It is not difficult to see that such a process does enable the distribution of clusters and single galaxies to adjust itself to produce a stable, steady state. For if by a fluctuation there are fewer clusters than usual in a region, the number of single galaxies captured in that region will decrease. Consequently, the number of single galaxies breeding new ones will increase. This in turn will increase the number of clusters. If, on the other hand, the number of clusters increases, this will initiate the reverse process leading to the escape of a larger number of single galaxies. In this way the system adjusts itself to a stable state which would persist indefinitely.

Another interesting feature of the theory is that the galaxies that arise together, whether in a cluster or alone, keep on drifting from one another. But as they continue to recede new ones are formed in the vast intervening spaces between them where the bulk of the new matter is being created. As a result, *young* galaxies are interspersed between the *old* so that the old veterans are far from one another with the parvenus continually emerging between them. Since the general behaviour of the universe never changes, it is even possible to compute the age pattern of the galaxies that remain within reach of any observer at any time. Thus the average age of a galaxy is computed to be about three billion years, but since this is an average it is only a statistical attribute and does not preclude any particular galaxy's being much older though not indefinitely so. For example, our own galaxy, the Milky Way, may well be ten billion years old as some observations seem to show even though it is a bit difficult to believe that it could exceed the average by a factor of three.

A more rewarding confrontation of the predictions of the theory with observation is possible by a more detailed specifica-cation of the age pattern of galaxies. If we correlate the age of a cluster with the number of galaxies it contains on the plausible assumption that the more numerous clusters are the older ones, having had longer time to acquire more galaxies by breeding and capture, the theory predicts the number of clusters containing a given number of galaxies, the latter number being taken as an index of its age. For example, it seems to show that in a sphere of about 100 million light years diameter the number of 'young' clusters containing 100 galaxies each is 100, as against only ten for the oldest clusters containing 10,000 galaxies. While these results are not unreasonable, a more detailed comparison of the prediction with observation cannot be said, at any rate for the present, to confirm the theory. We shall deal with this question later.

The steady-state theory has been violently criticized for its alleged atavistic regress into a search for *perfection* in celestial matters. In earlier times the search for knowledge was handicapped by the presumed perfection of the motion of heavenly bodies. For when these motions were actually observed to be imperfect, that is, non-circular (circular motions alone being considered perfect), an incredible amount of ingenuity was expended on inventing a theory of heavenly motions designed to reconcile observation with the preconceived principle of perfect motion. It is contended that the steady-state theory proposes to do exactly the same. It prefers to hold a preconceived principle, the perfect cosmological principle, inviolable even though observation shows clearly that the density of matter in the universe is continually thinning.

To make this density constant in spite of contrary observational evidence, it junks one of the best-established physical laws, the conservation of matter, and calls upon us to perform an act of *credo quia absurdum* like those metaphysical leaps by which some theologians choose to solve their ontological conundrums. For this reason there has sprung up a suspicion that the upholders of continuous creation are a band of blue-domers like Gurdjieff and Ouspensky seeking to make respectable a species of mystagoguism. But it seems to me that the suspicion is unwarranted. Quite innocent of theology and metaphysics, they were driven to it by what seemed at the time an almost irresistible argument, as we shall presently see.

If we examine all possible hypotheses about the duration of matter in the past, we find that there are three and only three possibilities:

(i) All matter has existed for all time, that is, during an infinite past.

(ii) All matter has existed for the *same finite* time which has elapsed since the date of creation. In other words, all extant matter was created at the same time, the epoch of creation.

(iii) Any particle of matter may have existed for any time. This means that matter is originating (or being created) without cause, continuously and at random everywhere from nowhere.

Now of these three possible hypotheses the first has to be rejected for the reason that the material universe shows no sign of being infinitely old. The bulk of material in the universe is still primeval hydrogen from which other elements are continually being synthesized in the interiors of stars. It is reckoned that almost half the material of the universe is even now primordial hydrogen in interstellar space. And it may actually turn out to be even more as explorations of the universe by the new radio telescopes and other means proceed apace. If all this hydrogen has existed for an infinite past, its continued existence at present is an enigma. For it should long ago have been transmuted into heavier elements. Not only that, the nuclear transformations and other processes that we see still at work in the universe should have reached their respective blind alleys, that is, gone on so far as they could with no possibility of further developments aeons ago.

There are only two ways of escape from this dilemma. First, primordial hydrogen might have hibernated for an eternity and started on its career of nuclear transformation a finite time ago – some ten billion years. This is hardly credible. The second alternative commands no greater credence, for it requires that the heavier elements transmuted from hydrogen in stellar interiors be broken down into hydrogen again to keep its proportion to the level observed at present. Now while transmutation of hydrogen into helium and higher elements releases energy, the decomposition of higher elements into hydrogen on the contrary *absorbs* it. There is no possible source of the prodigious energies required to break heavier elements back into hydrogen. If for this reason we reject the eternal existence of matter in the infinite past – hypothesis (i) – we are left with only a choice between hypotheses (ii) and (iii). That is, either all matter was created at one blow some finite time ago, as for example in the big-bang theory of Lemaître and Gamow, or it is being created continuously in the manner envisaged by the steady-state theory.

Now at the time Bondi, Gold and Hoyle suggested the adoption of the continuous-creation hypothesis (iii), there was very serious difficulty in accepting hypothesis (ii). For the big-bang theory of creation at the time allowed the universe barely two billion years of existence as against a minimum of four to five billion years ascribed to the earth itself by independent geological evidence. It is true that with Baade's revision of the galactic-distance scale and the consequent disappearance of the age difficulty the argument in favour of continuous creation is now no longer as strong as it was at the time of its original formulation. Nevertheless, the difficulty still persists. As we saw in Chapter 7, the putative age of the universe according to relativistic cosmological models is still ten or at most thirteen billion years, whereas some of the older galaxies are believed to be fifteen to twenty-five billion years old. Consequently some cosmologists still prefer the continuous-creation hypothesis to that of all-at-once creation as it neatly resolves the 'age' paradox of some of the constituents of the universe being older than itself. Moreover 'creationist' cosmologies are obliged to put the creation process itself out of bounds to scientific inquiry. In other words, it makes of initial creation what Milne called a grand irrationality. It seems to critics of this hypothesis unscientific to assume that physical laws have held in all rigour all along except at the epoch of creation itself when anything could have happened.

If the continuous-creation hypothesis owed its origin to what at the time must have looked like an almost impregnable argument, why are its votaries suspected of importing metaphysics into their reasoning? Apparently this suspicion arises from the manner in which the perfect cosmological principle has been combined with Hubble's observation of expanding galaxies to provide a logical underpinning of the continuous creation of matter from nowhere. For clearly such a queer-looking hypothesis cannot be justified merely by the need to preserve any preconceived cosmological principle, perfect or otherwise. This is particularly so here, as one cannot claim these principles to be true *a priori* as one might do for some of Euclid's axioms (for example, that of the equality of two equal terms when equals are added to both). In other words, they are not sheer necessities of thought. Consequently, they must depend for their truth on an appeal to experience. But this signally fails to establish their truth.

It is well known that even the narrow cosmological principle is not satisfied in any small neighbourhood such as our own solar system or of our galaxy, the Milky Way. Nor is it satisfied in a larger region containing some hundreds of galaxies. All that observation on the basis of counts of very large numbers of galaxies warrants so far is that it may hold statistically for the universe on the very large scale. Consequently the utmost that experience can justify for the present is the assumption of *only* the narrow principle as adopted in the treatment of the smoothed-out models of the universe. It is therefore obvious that the attempted deduction of the creation hypothesis from the *perfect* cosmological principle in conjunction with Hubble's observation of dispersing galaxies is gratuitous. Nor does the invocation of the perfect principle really end the arbitrariness, as Bondi claims, in the choice of physical laws adopted as a basis for cosmology. For, as Milton K. Munitz has pointed out, *any* theory by claiming certain relationships to hold and not others that are logically possible is arbitrarily selective. The perfect cosmological principle, by straight-jacketing the universe into a steady state, is in effect as arbitrary as an evolutionary theory.

Realizing the weakness of a cosmology reared solely on a cosmological principle no matter how 'perfect', Hoyle and Narlikar have recently attempted to produce a new theory of gravitation to support the creation hypothesis. In formulating it they accept the basic concepts and philosophy of Einstein but abandon the field point of view he had adopted. As will be recalled, Einsteinian reformulation of classical mechanics occurred by a switch from the discrete particle to the space-time continuum or continuous 'field' as the more fundamental concept. The enormous success of Einstein's field theory gave rise to the hope that everything in physics henceforward would be interpreted as a field without recourse to the particle point of view. Nor was it an unreasonable hope at the time. For if the gravitational attraction of discrete particles does not occur by action at a distance but stems from the structure of the space-time in which they are embedded, may not other kinds of forces – electromagnetic and nuclear – arise in the same way? Consequently numerous unified field theories were devised during the past four decades in order to extract out of space-time a geometrical structure from which the behaviour of other forces of nature, like electromagnetic and nuclear forces, could follow. Unfortunately none of them, including one of Einstein himself, came up to expectations largely because of two difficulties. First, we do not have any simple suggestive relation between gravitational forces and electromagnetic and nuclear forces like the one between the fictitious and gravitational forces (the equivalence of gravity and acceleration) that guided Einstein to his general theory. Second, it is likely that unification cannot be achieved in this fashion because the general theory of relativity does not embody quantum aspects so necessary in the description of the microscopic properties of matter and its interaction with radiation. This is why there has been a swing back to particle-type theories. Indeed, during the past three decades it has been demonstrated that even such a phenomenon as electromagnetism, where the introduction of the field concept secured its most spectacular success (prediction of radio waves), can be expressed in a suitably refined action-at-a-distance theory fully equivalent to Maxwell's field formulation. It therefore means that field and action-at-a-distance types of physical theories are alternatives with their own advantages and disadvantages.

Deterred perhaps by the persistent failure of unified field theories but encouraged by the demonstrated amenability of even classical field theories like Maxwell's electromagnetic theory to the particle point of view, Hoyle and Narlikar abandoned the field approach of Einstein in favour of a direct-action theory. They were further encouraged to adopt it because the epistemological objection to a direct-action theory appears in their eyes to have lost its edge now. As they remark,

The success of field theories has overshadowed the 'action-at-a-distance' theories, although, ironically, we nowadays have no difficulty with the problem that seemed so worrying to Newton and his followers, namely the mystery of how particles manage to act on each other when they are at a distance apart. We now know that particle couplings are propagated essentially along null geodesics – i.e. at no distance in the four-dimensional sense. Strictly the phrase 'action-at-a-distance' should be changed to 'action-at-no-distance'.

This way of overcoming the action-at-a-distance difficulty is better understood if we recall that by 'distance' in this quotation Hoyle and Narlikar mean not the distance between two locations in our ordinary perceptible space, which is what we mean when we use this term in daily parlance, but its analogue in the four-dimensional space-time continuum, or what we earlier called the 'interval'* between two events. With this extension 'null geodesics' become virtually the paths of light rays or electro-magnetic radiation. It merely means that action at a distance propagates itself in the same way as light along a world line in the space-time continuum, any two points of which are at zero interval – the 'null geodesic'. It therefore results that action at a distance between particles, occurring as it does via lines of null interval, boils down to action at *no* 'distance' i.e. at zero interval.

With the action-at-a-distance difficulty having been conjured away even though only formally by a sort of mathematical jugglery, Hoyle and Narlikar proceeded to incorporate fully into their theory of gravitation Mach's principle, for which there is now growing observational support, which Einstein never did. As a result, mass or inertia of matter, according to Hoyle and Narlikar, is not its inherent property but is deemed to stem completely from the long-range interaction of distant masses. Indeed, they have produced an equation that neatly ties up the mass of any body to the total mass in the observable universe – a universe that is moreover assumed to be a conglomeration of discrete particles that any particular observer can possibly include within his observational ken. Furthermore the equations

* See page 140.

imply creation of new matter not by an *ad hoc* insertion of new terms in Einstein's field equations as in Hoyle's original version of the steady-state theory but as a natural consequence of the new theory. For according to the new theory matter is created not *ex nihilo* but out of the very energy of the universe's expansion. This link between energy and matter is secured by recourse to Einstein's mass-energy relation $E = mc^2$.

The expansion of the universe then provides the reservoir of energy whence arises newly created matter – the famous Creation Field or C-field of Hoyle and Narlikar although it is not a 'field' in the normal sense used by physicists. On the other hand, C-field is an abstruse mathematical concept designed to balance the cosmic energy books disturbed by the creation process. Since the creation of matter of any mass m means virtually the addition of mc^2 ergs of energy in the universe, its creation leads to a recoil effect in the C-field which receives the equivalent debit, that is, mc^2 of negative energy. In this way C-field becomes a peculiar kind of reservoir in that it is a store of negative energy with the result that the more it yields the greater it grows. The situation is not unlike that of a man with a vast resource of negative credit, the more he spends it (this negative credit) the richer (in the negative sense) he becomes.

The notion of C-field as a reservoir of negative energy no doubt looks queer since energy, as ordinarily understood, is essentially a positive quantity. But the idea of negative energy was first mooted by Dirac thirty-seven years ago when he proposed a new wave equation to describe the behaviour of an electron in a given electromagnetic field. Although the idea appeared bizarre and was regarded at the time with a good deal of suspicion, it paid rich dividends in predicting the existence of positively charged electrons or positrons later observed by Anderson, as well as other phenomena like pair creation and annihilation, that is, creation of a pair of positron and electron out of a photon of gamma radiation or conversely the annihilation of the pair by their mutual collision into a photon of gamma radiation. It is well known that Dirac's mathematical 'excesses', like those of Maxwell, paid off in that his innovations sparked a new era of developments in physics. It is a rule of mathematical physics that nothing succeeds like excess *if* it 'works'. Indeed, Dirac's mathematical 'excess' of empty space as a virtual 'sea' of electrons of negative energy wherein 'holes' can appear worked so well that it has already given rise to a new legend in Cambridge similar to the well-known legend of Newton's falling apple.

According to the legend, Dirac was led to interpret the 'holes' in a background of negative energy as positrons as a result of a dream he had after participation in a competition organized by the Cambridge Students' Mathematical Union. One of the problems posed was this:

Three fishermen caught a haul of fish and lay down on the shore to sleep. One of the fishermen awoke and decided to leave, taking his share of the catch. He discovered that he could divide the catch into three if he threw one fish into the sea. This he did, and taking his third share, he went home. Soon the second fisherman woke up, and not suspecting that the first one had gone, again began to divide the catch. Just like the first fisherman, he divided the whole catch into three parts and also discovered that one fish remained over. Having thrown it away he took his third share of the catch and went home. The third fisherman went through exactly the same operation discovering that as a result of dividing up the catch into three parts there was again one fish left over. It was required to determine how many fish there were originally, indicating as a result of this the minimum permissible number.

A simple calculation will show that the minimum number is twenty-five. But Dirac suggested a queer answer, namely *minus two*. Indeed this is a *solution* in a formal sort of way. For if one fish is thrown away, then minus three fish remain, which can be divided into three and for the next fisherman once again there remains minus two fish. It therefore results that with an initial haul of minus two fish the three fishermen can play their game not once but over and over again indefinitely. Hoyle and Narlikar's C-field of negative energy is not unlike the negative haul of fishermen. It can sustain the game of creation indefinitely at least in the same formal way.

However, the main point in relating the fisherman's problem is quite different. It is that mathematization of a physical problem sometimes throws up spurious solutions as every school boy attempting to solve his arithmetical problems by recourse to algebra knows. He knows that the transcription of his problem into the language of mathematics leads to one or more algebraic equations. But *all* roots of the equations are not necessarily admissible solutions of his problem. Consequently he rejects those he regards as inadmissible on various grounds, sometimes because they are negative, at others because they are imaginary. This is because the mathematical apparatus devised is overpowered. It not only grinds out the solutions to the real physical problem in hand but throws up in addition solutions of certain pseudo or other purely formal problems apparently without physical meaning. It is often the same with equations of mathematical physics. But once in a *rare* while by a piece of good fortune and/or inspired intuition a system of physical equations may open up an unsuspected gold mine. It does so by yielding some apparently bizarre or queer solutions, which, though without physical meaning at first encounter, do reveal some novel aspects of reality not previously understood or even suspected. Cases in point are Maxwell's equations of electrodynamics and Dirac's wave equations of electron behaviour. If the Hoyle-Narlikar equations with their accompanying queer-looking C-field of negative energy happen to provide new physical insights not given us before, it would be worth while spinning a legend or two around the Hoyle-Narlikar axis too. But unfortunately no fresh physical insight has *yet* been vouchsafed by their theory. All that they can claim in its favour so far is that it solves some awkward puzzles that have worried theorists for a long time besides providing a sophisticated mathematical rationale of some sort of their own steady-state cosmology since radically revised under the duress of mounting observational evidence against the original version.

The intellectual puzzles which Hoyle and Narlikar claim to have resolved are mainly three. First, their theory obviates the existence of the worrisome singular state of our universe at the epoch of its putative creation so that it does not lead to a state of affairs where the theory ceases to apply as is the case with the rival big-bang cosmology. Second, by accounting for gravitation in terms of particle interactions and summing these up for the whole universe they implement Mach's principle completely for

Plate 1. Coal Sack Nebula, a huge cloud of dust and gas in the Southern Cross. (Yerkes Observatory.)

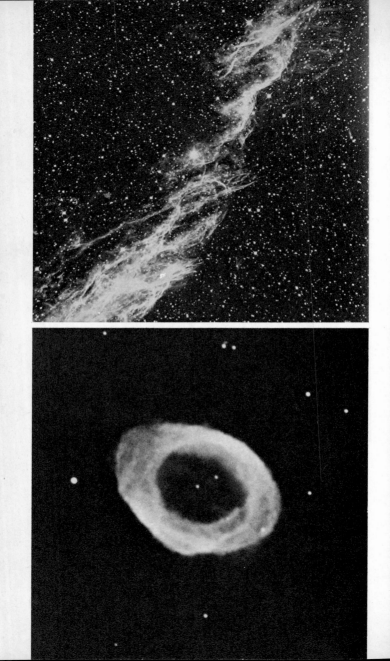

(ABOVE LEFT) *Plate 2*. Network Nebula in Cygnus.
(BELOW LEFT) *Plate 3*. NGC 6270 Ring Nebula in Lyra.
(BELOW) *Plate 4*. Radio galaxy M 82. Red (Hα) image with photovisual image photographically subtracted.

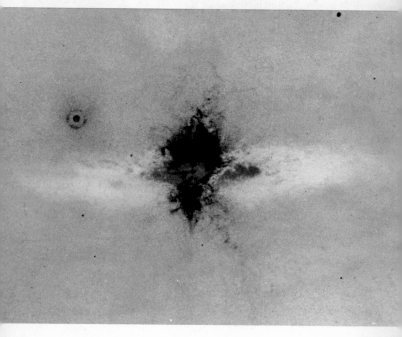

(ABOVE) *Plate 5*. Quasi-stellar radio source 3C 273.
(ABOVE RIGHT) *Plate 6*. Crab Nebula, the remains of a supernova, photographed in red light.
(BELOW RIGHT) *Plate 7*. Elliptical galaxies. The photographs show four galaxies with increasing degree of ellipticity from E0 in NGC 3379 to E7 in NGC 3115.

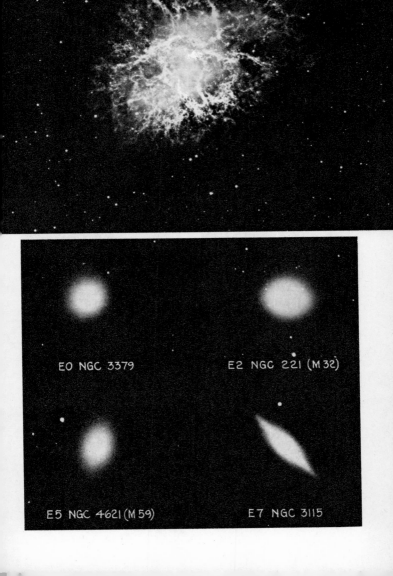

EO NGC 3379

E2 NGC 221 (M 32)

E5 NGC 4621 (M 59)

E7 NGC 3115

Sa NGC 4594

SBa NGC 2859

Sb NGC 2841

SBb NGC 5850

Sc NGC 5457 (M101)

SBc NGC 7479

(LEFT) *Plate 8*. Spiral galaxies. The photographs show three types of spiral galaxies. The nucleus is dominant in Sa types while the arms predominate in Sc types.

(RIGHT) *Plate 9*. Barred galaxies. The photographs show three types of barred spiral galaxies.

(ABOVE) *Plate 10.* Rosette Nebula. Small dark globules can be seen projected against the luminous background of the nebulosity.

(BELOW) *Plate 11.* Photograph of a very young unstable star of the T Tauri class, taken with the 120-inch reflecting telescope at the Lick Observatory. Two wisps of the small luminous nebula (Herbig-Haro Object) in which the star is embedded can be seen protruding above and below the overexposed image of the star. (G. H. Herbig, Lick Observatory, Mount Hamilton, California.)

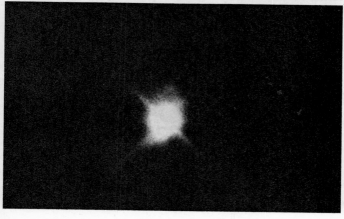

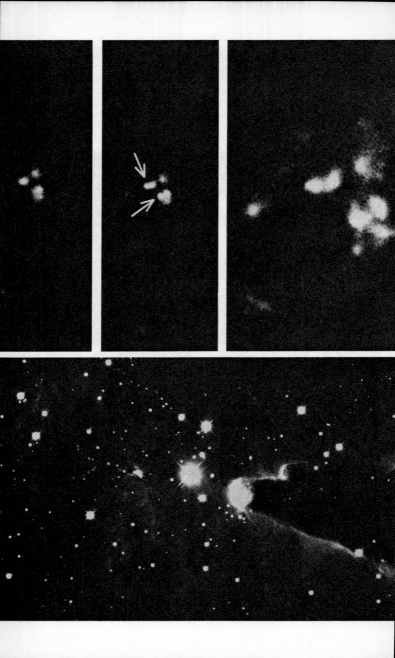

(ABOVE LEFT) *Plate 12.* Direct photographs of the Object Haro 12 = Herbig 2, in Orion. The two plates at left were obtained on 20 January 1947 and 20 December 1954 with the 36-inch Crossley reflector and blue-sensitive emulsions; that on the right in 1959 with a 120-inch reflector and red light. North is at the bottom and east at the left. The angle can be obtained from the fact that the two brightest stars on the 1947 photograph are 8″ apart. (G. H. Herbig, Lick Observatory, Mount Hamilton, California.)

(BELOW LEFT) *Plate 13.* T Tauri stars in the young star cluster NGC 2264. (G. H. Herbig, Lick Observatory, Mount Hamilton, California.)

(BELOW) *Plate 14.* Hubble's relation between red shift and distance for extra-galactic nebulae.

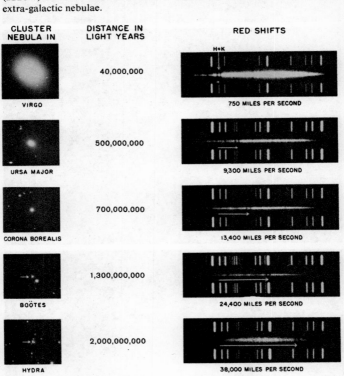

CLUSTER NEBULA IN	DISTANCE IN LIGHT YEARS	RED SHIFTS
VIRGO	40,000,000	750 MILES PER SECOND
URSA MAJOR	500,000,000	9,300 MILES PER SECOND
CORONA BOREALIS	700,000,000	13,400 MILES PER SECOND
BOÖTES	1,300,000,000	24,400 MILES PER SECOND
HYDRA	2,000,000,000	38,000 MILES PER SECOND

Red shifts are expressed as velocities, $c\, d\lambda/\lambda$.
Arrows indicate shift for calcium lines H and K.
One light year equals about 6 trillion miles,
or 6×10^{12} miles

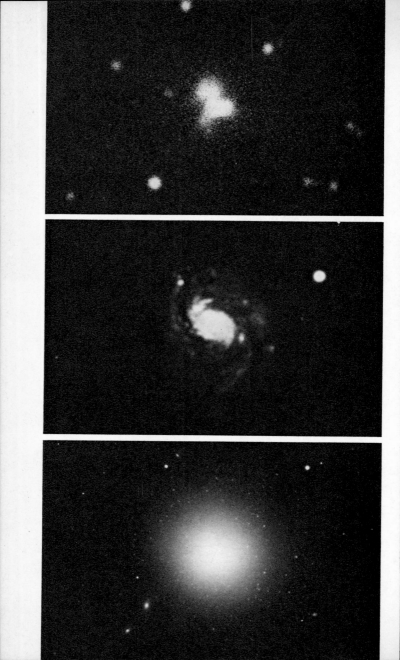

(ABOVE LEFT) *Plate 15.* The strongest extra-galactic radio source is the peculiar galaxy or pair of galaxies known as Cygnus A. In this photograph, made with the 200-inch telescope by Walter Baade, Cygnus A is the object in the centre that looks like two galactic nuclei in contact. The radio-emitting regions lie some 150,000 light years to each side of the visible nuclei. Cygnus A itself is about 700 million light years from the earth.

(CENTRE LEFT) *Plate 16.* NGC 1068, a late spiral forty million light years away, emits 10^{10} ergs per second of radio energy.

(BELOW LEFT) *Plate 17.* NGC 4486. Globular nebula in Virgo. Messier 87.

(ABOVE) *Plate 18.* NGC 4486. M 87 jet.

Plate 19. NGC 5128. This object is a strong radio source. 200-inch photograph.

the first time since its formulation by Mach ninety years ago. Third, they show why gravitation must always be attractive and unlike electromagnetic forces never repulsive, although McCrea in a recent issue of *Nature* has shown that in this respect the Hoyle-Narlikar claim of superiority over standard relativity and even classical theory is inadmissible. For their theory too makes the same tacit assumption that mass must always be positive as other theories do. McCrea has indeed tried to show that no improvement on existing theory is required in this particular regard, nor has been brought about by Hoyle and Narlikar.

But however gratifying the solution of such puzzles may be and however impressive the mathematical apparatus they have invented, the acceptance of a *physical* theory depends on its ability to suggest a way in which it can be tested experimentally. Unfortunately it has not yet been able to suggest any crucial test, the kind of observable *experimentum crucis* that led to the acceptance of the unorthodox theories of Maxwell, Einstein and Dirac and justified their mathematical 'excesses'. In fairness to Hoyle and Narlikar, however, it must be conceded that the theory perhaps is not wholly to blame for this failure although they hope to remedy even this defect in future. It is a failure inevitable in almost *any* new theory of gravitation. For although the force of gravity which holds us to the surface of the earth is so obvious a fact of our life that we dare not ignore it, it is relatively to other forces of nature the feeblest. Thus the electrostatic force between two charged elementary particles is 10^{40} times their gravitational attraction. Because of this weakness of gravitational interaction it is extremely difficult to perform worth-while laboratory experiments concerning gravitation. Consequently with very few exceptions, like the sophisticated experiments of Eötvös and Mössbauer, we have to rely merely on observations of large bodies of astronomical size like the earth and sun which alone give rise to appreciable gravitational effects. And it is notorious that we cannot experiment with them – merely observe them. If we could experiment with them – perish the thought – we could have carried out the imaginary experiment suggested by Hoyle and Narlikar to show in a graphic manner the difference between their theory and its Einsteinian or classical predecessors: abolish one half of the distant galaxies in the universe and observe what happens to the solar system. If Newton's or even Einstein's theory is true, nothing will happen here. The cosmic catastrophe imagined will not cause even a ripple. But if the Hoyle-Narlikar view is right, the sun will grow to be 100 times brighter and 'fry the earth to a crisp'.

Since such frying-the-earth-to-a-crisp kind of experiments cannot fortunately be performed but merely imagined, the validity of their theory can only be tested indirectly by investigating its astrophysical consequences and putting them to observational tests. It is fruitless to investigate its local effects, say, in our own solar system for they are the same as those deduced from Einstein's theory, as indeed they must be. Consequently we must deduce its *cosmological* consequences which alone can be different from its rivals. One such consequence is a precise tie-up between the rate of recession of galaxies, rate of creation of new matter and the mean density of matter in the universe as a whole. All three are tied up in terms of a single constant f on which the strength of C-field depends. The theory by itself does not suffice to determine f. But if the value of Hubble's constant (H) is assumed to be known, we can fix f and hence both the mean density of matter in the universe as well as the rate of creation of new matter. Adopting the current value of H, the rate of creation of matter is found to be about one new atom of hydrogen per litre of volume per 100,000 years and mean density of matter one hydrogen atom per 100 litres or $3 \cdot 10^{-29}$ gm./cm.^{3}. While the calculated mean density of matter is of the right order of magnitude, it is not possible to observe directly the feeble rate at which new matter is supposedly being created. The only way therefore to prove its validity is to investigate its cosmological consequences and put them to observational tests.

In recent years two such tests have been applied. Unfortunately for the steady-state theory both have tended to disprove it. First is the correlation between red shift or the recessional velocities of the most distant galaxies and their luminosity or distance graphed in Figure 26 on page 156. The observed scatter of galaxies on the red-shift luminosity diagram cannot be reconciled with the steady-state theory. Nor are the observations of radio astronomers in any better accord with it. Thus Ryle and his collaborators have shown that the spatial distribution of radio sources is also inconsistent with the requirements of the steady-state theory.

What Ryle and his team did in effect was to count the number (N) of radio sources within a sphere of radius r. By measuring N for a series of pre-selected r's, they obtained N as a function of r. Now we can also calculate the form of such a function given a cosmological model. By comparing the observed with the computed forms of N we can test the validity of the model assumed.

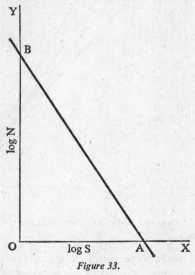

Figure 33.

Consider, for instance, a model universe consisting of radio sources, all equally powerful, at rest, and evenly distributed throughout space assumed to be Euclidean. Then the number of sources, N, inside a sphere of radius r will be proportional to its volume, $\frac{4}{3}\pi r^3$. The flux density S, of a galaxy G on the surface of the sphere, like the apparent brightness of a star, varies as $\frac{1}{r^2}$. Obviously all galaxies inside the sphere appear brighter than

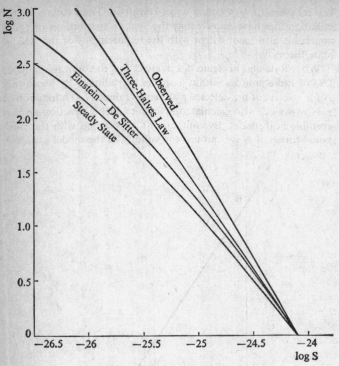

Figure 34. Ryle's radio observations.

G, being closer, while those outside it appear fainter. It follows that if N is the number of galaxies whose power is *greater* than a prescribed amount, S, then N is proportional to r^3, while S is proportional to $\frac{1}{r^2}$. Consequently N is proportional to $\frac{1}{S^{\frac{3}{2}}}$ so that

$$\log N = -\tfrac{3}{2}\log S + \text{constant}.$$

A plot of $\log N$ and $\log S$ will therefore be a straight line AB whose slope is $-\frac{3}{2}$ as shown in Figure 33. This is the famous 'three-halves' law.

The actual log N versus log S plot obtained by the Ryle team is shown in Figure 34 wherein are also drawn the plots according to

(i) the three-halves law just derived,

(ii) the ever-expanding relativistic model of Einstein and de Sitter in Euclidean space described in Chapter 7; and

(iii) the steady-state theory.

It will be seen that the observed plot departs significantly from all three. The departure from plot (i) need not surprise us as we know that the universe is *not* static as the model assumes but is expanding. The departure from plot (ii) is not difficult to explain if we allow for possible evolutionary changes. Suppose, for example, the radio sources become weaker as they grow older. Then the sources we observe at great distances – some of which are as far as four billion light years – are stronger than those near us because they are younger. Now the curve for the Einstein–de Sitter model in Figure 34 was calculated on the assumption that the sources do not change with time. If we re-calculate it on the basis of our new assumption, this has the effect of raising the top left-hand portion, and, by supposing a suitable rate of weakening with age, it can be brought into coincidence or at any rate greater concordance with the observed Ryle line. Thus possible evolution of radio sources over billions of years since the start of the expansion can save relativistic models. But no such course is available to the steady-state theory. For by hypothesis, radio sources, like everything else, must on the average stay always the same so that no systematic evolution is possible.

Hoyle has, however, objected to Ryle's conclusion on the ground that counting of a class of objects whose physical nature is unknown is no longer a satisfactory procedure now that a new species of celestial objects, the quasars, has been discovered. Before their discovery it could be argued that all radio galaxies were more or less alike and they could be counted as objects of one class. It is no longer permissible to lump together radio galaxies and quasars as we do when plotting log N against log S. If the log N versus log S curve is redrawn separately for radio galaxies and quasars as has recently been done by Philippe Veron, the radio galaxies essentially fit a slope of $-1 \cdot 5$ and the quasars a slope of $-2\cdot2$. This is why the combined curve has a slope of $-1\cdot8$. We shall deal with the cosmological significance of quasar red shifts more fully in Chapter 12. Suffice it to mention here that if quasars are nearly as distant* as their large red shifts seem to suggest, they are very likely remnants of a very old and transient phase of our evolving universe that prevailed several billion years ago close to the epoch of the supposed big bang with which our universe is believed to have originated according to the rival creationist cosmology. It has taken light and radio signals we are now observing all these billions of years to reach us so that what we are witnessing at present is a state of affairs that prevailed long ago soon after the first dawn of creation. Obviously the discovery of quasars does not accord with the steady-state theory that requires the universe to stay put in the same state for all time.

Realizing the adverse implications of mounting observational evidence against the steady-state cosmology, Hoyle and Narlikar re-examined their stance in 1965. They then conceded – though only to withdraw the concession three years later – that the existence of quasars, particularly the 'quiet' ones then newly discovered by Sandage, would seem to imply that the universe has expanded from a state of higher density. They also agreed that there was mounting observational evidence such as Ryle's number counts of radio sources, $3\cdot2$ cm. microwave background radiation reported by Penzias and Wilson and the observed helium-to-hydrogen ratio in stars and galaxies reinforcing the same conclusion that the universe was much denser in the past than now. They therefore abandoned the original 'simple' version of the steady-state theory. They proposed that their earlier picture of a homogeneous and isotropic universe in which gentle creation went on everywhere was no longer tenable. According to the revised model offered by them, 'the universe is inhomogeneous, made up of galaxies and clusters of galaxies and with a non-uniform rate of creation'. In such a model the creation of matter took place, not all over the universe, but in certain isolated pockets. To permit this uneven creation of

* But if not, their slope of $-2\cdot2$ has no relevance to the cosmological problem.

214

matter the C-field strength at any point was made to depend on the pre-existing particles. The more massive particles gave rise to more powerful C-field in their vicinity than less massive ones. Consequently the rate of creation of new matter was given a strong anti-have bias in that the emptier a region the less the creation there. In this way by concentrating the creation of new matter in the neighbourhood of pre-existing massive objects it was possible to have pockets of intense creation in the proximity of strong gravitational fields. They therefore suggested that the massive quasars might well be such pockets of creation.

Linking the creation rate of new matter in a region with the magnitude of pre-existing matter in its neighbourhood, they produced a cosmology that was a cross between the original steady-state and relativistic oscillating models already described. Thus according to the view Hoyle and Narlikar held three years ago, we seemed to be living in a rarefied 'bubble', within a much denser more rapidly expanding universe. The 'bubble', that is, the sample of the larger infinite universe that we could observe, was believed to have expanded not from a state of infinite density as in oscillating models but from a state of very high though finite density ($3 \cdot 10^{-9}$ gm./cm.^{3}) which is 10^{21} times denser than the present density. Further, since, the expansion of the steady-state universe was due to pockets of creation and the pockets in a bubble would begin to die out as the bubble expanded, we should be able to see more pockets in the form of visible objects in the past than at present as we actually do.

All in all the revised steady state theory required that although the universe as a whole expands as in the original steady state, it may occasionally develop bubbles which expand and contract without creation. Hoyle and Narlikar suggested that we live in such a bubble. In other words, the portion of the universe we inhabit is merely a statistical 'fluctuation' from the steady state of the infinite universe as a whole. Obviously such concordance with observation as Hoyle and Narlikar managed to secure by treating the 'bubble' we live in as a statistical deviation from the steady-state norm is more contrived than natural. This is why steady-state cosmology even in its modified form fails to convince.

Apparently it failed in the end to convince the authors themselves, which is why they have now reverted to Hoyle's earlier stand relying on their enormous intellectual resources to revive it by a fresh two-pronged attack on its rival big-bang cosmology. They now argue that while earlier inferences drawn from the new observational evidence are not warranted, certain theoretical considerations point to a deep link between local electrodynamics and cosmology supporting only Hoyle's old-style steady-state theory and ruling out all other cosmologies. We shall comment more fully on the latter in Chapter 14. Suffice it to say here that the support sought from electrodynamical theory is illusory. Their attenuation of the observational evidence in favour of the big-bang rival is equally unavailing. Thus the universal black-body 3·2 cm. microwave radiation, the supposed relic of the big bang, is now given another interpretation on the ground that its energy density $(4 \times 10^{-13}$ erg/cm.$^3)$ is about the same $(6 \times 10^{-13}$ erg/cm.$^3)$ as would ensue as a by-product of the conversion of hydrogen into helium since its putative 'creation'. They consider this coincidence on the basis of the usual explanation quite disconcerting, to say the least. For according to the big-bang theory the energy density of the background radiation decreases as the universe expands so that the correspondence with the energy density of the conversion of hydrogen into helium would have to be an accident which just happens to occur at the epoch we are living in. The objection is by no means lethal as the estimate of the amount of helium in the universe may well be wrong by a factor of 100. And even if it were right, their near equality could well be a coincidence. Nevertheless, Hoyle and his collaborators consider it sufficiently decisive to warrant a series of new hypotheses to explain their supposed coincidence. Accordingly they argue that the energy densities are alike because of the way vast tracts of dust seen to be present in galaxies control the evolution of the universe much as the rate of reactions in a nuclear reactor is controlled by cadmium rods. They point out that much interstellar dust is now believed to consist of graphite flakes whereon can condense solid hydrogen under certain conditions. They have tried to specify these conditions by working out the density of hydrogen gas in equilibrium with solid hydrogen for a range of temperatures believed to be characteristic of interstellar space. The temperatures, at which the densities of hydrogen gas corresponding to galaxy formation and to the formation of stars within a galaxy are in equilibrium with solid hydrogen are remarkably close to the 3°K. of microwave background. Consequently the coincidence between the two energy densities is explained by the assumption that the formation of galaxies, and the formation of stars within galaxies, goes on readily when hydrogen is condensed on the grains as a solid, but not when it is gaseous. This means that the formation of galaxies and stars proceeds until the energy from the conversion of hydrogen into helium raises the background temperature enough to evaporate the mantle of solid hydrogen on the graphite grain (about 3°K.) thereby cutting off the process of galaxy formation and star formation within galaxies. Thus a feedback system is set up which keeps the background temperature at the level corresponding to condensation of hydrogen on dust grains very much as a thermostat does in a refrigerator automatically ensuring the equality of the two energy densities.

The question then arises as to why galaxy and star formation go on when solid hydrogen is condensed on the dust grains but not when the hydrogen is gaseous. This leads Hoyle *et al.* to the further supposition that the intergalactic magnetic field may be too high to allow galaxies to form from hydrogen gas. When hydrogen is condensed on grains it can slip across magnetic fields. While it is true that the way solid-state processes invoked by Hoyle *et al.* – the condensation of hydrogen on graphite – control the evolution of the universe may find a place in both the steady-state and the big-bang cosmologies, the big-bang explanation of the microwave radiation seems more natural. In fact, it was actually predicted from theory before it was observed.

Nor is Hoyle's rescue of his much beleaguered steady-state theory against the evidence of counts of radio sources any more effective. As we saw, the log N versus log S curve based on Ryle's count of radio sources made by his Cambridge team favoured the evolutionary big-bang cosmology. In other words, the observations of the Cambridge radio astronomers seemed to show that radio sources are more densely packed at distances corresponding to the earlier days in the history of the universe as predicted by the big-bang theory. But more recent counts of radio sources reported from Australia by a group using the 210-foot Parkes telescope have thrown the issue into the melting pot all over again. For the Australians claim to find no evidence that counts of distant radio sources are in conflict with the requirements of the steady-state theory. There is, however, a crucial difference between the recent Australian and the earlier Cambridge counts. Because the Australian team carried out their survey at a frequency of 2,700 megacycles per second against the much lower frequencies of 178 and 407 megacycles per second of the Cambridge measurements, the Australians suggested that the reason for the discrepancy may be in spectra of the radio sources. The Cambridge reply is that there is, in fact, little discrepancy between the two counts. What there is can be explained by statistical uncertainties caused by the small number of sources observed in the high frequency survey. As G. G. Pooley has pointed out, in the limited area of sky so far observed at 2,700 megacycles per second frequency there are only four or five sources with unusual spectra, which could account for the whole of the discrepancy. However, the present flurry of activity of radio astronomers is churning up more observations than can be brought under the aegis of current theories. A case in point is the latest survey from Cambridge, at 408 megacycles per second, which shows a cut-off in the number of very weak sources. The Cambridge team says it is clear from this that at an early epoch the sources underwent a change which may have had something to do with galaxy formation. While further research on radio sources and galaxy formation will no doubt clarify the situation during the next few years, it does seem that Hoyle's steady-state theory in either of its two versions is not likely to survive the mounting adverse observational evidence piling up against it.

CHAPTER 10

A New Cosmological Principle

THE cosmological principle, perfect or otherwise, is not the only general principle on which a cosmology may be based. A radically different principle has been suggested by Dirac as a new basis for cosmology. To understand it, it is necessary to digress a little into what is usually called dimensional analysis. The essence of dimensional analysis is the well-known second theorem of Buckingham which, shorn of its mathematical technicalities, simply means that any complete physical equation (i.e., any equation in which no relevant quantity is overlooked) can be expressed in a dimensionally homogeneous form. This is a way of saying that the equation can be put in a form which remains valid no matter in what units we may decide to measure the physical quantities appearing therein. Now the simplest condition under which this can happen is that the equation continue to hold when each term in it is multiplied by the *same conversion* factor required to effect the change of units. But what is a conversion factor?

To measure any physical magnitude such as distance, we require first of all a scale or unit on which to measure it. Obviously any given length may be measured in any unit – miles, yards, feet, inches, kilometres, kos, versts, leagues, light years, parsecs, or any of their infinite aliquot parts, fractions, or multiples. We can easily transform a length measured in any unit into any other by multiplying it by an appropriate conversion factor. Thus if 5 is its measure in one unit (say, yards) its magnitude in another unit (say, feet) may be derived from 5 by multiplying it by the conversion factor 3. More generally, if the magnitude of a length is x in one system of units (say, miles) and x' in another (say, millimetres) and a the conversion factor,* then clearly we may pass from miles to millimetres by means of the equation $x' = ax$.

If we think systematically about the conversion factors needed

---

* That is, one mile = a millimetres.

to transform physical quantities from one system of units to another, we find that every physical magnitude has certain 'dimensions' to be written as exponents. Take, for instance, velocity. It is simply the quotient of distance divided by time. If we represent the dimension length by L and the dimension time by T, the dimensions of velocity are L/T, which may also be written LT^{-1}. This symbolism is a device to indicate the power to which the conversion factor in question must be raised in any given case. Thus in the case of velocity, if we change the unit of length by the conversion factor a and the unit of time by the factor b, then the measure of velocity will have to be changed by the conversion factor a/b or ab^{-1}. Similarly density, which is the ratio of mass and volume, will have the dimensions of mass M divided by that of volume L^3, that is, M/L^3, or ML^{-3}. This merely means that if we change the unit of mass by a conversion factor c and that of length by the same factor a as before, the measure of density will have to be changed by the conversion factor ca^{-3}.

Dimensional homogeneity of a physical equation simply requires that the *same* conversion factor will suffice to transform the units of every one of its terms to any other system of units. Clearly if the conversion factor required to transform the system of units is unity, all the terms occurring therein remain unaffected by any change of units. In other words, the terms are pure numbers, or numbers with no dimensions. Such numbers are known as dimensionless. A simple example of a dimensionless number is the ratio of the age of two men, for clearly this ratio remains the same no matter in what units – years, weeks, months, days or any other standard duration – we may choose to measure their ages. Such dimensionless numbers are obviously useful and indeed indispensable if we do not want the arbitrary character of the units of measurement to obtrude itself too much in our statements of physical laws. In fact, Buckingham's theorem referred to earlier as the hard core of dimensional analysis enables us to do exactly this. For the theorem states that not only can a complete physical equation be reduced to a dimensionally homogeneous form but that this form involves *only dimensionless numbers*. In other words, every complete physical equation can be written down as a relation between dimensionless numbers independent of any particular choice of the units of measurement.

Dirac proceeds to construct two such dimensionless numbers, one from atomic and the other from cosmic physics. The one from atomic physics is what is called the force constant. As is well known, there are two different kinds of forces acting between a proton and an electron. First, there is the electric force which, by Coulomb's law, is directly proportional to the product of the charges and inversely proportional to the square of the distance between them. The second force is the gravitational attraction which, by Newton's law, is proportional to the product of the masses and inversely proportional to the square of the distance. In symbols, if e is the charge on a proton or an electron and r is the distance between them, the former by Coulomb's law is e^2/r^2. Likewise, if m_e and m_p are the masses of an electron and proton, the latter by Newton's law is $\gamma m_e m_p/r^2$, where γ is Newton's constant of gravitation. The ratio of these two forces, namely, $e^2/\gamma m_e m_p$, the force constant, is evidently independent of our choice of units of mass, length, and time. It is therefore a dimensionless number or pure constant. From experiment its value is found to be $2\cdot3 \times 10^{39}$.

This is truly a giant among numbers even in cosmology, a realm where such giants abound. Written in the usual decimal notation instead of the compact algebraic notation we have employed, it requires forty digits to be spelled out. Even this may not indicate fully its truly cosmic size. To appreciate its magnitude better, it is necessary only to remark that the very large number of grains of wheat poor King Shirman* was inveigled into promising his sly grand vizier as a reward for his invention of chess is only a twenty-digit number.

Dirac's second dimensionless number, chosen from cosmic physics, is equally gargantuan. If we assume with Milne or Lemaître that the galaxies started receding from a common point

* The allusion here is to the well-known legend of the grand vizier asking for one grain of wheat in the first square of the chessboard, two in the second four in the third, eight in the fourth, sixteen in the fifth, and so on till the sixty-fourth square. The poor king never suspected till it was too late that the total number of grains required to fill the board in this manner would exceed the total world production of wheat during two millennia at its *present* rate of production.

of coincidence at the epoch of creation at cosmic time $t = 0$, the present epoch or age of the universe would, according to Hubble's revised recession law, be about 10^{10} years. Now consider another period of time on the atomic scale, the time taken by light to traverse the radius of an atomic nucleus. This has a value about 10^{-22} seconds. The ratio of these two periods, that is, the present age of the universe reckoned in atomic units of time, is obviously a pure number independent of the units employed. A simple calculation shows that its value is $3 \cdot 2 \times 10^{39}$.

What is the significance of the numerical concordance of these two very large dimensionless numbers? Before we answer the question, it is necessary to consider the objection that this concordance has been deliberately manufactured by an arbitrary selection of the atomic units. If we used slightly different units, the concordance would vanish. For instance, in constructing the force constant we could as well take two protons or two electrons instead of an electron and a proton as we have done and measure the ratio of their electrical and gravitational interactions. But this would not change the value of the force constant by more than a few powers of ten. Likewise, we could decide to reckon the present epoch in terms of a unit of time required to describe a length equal to the radius of an electron instead of an atomic nucleus. This again would not alter its value by more than a factor of similar magnitude. In view of the enormous magnitudes of these numbers, a slight arbitrariness in the choice of atomic units of length and time (which in any case does not alter their value by more than the addition of a few zeros) seems hardly material. We may therefore conclude that the concordance between their values is real and not a product of the arbitrary choice of atomic units, at any rate within the limits of small variations of the order of a few powers of ten.

Dirac considers that such coincidence is presumably due to some deep connexion between cosmology and atomic theory. Accordingly, he surmises that this coincidence may hold not only at the present epoch but for all time, so that, for example, in the distant future when the age of the universe is 10^{50}, measured in atomic units of time, we may expect the force constant to be of the order of 10^{50}. He thus infers that a quantity like the force constant, usually considered a universal constant, must vary with the passage of great intervals of time.

Besides the afore-mentioned two dimensionless numbers, the study of cosmology throws up a class of other very large dimensionless numbers. Such, for instance, is the dimensionless number called the cosmical number N obtained by dividing the total mass M of the universe by the mass of an elementary particle (say, that of a hydrogen atom). Its magnitude is of the order 14×10^{78}. It happens that most of these large dimensionless numbers turn out to be of the order of 10^{39} or sometimes 10^{78}. Dirac then assumes by a natural extension of the foregoing ideas that all those numbers of the order 10^{39} increase proportionately to the epoch or age of the universe, and all those of the order 10^{78} proportionately to the square of the age, the latter number 10^{78} being the square of the former.

This leads him to formulate a new cosmological principle, namely, that 'all very large dimensionless numbers which can be constructed from the important constants of cosmology and atomic theory are simple powers of the epoch (or age) with coefficients of the order unity'. Thus, for instance, the force constant varies directly as the age of the universe at which it is reckoned.

But it is possible further to generalize the principle by discarding the assumption made at the outset that the velocity of recession of each galaxy remains constant. If we give up this assumption, we can no longer speak of a natural origin of time or the zero hour of creation, though we can still talk about the epoch of an event. This has the consequence that in the absence of a natural zero from which to date the epoch, only the difference of two epochs can enter into laws of nature. Consequently the formulation of Dirac's cosmological principle given earlier assumes the generalized form: 'All very large dimensionless numbers which can be constructed from the important natural constants of cosmology are connected by simple mathematical relations involving coefficients of the order of magnitude unity.'

Dirac now proceeds to investigate the main consequences of this assumption to build his theory of cosmology. He begins by expressing the distance between two neighbouring galaxies not in terms of light years, the usual scale of distance reckoning, but in terms of an atomic unit, the radius of an electron. The distance between the galaxies thus becomes a dimensionless number and by Dirac's cosmological principle is some unknown function $f(t)$ of the epoch t of its reckoning. He then brings into the argument the average density of matter in the universe and, assuming conservation of matter, he is able to prove that the age of the universe in these atomic units is only one-fifth of the value ascribed to it on the basis of Hubble's law, that is, only about 2×10^9 years.

This value may seem to land Dirac's theory into the same time-scale difficulty from which the earlier cosmological theories based on relativity were rescued by Baade's revision of the scale of galactic distances. Dirac, however, extricates himself from it by pointing out that a thorough application of his ideas would require the rate of radioactive decay on which the geological estimates of the earth's age are based also to vary with the epoch, so that it would have been a lot faster in the distant past than it is now. Estimates of the earth's age based on the present, much slower, rate of radioactive decay are not comparable.

Having derived the law of recession of galaxies from his principle, Dirac deduces from it the character of our space. As we saw in Chapter 7, space is of three main types according to whether its curvature is positive, zero, or negative. Dirac eliminates the two possibilities that space would have positive or negative curvature by an ingenious type of *reductio ad absurdum* argument. For if space curvature is positive, it is a closed universe – finite but unbounded like the Einstein world we considered in Chapter 7. Consequently the total mass of the universe is finite, and, expressed in an atomic unit such as in terms of the mass of a hydrogen atom, it is again a very large dimensionless number N of the order 10^{78}. According to Dirac's principle, this number should therefore increase as the square of the age of the universe. But if so, the law of conservation of mass would be violated, which is impossible. Dirac therefore concludes that space curvature cannot be positive. The argument ruling out the possibility of negative curvature of space is similar, though more involved. Dirac is thus left with the case of zero curvature or flat space as the only possibility consistent with his fundamental principle and the law of conservation of mass.

Finally Dirac attempts to determine the motion of a free particle under the action of the gravitational field of the universe as a whole in a manner similar to Milne's derivation described earlier in Chapter 8. Here he concedes that since general relativity explains so well the local gravitational phenomenon it must be expected to have some applicability to the universe as a whole. Consequently he is obliged to include orthodox relativity in the framework of his theory. But this raises a difficulty.

As we have seen, the force constant, that is, the ratio

$$\frac{\text{electric force}}{\text{gravitational force}}$$

varies, according to Dirac's principle, directly as the age of the universe. But as with atomic units of time, distance and mass, the electric force between an electron and proton at a constant distance remains constant, only the gravitational force between them can vary. Since this latter force appears in the denominator of the ratio we have called force constant, it must vary inversely as the age of the universe. It therefore follows that the constant of gravitation must be inversely proportional to age. But the gravitational constant in general relativity does not change. To keep it constant, Dirac has to set up a new system of units whose ratio to the old units may vary with the age, so that with respect to the new units the gravitational constant does not vary with the age of the universe and general relativity may be expected to apply. Further, the new unit of mass must not be such that its ratio to the old one varies with age, as in that case we should have the same mass of proton or neutron varying with age and general relativity forbids any variation in the mass of an isolated particle. Consequently the changes in units must be confined to distance and time in such a way as to keep the velocity of light unity. He is thus led to two distinct measures of distance and time, one for atomic phenomena and the other for ordinary mechanical phenomena included under general relativity. The situation is exactly the same as in Milne's theory of two time scales, the t- and τ-times, except that the ratio of the two measures is just the inverse of what it is in Milne's theory.

Dirac's attempt to incorporate general relativity as part of his theory is not as impeccable logically as the preceding part concerned with the deduction of the law of galactic recession and space curvature. The main weakness of this part of his argument is that since the force constant is taken to be a variable, the charge of an electron has to vary if the mass and constant of gravitation in the new units are kept constant. But conservation of charge is as integral a part of relativity theory as conservation of mass. Consequently no theory treating the force constant as a variable can be reconciled with orthodox general relativity.

Dirac's principle is not based on any definite evidence, observational or otherwise. The only reason for assuming it is that it seems a simple way (even though a purely hypothetical one) of explaining the remarkable concordance between large dimensionless numbers of the order of magnitude 10^{39} and 10^{78}. For to do so the observed coincidence of these large numbers need have no intrinsic significance except the purely fortuitous one of being observed together at the same time. That is, the sole common feature of these numbers that need be assumed is the known identity of the date of their observation. Dirac therefore rejects the validity of epistemological attempts* like that of Eddington which seek to discover a 'deeper' reason for their coincidence. Instead, he is content to infer that they vary with the epoch of their observation.

But this assumed variation leads to a difficulty. For example, as we have seen, Dirac's principle requires that the constant of gravitation vary inversely as the age of the universe. Now this variation, as Dr Edward Teller has pointed out, may easily come into conflict with geological evidence. For the well-known relation which connects the masses with the luminosities of stars involves the constant of gravitation. Consequently the luminosity of a star like that of our own sun would also vary with the age of the universe. Dr Teller has calculated that on the basis of Dirac's principle the value of the constant of gravitation some 200 to 300 million years ago must have been about 10 per cent higher

* See Chapter 11.

than now. Corresponding to this higher value of the gravitational constant, the mass-luminosity law yields such a fierce rate of solar radiation as to make the oceans boil and the earth unfit for habitation in any form.

But 200 to 300 million years ago the earth corresponded to what the geologists call the Palaeozoic period, when life had already appeared. It is true that the assumed changes in the gravitational constant may be compensated by corresponding (assumed) changes in other factors, such as the opacity of the sun by changing chemical composition, thus saving Dirac's principle from complete refutation. But if we are to avoid multiplication of factors not warranted by any observation merely to preserve a principle, it would seem simpler to assume that physical constants like the gravitational constant are true constants and do not vary with time so as to ensure a reasonably steady temperature required for the existence of life on earth during the last 500 million years.

Nor can Dirac's principle be justified as a simplifying assumption in cosmology, as was attempted with the perfect cosmological principle of Bondi and Gold. In this respect it is a complete antithesis of the perfect principle. For while the perfect principle allows no evolution to the universe at all and strait-jackets it in the same steady state for all eternity. Dirac's principle postulates as much evolution as it can have. For it means that not only does the universe evolve as a whole but that this evolution proceeds apace according to an ever-evolving scheme of fundamental laws. By thus introducing what is virtually a new dimension in the field of possibilities of change in the pattern of cosmic evolution Dirac's principle greatly adds to the complication of cosmology. Because it makes it infinitely harder for cosmology to sheet-anchor itself, cast as it is in a sea of limitless possibilities, Dirac's principle has seemed to some a counsel of despair.

In spite of Teller's tentative evidence to the contrary and the difficulties that it adds to cosmology, some authorities nevertheless feel that Dirac's principle points to the close connexion between the atomic world and the cosmos which has been dawning on scientists ever since the emergence of universal dimensionless numbers like the force constant and cosmical number. But they think that its full significance will not be apparent till atomic theory is reformulated in a new way in which all variables, and particularly those of length, time, and mass, are expressed in dimensionless form somewhat on the lines suggested by L. L. Whyte. That such a dimensionless reformulation of atomic theory is already overdue seems clear from the inability of the older dimensional methods to treat most of the fundamental physical problems that have been piling up unresolved during the past twenty-five years.

One cause of the difficulty seems to be a confusion between several essentially different modes of interlocking between the different fundamental units, now recognized to be an inherent feature of the normal dimensional methods hitherto employed. For example, if we employ the velocity of light c in the usual dimensional form as so many miles per second, we assume a connexion or interlocking of one sort between space-unit (mile) and time-unit (second). Now if we employ another expression like e^2/h, where e is the charge of a proton and h Planck's constant, we imply a connexion of a second, different kind between the same two types of units. That is why Dirac, as we saw, was led to two distinct measures of time and distance, one for atomic phenomena and the other for macroscopic phenomena.

In many cases such duplicity multiplies several times over into an embarrassing ambiguity. It has been suggested that the use of dimensionless methods alone, whereby the role of fundamental units is greatly minimized, can successfully avoid such confusion. Only when atomic physics learns to speak in dimensionless language will it be possible to understand the message that cosmology is already broadcasting to us in this language. Dirac's principle, the first crude translation of the cosmological message, interprets cosmic history as the evolution or unfolding in time of certain dimensionless ratios of significant physical quantities. These authorities therefore expect to hear more of it if and when atomic physics becomes truly dimensionless on the lines envisaged by them.

As we saw, Dirac was led to formulate his new cosmological principle by the coincidence of large dimensionless numbers such as the force constant and the age of the universe reckoned in atomic units. Jordan's approach is more cautious, but only at the beginning. Instead of trying to derive his model of the universe deductively from certain physical principles such as the field equations of relativity, he adopts what he calls an essentially inductive approach – dimensionless analysis of certain empirically derived physical quantities. These are primarily six magnitudes which comprise the sum of our knowledge of the structure of the

TABLE 5

Item No.	Physical Quantity	Symbol	Observed Value in C.G.S.	Dimensions	Remarks
1	Velocity of light	c	3×10^{10}	LT^{-1}	
2	Constant of gravitation in general relativity	$f = \dfrac{8\pi\gamma}{c^2}$	2×10^{-27}	LM^{-1}	Symbol γ is the Newtonian gravitational constant and f the relativity gravitational constant
3	Maximum age of the oldest heavenly body	A	10^{17}	T	Jordan assumes that all the known stars and star systems could not have existed much longer than a certain maximum time which is apparently four billion years
4	Density of matter in the universe	$\rho*$	10^{-28}	ML^{-3}	This is based on the estimated masses of a cross-section of typical galaxies. The actual value is uncertain by a few powers of ten, but Jordan adopts that given here
5	Hubble's constant	$H*$	10^{-17}	T^{-1}	
6	A constant in galaxy counts	$R*$	3×10^{27}	L	For an explanation of this constant see the text

* The older values of ρ, H and R shown in Table 5 have not been changed as Jordan's order-of-magnitude argument remains unaffected by any subsequent revisions in values of ρ, H and R which are in any case still uncertain by a few powers of ten.

universe on the large scale. Table 5 lists them as also their observed values and physical dimensions.

The significance of the first five constants listed in Table 5 is fairly clear. To understand the meaning of the sixth we may remark that if the galaxies are uniformly distributed throughout a flat Euclidean space of zero curvature, the number of galaxies in a thin spherical shell at any distance r can be shown to be proportional to $4\pi r^2$. But Jordan, in order to avoid Olbers's paradox, assumes a curvature of space which reveals itself in a thinning out of galaxies with increasing distance.

This means that the number of galaxies in a thin spherical shell at distance r will not be proportional to $4\pi r^2$ as in a flat Euclidean space but to a slightly smaller quantity instead, namely $4\pi r^2$ $[1 - (r^2/R^2)]$, where R defines the constant in galaxy counts listed in Table 5. It is therefore a measure of the thinning out of galaxies with increasing distance resulting from the curvature of space. The value given for this coefficient of rarefaction of galaxies at increasing distances in our table is the one adopted by Jordan based on the early counts of Hubble. Its actual value is still unknown. A large margin of error in the value adopted by Jordan is therefore by no means unlikely.

With this assembly of cosmological data, Jordan proceeds to show that three and only three dimensionless numbers can be built out of them. As may be readily verified, they are AH, R/cA and $f\rho c^2 A^2$. Their values are shown below:

$$(a) \quad AH \quad\;\; = 1,$$
$$(b) \quad R/cA \;\;\; = 1,$$
$$(c) \quad f\rho c^2 A^2 = 1\cdot8.$$

In view of the uncertainties of the data we may consider the values of all three of these dimensionless numbers to be about 1. Jordan considers it an extraordinary thing that all of them should be of the same order 1. This coincidence is his justification for interpreting them in the following way: R, the constant of nebular rarefaction in galaxy counts, is interpreted as the radius of curvature of a closed Riemannian space. In the same intuitive way the Hubble effect is interpreted as the real recession of galaxies in spite of attempts to explain it in other ways. With this interpretation, the empirical equation (b), which equates R and cA, means in effect that the radius of the universe increases (or the periphery of the universe expands) with the velocity of light. The equation (a), namely $AH = 1$, implies that space has been expanding with the velocity of light ever since its origin, when all the galaxies were in coincidence and the universe confined within a mere point something like the eye of a needle. It therefore reiterates what is said in equation (b).

To interpret the third equation (c), Jordan resorts to a little algebraic manipulation whereby he is able to show that it can be transformed into a seemingly new one: $fM = R$,* where ρ, the density of the universe in equation (c), is replaced by its total mass M. But in this guise it leads to what Jordan himself acknowledged to be an astonishing conclusion when he first came to it. For if R is continually increasing in a universe whose radius is expanding with the velocity of light, then fM can no longer be regarded as invariable.

Of the two factors, the gravitational constant f and mass M of the universe, at least one must vary with time. But if we replace f by its value $8\pi\gamma/c^2$, shown in the above table in the equation $fM \approx R$, we find $8\pi\gamma M/R = c^2$, or

$$\frac{8\pi\gamma M^2}{R} = Mc^2. \tag{1}$$

Now Einstein has shown the equivalence of mass and energy, the equivalence relation being $E = mc^2$, where E is the rest energy of a particle of mass m. Consequently the term Mc^2 on the right-hand side of (1) may be interpreted as the sum total of the rest energies of the masses of the stars in the universe. The term on the left of (1) is also simple to interpret. It is merely the potential energy of gravitation for the whole universe, which is known to be negative.† Equation (1) then can be taken to mean that the total energy of the universe is exactly zero, the positive rest energy of the stellar masses Mc^2 cancelling the negative

* Since volume of the universe of radius R is proportional to R^3, its mass M will be proportional to ρR^3, ρ being the density. Substituting M/R^3 for ρ and R for cA in equation (c) we have $fM/R^3 \cdot R^2 \approx 1$ or $fM \approx R$.

† Except for the numerical constant 8π.

gravitational energy $\gamma M^2/R$ due to them. The evolving universe thus passes through a sequence of states in each of which its total energy remains zero. It is in this way that Jordan secures the conservation of energy by permanently pegging it at zero!

To proceed further Jordan requires some more dimensionless numbers, as those which can be produced from the six cosmological quantities c, f, A, H and R have all been used up. This he does by borrowing the dimensionless numbers with which Dirac began, namely the force constant, the present age of the universe A in terms of atomic units, and the total number N of particles in the universe.

We found that the force constant was of the order $2 \cdot 3 \times 10^{39}$, A in terms of atomic units $3 \cdot 2 \times 10^{39}$, and N, 10^{79}. In accordance with Dirac's principle, the force constant would be directly proportional to age A, and N to its square A^2. We saw earlier that Dirac accepted the former while rejecting the latter as being inconsistent with the law of conservation of mass. That was also Dirac's reason for rejecting curved space in his cosmology. Jordan is of the view that Dirac was led away from such a view by the fear of contradicting the principle of conservation of energy. But since this difficulty, according to him, has now been resolved by his equation (1) whereby the total energy of the universe remains conserved at zero, he thinks he is free to deduce from Dirac's principle that the number N of particles in the universe increases as the square of its age A. This means that there is continual creation of matter in the depths of the universe.

To answer the question as to how and where this creation of matter takes place, Jordan abandons the universe and proceeds to subject the individual stars to dimensional analysis in the same way as he did with the whole universe earlier. The maximum mass of a star is known to be about fifty solar masses, or 10^{35} gm. Expressed in atomic units (say, the mass of a hydrogen atom), it yields another large dimensionless number of the order 10^{60}. By Dirac's principle this must be a simple power of the age of the universe, at present a number of the order 10^{39}. It therefore follows that the present value of the upper limit of stellar mass (10^{60}) must be a simple power of A, namely $A^{3/2}$. But we saw earlier that Dirac's principle also requires the gravitational constant f to vary inversely as the age A. Consequently stellar mass must be proportional to $f^{-3/2}$. Jordan takes as a confirmation of his conjecture the fact that astrophysical theories of the internal constitution of stars also lead to stellar masses being proportional to $f^{-3/2}$.

Jordan now examines the exact *modus operandi* of the creation of new matter. He considers that the principle of conservation of energy – its permanent pegging at zero level – must hold rigorously not only for the universe as a whole but also for any finite region where this creation occurs. If then we consider a homogeneous sphere of radius R_0 and mass M_0 in an approximately flat Euclidean space, and if the analogue of equation (1) for this region (i.e., $fM_0 = R_0$) holds, then its total energy, like that of the universe as a whole, also remains zero. That is, a dispersal of the material of this region into a gas cloud spread over infinitely remote regions would require as much energy to overcome their gravitational force as the sum of their rest energies.

This leads Jordan to conjecture that cosmic creation of new matter does not take place piecemeal fashion in diffuse hydrogen atoms scattered everywhere, as in the continuous-creation theories of Hoyle and Bondi, but by the sudden, spontaneous and instantaneous appearance of large lumps of new matter so big that the principle of conservation in the sense that the total energy of the lump retains the value zero holds for the entire lump as a whole. Jordan considers that the density of each lump can only have one value, that of the atomic nucleus. That is, it is of the order of a proton mass spread through the interior of a hollow shell of atomic dimensions. This requires that the size of the lump appearing at any given time be a function of the gravitational constant at that time. Since drops are obtained at present containing 10^{60} particles, their mass would be proportional to $A^{3/2}$, or $f^{-3/2}$. Each lump thus appears in the space of our universe as a star in its own right. This means rejecting the accretion process of Hoyle and Lyttleton as the origin of stars by the gravitational condensation of interstellar material.

In Jordan's cosmology stars come literally out of the blue like Athena leaping forth from Zeus's brain mature and in complete armour. The supernovae that we see in the sky suddenly flaring up in astounding luminescence once in a few centuries in each galaxy are these celestial Athenas – the newly created lumps of matter. To give the supernova origin of stars a semblance of physical reality, Jordan suggests that at any time there may be within our larger universe several smaller three-dimensional spaces totally unconnected with the former. The justification for this seemingly strange assumption is that if we make a 'cut' of

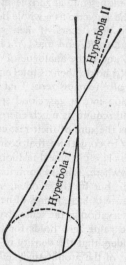

Figure 35. The figure shows the two unconnected branches of the hyperbola I and II obtained by a plane cutting a cone.

our four-dimensional space-time manifold at any time, we may obtain several unconnected slices of three-dimensional space, each isolated from the main universe.*

Now consider any such isolated 'cut' of three-dimensional space. Such a slice, existing quite apart from the main universe,

* A three-dimensional analogy will make Jordan's meaning clearer. If we make a cut of three-dimensional cone by a plane, we may obtain two isolated sections of a curve known as hyperbola (see Fig. 35). The two unconnected sections or branches of the hyperbola are the two-dimensional analogues of unconnected three-dimensional slices of space.

can coalesce with it through sheer unfolding of time. If we further suppose that in this small world the mass density remains constant and the gravitational constant f varies inversely as the square of its age A, then by Jordan's equation (c) the conservation of energy will hold in this small world as well. But during the course of its evolution this small world may reach a stage when its value of f becomes identical with that in the larger universe. If this happens when the small slice of three-dimensional space coalesces with the main universe, the material content of this slice appears as an embryonic star in the larger universe.

With this theory it never rains but it pours. The newly created material does not appear as a sort of imperceptible cosmic drizzle as in the earlier version of the continuous-creation hypothesis of Hoyle and Bondi but in large cosmic hails of supernovae. Stars come on the celestial scene as fully formed lumps of matter in the shape of supernovae as though they floated into our ken from some extraneous spatial dimension in the manner first conjectured by Jeans when he described spiral arms of galaxies as 'singular points' through which matter is poured into ours from another part with which we are not connected. This provides an opportunity to confront the theory with observation, as the rate of supernova emergence can be calculated on its basis. The number of stars per galaxy and the number of galaxies in the universe are interpreted as proportional to $A^{1/4}$. The average rate of occurrence of supernovae can then be shown to be about one per galaxy per year. As we noted earlier, the actual rate is much slower – about one to one-fifth per galaxy per century.

Jordan's development of Dirac's cosmological principle depends on two main additional assumptions. First, he uses Einstein's formula $E = mc^2$ relating the mass m with its energy E to *identify energy with matter*. On this basis he postulates in effect that the loss of gravitational potential energy of dispersing galaxies leads to the *creation* of new matter. But gravitational potential energy is simply a mathematical invention devised by Lagrange to make calculations from Newton's law of gravitation easier. Newton himself was quite unaware of it. Later the term *kinetic energy* was added so that mechanical problems could be dealt with in a very general way without having to take account of all the minute details of forces at play.

In other words, the original *raison d'être* of energy was a simplified mathematical treatment for avoiding considerations of details of physical processes. It is thus no more than a mathematical construct obtained by multiplying some numbers obtained by measurement – a merely metrical feature of a physical process exactly as ton-miles is a metrical feature of the process of carrying goods. It is certainly useful to express the capacity of a transport system in terms of ton-miles. It may in theory be even possible to envisage transport systems in which the ton-miles of freight carried are conserved. A simple instance of such a system is the case of a three-ton lorry that could make one trip 100 miles long, two trips each fifty miles long, or three trips each 100/3 miles long, and so on. Such a transport system would be subject to the conservation law of ton-miles as its total haulage potential is permanently fixed at 300 ton-miles no matter what number of trips the lorry made.

The utility of this conservation law for the transport system is that it enables us to calculate the length of the haul given the tonnage we have to carry, or inversely the tonnage we may carry given the length of its haul. But if we make it the basis for speculating that the transport system could in some magic way transform three tons hauled 100 miles into six tons by *merely* halving the length of haul, we should certainly be making the inference at our own peril. There is no justification in the conservation law of ton-miles for inferring that somehow we may conjure tons by paying out in miles or vice versa. It is no answer to say that the analogy is not applicable because the distinction between mass and energy is obsolete in modern physics and that the two are absolutely synonymous. For the only reason for the proposed abolition of this distinction is that the *disappearance* of matter is observed to be accompanied by radiation or emergence of energy, the two being related by Einstein's equation. It seems that the root cause of the difficulty here is inability to distinguish 'mass' from 'matter' which is characteristic of all creationist cosmologies, whether continuous like that of Bondi and Gold or spasmodic like that of Jordan.

The second assumption of Jordan is even more naïve. To assign a definite physical reality to a purely mathematical construct such as space-time is to press a useful description to the breaking point. There can be no meaning in the picture Jordan tries to paint of supernovae in independent spatial worlds suddenly floating into our view as if from nowhere. That these pictures are painted at all – of material lumps falling out of abstract mathematical inventions like gravitational energy or supernovae bursting into our ken by the coalescence of isolated three-dimensional slices of space-time with our own universe – is the result of our cosmological Odysseuses listening too intently to the siren song of metrical symbols without tying themselves firmly to the bark of empirical observation.

A more plausible rationalization of Dirac's principle that is free from the charge of neglecting observation is that recently provided by Brans-Dicke cosmology reared on a new theory of gravitation. Their theory, in essence, is a blend of two of the only theories of gravitation that have won universal acceptance, namely those of Newton and Einstein. As we know, in Newton's theory a body is accelerated in an absolute space by the gravitational forces exerted by other bodies. In Einstein's theory there is no gravitational force. Only the four-dimensional space-time is distorted or curved by gravitating bodies and a small test object moves along the 'shortest path' in this curved space-time continuum. The Brans-Dicke theory of gravitation is far more complex. But it boils down to a description of gravitation as partly due to Newtonian attraction treated relativistically within the framework of Einstein's general relativity and partly due to the 'curvature' of space.

While the predictions made by the theories of Einstein and Brans-Dicke agree in most respects, such as planetary motions, gravitational red shift, etc., they do differ in some others. The most important of these differences from our present point of view is the fact that Brans-Dicke theory requires a steady weakening of gravitation in relation to other, e.g. nuclear and electromagnetic, interactions. Their calculation shows that the constant of gravitation decreases at the rate of one part in 10^{11} per year.

This is about ten times slower than that under the Dirac-Jordan cosmology described earlier.

R. H. Dicke has suggested a neat geophysical test of the theory. Adopting atomic units of length and time a slow decrease in the strength of gravitation relative to the nuclear and electromagnetic interactions leads to a lengthening of the planetary orbital periods and an increase in their orbital radii. Telescopic observations over the past two centuries would be sufficiently precise to reveal a change in the planetary period induced by a gravitational weakening as small as 10^{-11} parts per year provided a suitable atomic clock were available for this period. Since real atomic clocks have not been available nearly long enough to be useful, recourse has to be had to a somewhat inferior substitute – the earth spinning around its axis. This raises a number of ticklish problems. But taking them all into account, Dicke finds observational evidence of a fractional weakening of gravitation at a rate from $2 \cdot 1$ to $4 \cdot 2 \times 10^{-11}$ parts per year. Unfortunately, neither this test nor some others he has suggested are sufficiently compelling to enforce acceptance of their theory. But he has suggested some still more crucial experiments capable of deciding between the two rival theories as, for instance, measurement of the precession of the axis of a top spinning in a sputnik.

If future observations do prove the validity of Brans-Dicke cosmology, it will neatly resolve the age discrepancy between the ages of stars and galaxies believed to be twelve to twenty-five billion years in a universe whose putative age is only ten to thirteen billion years. For their theory shows that the strong gravitational interaction in the past would brighten the stars and shorten their lives. Dicke has shown that with the gravity weakening $1-2 \times 10^{-11}$ parts per year a consistent age pattern can be obtained in this way, a star apparently fifteen to twenty-five billion years old being actually six to eight billion years of age.

CHAPTER 11

Cosmology a priori

ALTHOUGH historically Eddington's cosmology preceded that of Dirac and Jordan, its starting point is similar to theirs. Like them he begins by constructing a series of dimensionless numbers from a number of primitive constants of physics so as to whittle down the role played by the arbitrary choice of fundamental units used in their measurement. But instead of the six magnitudes adopted by Jordan, Eddington starts with the six primitive constants of physics shown in Table 6.

TABLE 6

Item No.	Physical Quantity	Symbol	Observed Value in C.G.S.	Dimensions
1.	Velocity of light	c	3×10^{10}	LT^{-1}
2.	Constant of gravitation	γ	$6 \cdot 67 \times 10^{-8}$	$M^{-1}L^3T^{-2}$
3.	Charge of an electron	e	$4 \cdot 8 \times 10^{-10}$	$M^{1/2}L^{3/2}T^{-1}$
4.	Mass of an electron	m_e	$9 \cdot 106 \times 10^{-28}$	M
5.	Mass of a proton	m_p	$1 \cdot 672 \times 10^{-24}$	M
6.	Planck's constant of action	h	$6 \cdot 6 \times 10^{-27}$	ML^2T^{-1}

From these six physical constants he constructs the following dimensionless numbers:

$$\text{(i)} \ \frac{m_p}{m_e}, \quad \text{(ii)} \ \frac{hc}{2\pi e^2}, \quad \text{(iii)} \ \frac{e^2}{\gamma m_p m_e}$$

Number (i) is the mass-ratio of a proton and electron. Its observed value is about 1,836, as may be computed from the values of m_e and m_p shown in Table 6.

Number (ii) is a mysterious dimensionless number called the *fine structure constant*, which first appeared in spectroscopy. That it is a true constant in the sense that its value does not vary in time seems to be assured by observational evidence. The reason is that but for this permanence the spectral lines in a receding galaxy would be shifted in a complicated way and not by the same uniform amount as actually observed in the Hubble effect. Its value is found to be close to 137.

Number (iii) is the *force constant* we have already encountered in Chapter 10. It is simply the ratio between the electrical and gravitational forces between an electron and proton. For at any distance r between them, the former, by Coulomb's law, is e^2/r^2 and the latter, by Newton's law, is $\gamma m_p m_e/r^2$, so that their ratio is $e^2/\gamma m_e m_p$. This is precisely Eddington's number (iii). Its observed value, as we saw before, is of the order of $2 \cdot 3 \times 10^{39}$.

To these three dimensionless numbers Eddington adds a fourth, the cosmical number N representing the total number of elementary particles in the universe. We found it to be of the order 14×10^{78} in the last chapter. These four numbers, in Eddington's view, furnish a 'natural and complete specification for constructing a universe'. For according to him their values are not what they are 'by the whim of Nature'. They are the inevitable consequence of certain basic assumptions underlying the very mode of our measurement so that to the all-wise observer of observers the exact value of these constants is implicit in his first glimpse of an experimental physicist: 'I lifted up mine eyes, and looked, and behold a man with a measuring line in his hand.' (Zechariah ii, 1.)

What then is the basic theory of measurement that leads inescapably to the exact determination of these four basic dimensionless constants which determine the structure of the universe? This theory is a development of the Machian principle that the behaviour of any part of the cosmos is determined by all its parts. Until recently physics as a whole, both molar as well as microscopic, developed independently of any considerations as to the behaviour of the remoter parts of the universe. Any physical field could be thought of as existing anywhere. But a theory concerned with the basic structure of the universe must take the whole range of physical systems from the nucleus to the cosmos in its stride. In order to do so Eddington divides the universe into two parts, the object system studied in atomic physics consisting, for example, of nuclear particles, and the background environment of such systems, which comprises everything not specifically included in the object system.

While the object system may be studied without any reference to the background environment, any such study which treats the system as though the rest of the universe did not exist is obviously meaningless. Thus, for instance, if the object system under study is the behaviour of a single atom, it would be futile to develop formulae for its behaviour in conditions which imply that like Crusoe it is the sole inhabitant of the entire universe. For 'we can no more contemplate an atom without a physical universe to put it in than we can contemplate a mountain without a planet to stand it on.' But if even the behaviour of a single atom is to be studied in relation to the rest of the universe, some simplifying assumption must be made to make the problem tractable at all. Eddington therefore begins with the consideration of simple object systems, such as a pair of electrons or protons in interaction in a very simple environment, such as a universe of uniformly distributed particles at zero* temperature so that they can be assumed to be at almost exact rest. He calls such an environment a uranoid.

In theory a uranoid is thus merely an ideally simplified universe exactly like a geoid in geodesy. In geodesy we conceive of a fictitious uniform sphere from which the actual earth is obtained by adding or subtracting material at various points so as to produce, for instance, the flattening at the poles, elevation of mountain peaks, depression of canyons and the like. In an analogous manner the uranoid is conceived of as a structure formed by imposing the condition that each particle in the universe contributes half 'its existence to the existence of the uranoid'. In order to see how each particle gives away, so to speak, its 'half' existence to create the universe in which it adventures, it is necessary to digress a little on Eddington's concept of existence.

Leaping for once clear of the metaphysical conundrums of existence which have plagued philosophy since Plato's *Theaetetus*, Eddington begins by mathematizing the primitive notion of existence, namely that an entity either exists or it does not exist. In other words, he denotes the structure of this concept of existence by a two valued mathematical yes-or-no symbol J

* The zero temperature on this scale is equivalent to $-273°$ C.

which may be assigned either the value 'yes' if the entity exists or the value 'no' if it does not. We may also designate the value 'yes' by the symbol 1 and the value 'no' by the symbol 0. Now we can never observe a single entity. Observation only discloses a relation between two entities, and the simplest relation we can consider is that between two elementary particles with simple existence symbols J_1 and J_2 respectively. This relation exists only if both particles exist, that is, if both J_1 and J_2 have the value 1. But if either or both do not exist, the relation too naturally cannot exist.

Thus the conceptual frame in which physical knowledge is expressed involves a simple existence symbol J for a single entity and a double existence symbol $J_1 \times J_2$ for relations between two entities. Now a relation between two relations will exist only if both the relations exist so that it should be assigned a quadruple existence symbol such as $(J_1 \times J_2) \cdot (J_3 \times J_4)$. We could conjure out of quadruple existence symbols octuple and higher symbols, but Eddington cautions us against unnecessarily ascending the inflationary spiral of an ever-expanding notation.

If the object is merely to construct a structure, the relation of relations can be represented by the same set of symbols as the relations themselves so that there is no need to go beyond the double existence symbol $J_1 \times J_2$ to the quadruple existence $(J_1 \times J_2) \cdot (J_3 \times J_4)$. But when we jump from the structure of the existence of individual entities to that of the entire universe, we have to give its structure an existence symbol independent of the existence symbol of its individual particles. For obviously the universe could exist without the existence of one or more of its constituent entities. To do this Eddington duplicates the existence of each element in order to enable it to play a dual role – one as a part creator of its own environment and the other as an individual self with its own life to live. Thus as a contributor to the structure of the universe it is assumed to exist continually, but as an independent entity it may or may not exist. This has the consequence that the existence symbol J for any entity is duplicated into $2J$ from which is then abstracted the part which represents its existence as a contributor to the structure.

In other words, the independently existing particle has the

existence symbol $K = 2J - 1$. If J has the value 1, that is, if the individual particle exists, then K has the value 1, which means a positive contribution to the structure of the universe. In other words, a particle existing in its own right makes its presence felt in the uranoid. But if $J = 0$, that is, if it does not exist, then K has the value -1, that is, it leads to a definite subtraction from the structure of the uranoid. In other words, the non-existence of the particle now is not a mere negation denoted by a zero but a gap (denoted by -1) occurring in the structure. In this way Eddington arrives at his conception of a uranoid, which is built up by including in it half the existence of each particle, leaving the other half for its own independent existence.

But with this change in our point of view the particle is now considered as an embodiment of a relation to the whole structure of the universe or uranoid. Instead of assigning to it a yes-no symbol J when we thought of it as an entity existing in its own right independently of the universe, we now assign to it an independent existence symbol K, whose value depends on whether a particle having that relation to the universe exists or not. This means that the relation of the particle to the universe is conceived of as a pre-existing state and the value of the existence symbol K depends on whether that state is 'occupied' or 'unoccupied'.

Now if it could be shown that within the framework of physical measurement there can arise only a certain number, say, Z of such independent 'occupation' states which particles could occupy, then the number N of particles in the universe must be $2Z$. For Z, the number of occupation states in the universe, provides for only the particles incorporated in the uranoid, which, as we saw, is built up by each particle's contributing only 'half' its existence. This is not, Eddington hastens to add, necessarily the precise number of particles actually out there in the universe. But it is the number which anyone who accepts certain elementary principles of measurement must think there are if he is to remain consistent. The question then arises as to what these principles are that lead us into thinking so.

To discover them Eddington recalls that an observation is merely a relation between two entities or particles each with an independent existence symbol J_1 and J_2 respectively. It therefore follows that the existence symbol for an observation is the double existence symbol (J_1, J_2) which may have any of the following four values:

$$(1, 1), \quad (0, 1), \quad (1, 0), \quad (0, 0).$$

The relation, of course, exists only if both particles exist. There is therefore only one mode of existence for an observation or relation and three modes for its non-existence. Let us denote these four possible modes of existence (and non-existence) of an observation by (x_1, x_2, x_3, x_4). If (y_1, y_2, y_3, y_4) is the corresponding symbol for the modes of existence of another observation, then four varieties of x's can combine with four varieties of y's in $4 \times 4 = 16$ ways.

Of these sixteen ways four are symmetrical combinations like $x_1 y_1$, $x_2 y_2$, $x_3 y_3$, $x_4 y_4$, and the remaining twelve are unsymmetrical combinations like $x_2 y_3$, $x_4 y_1$ and so on. But clearly to each unsymmetrical combination like $x_2 y_3$ there corresponds its counterpart $x_3 y_2$ obtained simply by interchanging the subscripts 3 and 2 between x and y. If both x and y modes are to be treated as indistinguishable, the distinction between $x_2 y_3$ and $x_3 y_2$ has to be obliterated. Eddington does so by postulating that of the twelve such unsymmetrical combinations only half are *active* and the other half remain dormant.

These dormant ones are only activated when electrical characteristics of a particle are manifested. In the neutral uranoid which is the standard environment for the present they have nothing to act on. Consequently, we are left with only six (active) unsymmetrical combinations which, along with four symmetrical ones yield in all ten combinations. This number 10 is basic in Eddington's theory of measurement, as it defines the number of specifications required to identify a standard particle in a neutral uranoid. It may be mentioned that in Eddington's theory the term 'particle' is defined merely as the *conceptual carrier of a set of specifications*, just as a rod may be defined as a vehicle for the transport of an abstract quality like length. We shall require this number 10 in the later development of Eddington's theory of physical constants.

Now let us pass from an observation to a measurement; the

number of entities involved now becomes four. For two entities are required to furnish an observation, and two more to furnish the comparison observable used as standard. For example, in a measurement of distance, the extension between two objects is compared with the extension between two others such as that between two graduation marks on a standard scale. Again each of these two observations whose comparison provides the measurement has its own double existence symbols, say, (J_1, J_2) and (J_3, J_4), wherein each J can take either of the two values 0 and 1. The existence symbol for a measurement is therefore the quadruple existence symbol (J_1, J_2, J_3, J_4) which may have any one of the sixteen possible values like

$$(0, 1, 1, 1), \quad (0, 0, 1, 1), \quad (1, 1, 1, 1), \quad \text{etc.}$$

The value $(0, 1, 1, 1)$ of the quadruple symbol means that the first entity does not exist but the other three do. Likewise $(0, 0, 1, 1)$ means that the first two do not exist but the last two do. It is obvious that measurement can exist only if all the four entities do, so that while there is one mode of existence of a measurement corresponding to the value $(1, 1, 1, 1)$, there are fifteen different modes of its non-existence. We may denote these sixteen different modes of existence of the quadruple existence symbol associated with a measurement by sixteen E-numbers, $E_1, E_2, E_3 \ldots E_{16}$. Another measurement yields in the same way another set of sixteen modes of existence denoted by the F-numbers $F_1, F_2, F_3 \ldots F_{16}$. Now sixteen varieties of E and sixteen of F yield $16 \times 16 = 256$ kinds of EF numbers. Of these 256 EF numbers sixteen are symmetrical types like $E_1F_1, E_2F_2, \ldots, E_{16}F_{16}$ and the remaining 240 unsymmetrical like, say, $E_7F_{10}, E_{12}F_{16}$ and so on. But to each unsymmetrical symbol such as E_7F_{10} there corresponds its analogue $E_{10}F_7$ obtained by simply transposing the subscripts for E and F. Eddington considers that the feature of quantum theory according to which the elementary particles are so much alike that it is impossible to tell which is which requires that no distinction be made between pairs of unsymmetrical symbols like E_7F_{10} and $E_{10}F_7$ for otherwise it would be possible to say which particle has E modes of existence and which has F. This means therefore that of the 240 unsymmetrical EF symbols only half, or 120, are *active* and the rest dormant. If we add to the 120 active symbols the sixteen symmetrical (active) symbols, we find the total set of 256 symbols divides itself into two groups, one of the active $120 + 16 = 136$ *EF* symbols and the other of 120 dormant symbols.

It is no doubt possible to extend the construction of a double frame of *EF* numbers to triple or more highly multiple frames, but Eddington considers that we must not let ourselves run into an orgy of multiple-frame building. The reason is that for all ordinary purposes of physics the double *EF* frame suffices to provide a base for all measurements such as the basal measure, 'the complete energy tensor' with 136 independent components, 'which includes the ordinary energy tensor as well as spin components and other consequential variates'. For this reason Eddington accords to this number 136 the same importance in his fundamental theory as the number 10 derived earlier in an analogous manner from the double existence symbol $J_1 J_2$.

In spite of Eddington's counsel against letting ourselves run riot with multiple-frame building schemes, he finds it necessary to do exactly this for one last occasion. We saw how we could stop at a double existence symbol evolved to denote an observation or relation between two entities. In that case we should have to denote the structure of a relation of relation by the same symbol. But this would not provide a sufficiently fine structure to provide a basis for measurement which involves four entities. We were thus obliged to employ a quadruple existence symbol to evolve the *EF* frame in which each measurement can have 136 different components. As we saw, this suffices for practically all purposes. But although measurement is primarily a process involving four entities, the conceptual interpretation of measurement postulates in addition the existence in the structure contemplated of something else.

This something is what Whitehead calls a background of 'sameness' or uniformity behind the changing pattern of the complex of processes which is essential to the emergence of the concepts of number and measurement. According to Eddington, in the actual universe there exists a basis which is uniform to a very high approximation and this serves for almost all purposes. But to the much higher approximation required in the calculation of N, the ideal exact basis of uniformity does not exist. The very finitude of N causes the breakdown of the approximation. It is in order to take account of this breakdown of approximation that Eddington permits himself one more last 'orgy' of multiple-frame building.

From the quadruple existence symbol (J_1, J_2, J_3, J_4) we may derive the octuple existence symbol $(J_1, J_2, J_3, J_4) \times (J_5, J_6, J_7, J_8)$, in which each J as usual can have either the value 1 or 0. This means that the total number of possible values of such an octuple existence symbol is 256 instead of sixteen as in the case of the quadruple symbol $(J_1, J_2)(J_3, J_4)$ from which we derived the double EF frame. The octuple existence symbol thus yields a multiple frame $EFGH$ which is superfine enough to take account of measures having 256×256 active and dormant components. In other words, the $EFGH$ frame is simply the result of splitting each of the sixteen components of the EF frame into sixteen components. Such a frame can be shown to have 2^{256} as its 'grid' number. This is a way of saying that it has room for only 2^{256} packets of what we earlier called independent 'occupation' states.

Now Eddington shows that each packet is a bundle of 136 – the revealing number we have already encountered before – independent occupation states. It therefore follows that the total number Z of independent occupation states possible in the universe is 136×2^{256}. Since, as we saw before, the cosmical number N is twice Z, the value of N is $2(136)(2^{256})$, a number of the order of 10^{79}. To appreciate the full extent of its enormity, imagine a super-chessboard obtained by piecing together four ordinary chessboards. Such a super-chessboard will naturally have $8 \times 2 = 16$ squares along each side so that the total number of squares in it will be $16 \times 16 = 256$. If we now fill its squares in the manner prescribed by King Shirman's sly grand vizier, that is, by placing one grain of wheat in the first square, 2 grains in the second, 2^2 in the third, and so on till 2^{255} grains in the last, 256th square, the total number of grains on such a super-board will have to be multiplied by 2×136 times to obtain Eddington's cosmical number N.

Having deduced (after a fashion) the cosmical number N,

Eddington shows that it lies at the basis of the architecture of the world. For example, as we may recall, relativity theory requires the presence of matter in space to manifest itself as curvature of space. But the precise extent of this curvature is determined by the cosmical number N. To derive the connexion between N and the curvature of space, he has to delve deeper into cosmological theory. Starting with the standard uranoid he finds that it has two fundamental units of length which, though interconnected, may be treated independently. The first one is connected with the impossibility even 'in principle' of performing the famous cosmic calculation posed by Laplace.

Laplace once remarked that given the initial position and velocity of every particle in the universe at any particular instant and given all the forces at work in nature, a super-intelligence could calculate with precision the entire past and future history of the cosmos. But position at a point in space and velocity at an instant of time depend on two abstractions, namely that of 'point' as a pure location devoid of length, and that of 'instant' as an extensionless duration.

The underlying assumption by which these abstractions are evolved is the hypothesis that any length or duration, no matter how short, can in principle be measured. The new developments in modern physics, however, show that this assumption is no longer valid in the subatomic regions. The only way whereby we can measure the length between two adjacent points is by means of light. But two points whose distance apart is less than a certain function of the wavelength of light by which they are observed form only one image and are therefore seen as coincident. Consequently, the physical means of measurement impose a natural limit to the precision of measurement. Whatever we may do we cannot surpass this limit.

Eddington introduces this natural limit to the precision of linear measurements as a fundamental unit of length and denotes it by the symbol σ. He then proceeds to show that this fundamental length provides a natural scale for measuring everything constructed in the universe 'whether it be a nucleus, an atom, a crystal or the whole extent of physical space'. This is done by investigating the way in which the extensions of these various structures are related to it. For example, by considering a volume extensive enough to include a large number of particles in a still larger assembly of N particles in the uranoid as a background environment, he shows that the curvature of space in which the particles are embedded is simply a consequence of the natural limit of the precision of linear measurements. More exactly,

$$2\sigma = \frac{R}{\sqrt{N}}, \tag{1}$$

where R is the radius of curvature of spherical space.

Now an important elementary length known to physics is what is usually called the classical radius of an electron. It may at first sight seem strange to attribute a radius to what must appear to be finer than a point considering that even the traditional pin's head contains something like 85 million million million (85×10^{18}) atoms of iron and each atom has several electrons. Nevertheless, an electron has to have a non-zero radius if we are to avoid one aspect of what is known as the 'disease of infinities'.

This disease arises partly because the electrostatic energy of a charged sphere of radius r is, apart from a numerical factor, e^2/r, e being its charge. A charged sphere of zero radius will therefore have infinite energy or mass according to Einstein's law. If an electron were conceived of as literally a geometrical point, it would be such a charged sphere of zero radius and thus have infinite mass contrary to what we actually find by experiment. It can be shown that if the entire mass m_e of an electron is of electromagnetic origin, then it must have a radius $r = e^2/2m_e c^2$. Eddington assumes that the natural limit of the precision of linear measurement (σ) is no other than this electronic radius r. In other words, $\sigma = r$. It therefore follows from (1) that

$$\frac{e^2}{m_e c^2} = \frac{R}{\sqrt{N}}. \tag{2}$$

Finally Eddington identifies the background environment of the assembly of N particles, the uranoid, with the Einstein universe considered earlier in Chapter 7. If M is its mass, it can be shown that

$$\frac{\gamma M}{c^2} = \tfrac{1}{2}\pi R. \tag{3}$$

But since N is the total number of particles, both protons and electrons, the number of hydrogen atoms in the universe (each atom consisting of a proton and electron) is obviously $N/2$ so that $M = \tfrac{1}{2}Nm_p$ where m_p as usual is the mass of a proton which is very nearly the same as that of a hydrogen atom. Substituting for M in (3) we have

$$\frac{\gamma \tfrac{1}{2}Nm_p}{c^2} = \tfrac{1}{2}\pi R,$$

or,

$$\frac{\gamma m_p}{\pi c^2} = \frac{R}{N}. \tag{4}$$

Dividing (2) by (4) we find

$$\frac{e^2}{\gamma m_e m_p} = \frac{\sqrt{N}}{\pi}, \tag{5}$$

which gives the value of the *force constant*. Since N is of the order of 10^{79}, the force constant is thus seen to be of the order of 10^{39}.

This is how Eddington deduces one of the three dimensionless constants listed at the beginning of this chapter. To deduce the other two, he relies upon the two apocalyptic numbers 10 and 136 we encountered earlier. Thus the reason why the value of the fine structure constant (ii) is close to 136 is this: There are two 'naturally' occurring units of action.* One of them occurs in the study of radiation. It is the well-known constant of Planck h divided by 2π. The second natural unit may be formed by considering two elementary particles such as two electrons or protons. They have a natural electrostatic energy because of the repulsion between them. The time naturally associated with this energy is the time that light takes to go from one to the other.

* Action, by the way, is simply the product of energy and time, that is, it is so much energy for so much time.

Multiplying the two, we get the action naturally associated with the pair of particles.

It happens that it is always the same whether the pair of particles are close together or wide apart. For if they are wide apart, the energy is small but the time light takes to travel from one to the other is longer, so that their product remains the same. In symbols, if r is the distance apart, the energy is e^2/r and the time to traverse the distance r is r/c, so that action is e^2/c. Now the fact that the unit of action in radiation theory, namely $h/2\pi$, is 136 times that of elementary particles, namely e^2/c, is the result of there being only 136 *active* symbols in the double frame of *EF* numbers which give all basal measures in physical theory. Consequently the ratio of these two units, which is the fine structure constant, must be exactly 136. But the actual experimental value today is found to be very near to 137 and not 136. This, however, does not deter Eddington, for having accounted for the bulk of its value, he manages to adapt his theory by adding a unit for some obscure reasons which are difficult fully to understand.

Finally, to deduce the mass ratio (i) of the proton and electron, Eddington analyses the motion of the two particles in a hydrogen atom, namely, an electron and a proton, in the same way as astronomers do that of a double-star system. The object of the analysis is to replace the composite particle that a hydrogen atom is with its two constituent particles which conform to the ordinary conception of a simple particle. This means that the standard carrier which is specified by 136 components is to be split into two simpler carriers requiring only ten specification numbers. The mass ratio of the two particles, proton and electron, is then obtained in terms of these two numbers, 10 and 136, as the ratio of the two roots of the quadratic equation

$$10x^2 - 136x + 1 = 0,$$

which is 1,847·9, near the experimental value 1,836.

If the above outline of Eddington's deduction of constants of nature appears at times obscure and unconvincing, the fault may not be wholly that of the exposition. For his own exposition has not been understood by some of the finest minds of our time.

There are two almost irreconcilable opinions about the value of Eddington's theory. To a few, such as Whittaker and Milne, Eddington is the modern Archimedes who seeks to deduce all quantitative propositions of physics, that is, the exact values of the pure numbers that are constants of science, by logical reasoning from certain self-evident principles without making any use of quantitative data derived from observation. They consider that he has had a fair measure of success in this search. That he made mistakes they concede, but in a field where none had ventured before. To others, and this is the majority, his rediscovery of the constants of nature by pure calculation carries no conviction as they simply do not understand its rationale. For example, his derivation of the fine structure constant is rejected because of the obscurity of the argument, apart from its failure to throw any light on electron-photon interaction.

That his Fundamental Theory, the central theme of which is the rationalization of the constants of nature, is an awe-inspiring magnificent work, we know; but whether the symbol rhapsodies that this cosmological Cyrano de Bergerac composes therein to enchant the elusive Roxane of reality will succeed in transfixing her after all appears to me very much in doubt. For, in spite of the startling wealth of numerical coincidences between calculated and observed values of several constants, his theory hardly made any true prediction as, for example, Maxwell's theory of displacement currents did when it predicted the existence of radio waves, or Einstein's relativity theory the bending of light rays in the presence of the sun. On the other hand, many of the phenomena actually observed subsequently which a fundamental theory of this kind might be expected to have forecast were never predicted. Thus neither the existence of any of the various types of mesons nor that of more complex nuclear particles collectively called hyperons was so much as even hinted at. What is perhaps still more serious is that it could give no premonition of the large error in Hubble's original value of the recessional velocity of galaxies recently revealed by Baade's correction of the scale of galactic distances. On the contrary, it yielded a value close to the earlier erroneous figure. This is why the theory has not been able entirely to ward off the suspicion of fudging reasons to fit the facts.

Eddington's search for a universal theory linking the entire gamut of physical experience, from the atomic nucleus to the whole cosmos, is a romantic quest which, like such earlier counterparts as the *perpetuum mobile*, elixir of life and philosopher's stone, may perhaps lead finally to a welcome outcome. It is likely, however, to be very much different from that intended by its original precursor. Already such executors of Eddington's scientific testament as Whittaker, for example, do not consider it unjustified to make use of any *qualitative* empirical facts in building up such deductive schemes instead of merely cogitating by looking at 'a man with a measuring line'.

Before long Eddington's programme may come to mean nothing more than the normal aim of all scientific theory, namely to include in its broad sweep the widest collection of known facts of empirical experience. But the facts of empirical experience both in nuclear physics and cosmology are at present piling up thick and fast.

Thus in the field of nuclear physics unsuspected new particles and new properties of old ones are being found with such a bewildering rapidity as to leave theory far behind experimentation. No generally acceptable theory of fundamental particles bringing together all the 'brute' facts of the empirical situation has yet been constructed. As a result, there is a feeling of widespread mystification pervading even our current knowledge of the nuclear particles. Obviously, therefore, neither physics nor cosmology has yet come of age for the wedlock that Eddington tries to accomplish in his Fundamental Theory. This is why Eddington does look like a premature Oedipus come to confront the sphinx of the universe long before his time.

CHAPTER 12

Quasars and Cosmology

ALTHOUGH the discovery of quasars has brought cosmology to the threshold of a new break-through, we are not likely to make it unless we can comprehend their mystery. As we have seen, the quasar mystery stems from a combination of enormous energy output, small size, rapid light fluctuation and apparent remoteness. Thus their radio power is 10^{44} ergs per second and optical power 10^{46} ergs per second, *which is 100 times the total energy output rate of a giant galaxy*. And yet the size of the region where most of their radiation originates must be very small. For they vary in optical brightness and radio output over periods as short as a few years, months or even weeks and at least in one case (that of quasar 3C 446) barely a day. The period of light and radio fluctuation sets an upper limit to the size of the emitting region, because unless the objects are in a state of continuous explosion at enormous speeds (which, as we shall presently see, is ruled out by their spectra), the period of variation is determined by the time taken by light to travel the diameter of the radiating object. Consequently their size can hardly be a few light years or even months and weeks. By comparison the diameter of our own Milky Way which is by no means a giant among galaxies is 100,000 light years. It therefore follows that quasar diameters are some 10^5 to 10^7 times *smaller*, while their power output 10^2 to 10^3 times *greater* than those of galaxies. Finally, their apparent remoteness is inferred from the large red shifts exhibited by their spectra by recourse to Hubble's law. Earlier we noted that the quasar 3C 91 with red shift ratio $z = 2$ must be some nine billion light years away as against the largest red shift of 0·46 for an optical galaxy (3C 295) corresponding to a recessional velocity of about 85,000 miles per second and a computed distance of only four billion light years.

Each component of this fourfold mystery of quasar behaviour – their enormous energy output, small size, rapid light fluctuation and immense remoteness – is so strange that several alternative explanations have been offered to account for the observations. Take, for instance, their red shifts. They are so large that the reality of the associated recessional velocities and the consequential distances seems hardly credible. There are only two alternative explanations available. First, the quasars are the fragments of a local explosion that occurred some ten to 100 million years ago. There are two variations of this theme. Either the quasars were shot out of the centre of our own galaxy only 30,000 light years away or somewhat further off near the radio galaxy NGC 5128 Centaurus A at a distance of thirteen million light years. In either case it is difficult to explain why all the quasars exhibit red shifts and none has been observed to show a blue shift. After all, an explosion will scatter its fragments in all directions more or less equally. As a result some of the fragments of even a local explosion may be expected to approach us and thereby show a blue shift. Actually 85 per cent of the quasars observed spectroscopically show red shifts and there is no evidence that any of the remaining 15 per cent have blue shifts.

The second alternative, namely that red shift is due to strong gravitational fields, fares no better. It is true that relativity theory requires light emanating from a massive object to be red-shifted. But if the quasar red shift is gravitational, the gravitational fields needed to produce the observed reddening would require incredibly superdense stars. Calculation shows that the red shift (z) of a self-gravitating body of radius R and mass M is of the order of $\dfrac{\gamma M}{Rc^2}$, where c is the velocity of light and γ the constant of gravitation. Substituting the values of c and γ and expressing M and R in solar units, the expression becomes $10^{-6} \dfrac{M}{R}$. In order then to have a red shift as low as that of quasar 3C 273 for which z is only 0·16, $\dfrac{M}{R}$ would have to be 16×10^4. If the mass M is unity, that is, equal to that of the sun, R would then be $\dfrac{1}{16 \times 10^4}$ times the solar radius or $\dfrac{7 \times 10^5}{16 \times 10^4} = 4·4$ kilometres!

The upshot is a superdense star of such great density as would be obtained by compressing a solar mass within a sphere barely nine kilometres across! Even so the upper limit for red shift is only 0·3. For it will be shown in the sequel that red shift of light from superdense stars cannot exceed 0·3. Since many quasars have considerably higher red shifts, they cannot be superdense stars nor their red shifts gravitational.

We do not succeed any better even if we assume that quasars have bigger masses and bigger radii than superdense stars. For example, assuming that 3C 273 is a star at a distance of 1,000 light years we can calculate its radius R from its energy output. We find that R has to be about 140 times that of the sun. The red shift then requires that its mass M be $(0·16)$ (10^6) (140) or $2·10^7$ times that of the sun. According to W. A. Fowler such massive stars cannot even burn hydrogen in their interiors, because they would suffer gravitational collapse – of which more later – before the temperature would be high enough to burn hydrogen!

Since no other rational interpretation of quasar red shifts is available at present, there seems no escape from the conclusion that they are cosmological. That is, they are due to the fact that quasars like the galaxies are receding from us and from one another as part of the general expansion of the universe. But there are three disturbing aspects of the observed red shifts that suggest that they may not be due to Doppler recession conforming to Hubble's law after all. First, there is a strange distribution of red shifts among the quasars. Of the hundred quasars whose red shifts have been measured, several have red shift values (z) equal to or more than 2, nearly five times the value of the largest red shift $(0·46)$ of any optical galaxy. Moreover, while the red shift ratio (z) of optical galaxies rises continuously to its maximum of 0·46, there is a relative paucity of z values between 1 and 2 in the case of quasars. Second, the absorption lines now being found in the optical spectra do not have the same z values as do the emission lines. Third, the expected absorption of light on the short wavelength side of the red-shifted Lyman alpha line of hydrogen is not observed. Since this absorption is an extremely sensitive test of the presence of neutral hydrogen between the source and the observer, its absence poses a serious dilemma.

If the red shift is a Hubble effect and the source as remote as the red shift seems to suggest, it is difficult to explain the absence of absorption unless we assume that there is virtually no hydrogen in all the immense intergalactic spaces intervening between the source and the observer. The latter assumption too leads to awkward complications of cosmology. To add to the prevailing confusion Thomas Mathews has very recently produced evidence which throws some doubt on the cosmological character of quasar red shifts. He has reported a fivefold widening of a luminous band joining the star-like nucleus of quasar 3C 287 and its fainter red companion in one year by comparing its photograph taken in 1966 with those taken in 1950 and 1965. The band has apparently expanded by 2″ of arc in one year. Even if it had expanded with the velocity of light, it could not be more than one million light years away. And yet its red shift is so large that its computed distance via Hubble's law works out to be five billion light years! The nature of quasar red shift is therefore still under debate even though there prevails a provisional consensus that it is cosmological.

The supposedly cosmological origin of quasar red shifts initially encouraged the hope that their discovery may enable us to decide between rival cosmologies. For it is only at longer distances that the predicted differences between rival cosmological theories become larger than the ghostly errors of observation and therefore significant. It will be recalled that the plot of luminosity versus red shift of eighteen known clusters of galaxies depicted in Figure 26 seemed to favour an oscillating relativistic model. A similar plot that included even more distant objects was naturally expected to be more decisive. Accordingly plots of quasar red shifts versus their apparent luminosities were drawn like the graph shown in Figure 36 for fifty quasars. It will be seen that the points exhibit a good deal of scatter even though in a rough sort of way the fainter and therefore more distant sources do have larger red shifts. The diagram also shows three lines through the quasar 3C 273 on which the points may be expected to lie according to three cosmological models. Lines (a) and (b) correspond respectively to the oscillating and exploding relativistic models and line (c) to the steady-state model. The points do

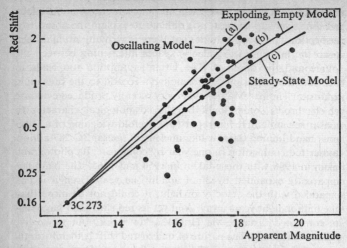

Figure 36. Fainter quasi-stellar sources (those with larger magnitudes) have larger red shifts. Plotted along the horizontal axis is the magnitude of the source, and vertically, on a logarithmic scale, is its red shift. Three lines have been drawn through the point representing 3C 273. They correspond to different cosmological models: (a) an oscillating universe, (b) an exploding, empty universe, and (c) a steady-state universe. Statistical studies suggest that the points probably fit line (a) or (b) better than (c).

not obviously fit any of these lines. The reason is that all the different quasars are probably not of the same *intrinsic* luminosity, thereby undermining one of the basic assumptions on which the theoretical model lines have been drawn. After all, the light from those quasars having red shift equal to 2 must have been emitted some eight to nine billion years ago which covers over 80 per cent of the putative age of the universe, if expansion started ten billion years ago. It is therefore too much to expect that all the quasars both relatively near and far have the same *intrinsic* optical or radio luminosity. And yet the objects must have the same average absolute magnitude to allow interpretation of the red shift versus luminosity diagram in terms of cosmological models. The difficulty can only be resolved if we know enough about quasars to tell how those which existed eight to nine billion years ago differ from those which are younger. It is therefore premature to use quasars as arbiters of rival cosmologies. We need to know more precisely their nature and, in particular, the source of their prodigious energy outpourings.

Various physical processes of their energy generation have been suggested since their discovery in 1960. But none of them is really satisfactory. It seems our theoretical understanding is not yet equal to the task. Thus if the energy source of the quasars were nuclear in nature as is the case with stars, it would be necessary to burn about 100 million suns in around 100 million years. It is unlikely that nuclear energy could be released in unison on such a large scale. Moreover, nuclear reactions usually occur in stellar interiors with gradual diffusion of energy to the surface. The spectrum of the radiation emitted is thermal with the surface electrons having low energy of only a few electron-volts. As we observed in Chapter 6, the radio spectrum of quasars is non-thermal synchrotron requiring relativistic electrons spiral-ling round magnetic lines with energies of the order of some bil-lion electron volts. It is difficult then to understand how the energy of the electrons can be elevated from a few electron volts to a billion.

If nuclear energy is ruled out, the only other candidate for consideration is gravitational energy. Although it is the weakest of interactions in nature, it is the only one that allows us at least in theory to extract up to 50 per cent of the rest-energy of matter against less than 1 per cent of nuclear source energy. This is why great attention has recently been bestowed on gravitational collapse as a possible source of the enormous energy output of the quasars.

The concept of gravitational collapse is nothing new. Already in the nineteenth century the equilibrium shapes of fluid masses were studied as a result of balance between the outward push of pressure and the inward pull of gravitation. The spherical con-figuration of equilibrium of any given mass of matter under some specified equation connecting pressure with density, the so-called equation of state, had been determined. But the question arose: how long can the equilibrium endure when the mass and con-sequently the resulting gravitational forces are increased more and more? Surprisingly the question remained unanswered till Landau's calculations in 1932.

To understand the idea underlying Landau's calculations, consider a large sphere of uncharged matter such as a spherical dust cloud. The cloud will begin to shrink under its own gravitation. The shrinkage will naturally continue till the elastic forces developed by the compression of the cloud begin to balance the inward pull of gravity. These elastic forces, which resist further gravitational compression of the cloud and are responsible for the 'equation of state' of the matter, arise from interactions of the immediate neighbours. Owing to its local origin, the equation of state, that is, the relation between pressure and density, is not significantly affected by the size of the spherical cloud. But the force of gravity is. Consequently if the radius is sufficiently large, the elastic forces of local origin will not be able to withstand gravity and the sphere will collapse. This is the elementary argument for gravitational collapse.

Recently it has been greatly refined by Archibald Wheeler *et al.*, who have shown that 'no equation of state compatible with casuality and with stability of matter against microscopic collapse can save a system from having a configuration which is unstable against collective gravitational collapse'. This is only a recondite, though highly cautious, way of saying that there exists a critical mass, Mcrit, such that if an object has a mass exceeding Mcrit it *must* collapse. Surprisingly Mcrit turns out to be quite low – about the mass of the sun in case of 'cold matter catalysed to the end of thermonuclear evolution'. But for hot gas of which stars are made Mcrit is much higher although according to Wheeler *et al.* there is reasonable probability of collapse even for masses below Mcrit.

It seems that under appropriate conditions of assembly there is no escape from eventual crushing by gravitational collapse of even the most elaborate molecular and atomic defences of matter – its electronic shells and nuclear structure. Gravitation, the feeblest of the four natural forces, wins in the end against all others whether electromagnetic (which holds atoms and molecules together) or nuclear (which holds atomic nuclei together) or 'weak' interactions involved in radioactive decays. The reason is again the same, namely whereas gravity is global the latter three are local. That is, while gravitational pulls of huge numbers of atoms in astrophysical objects (10^{57} in the case of the sun) can and do combine together to make one mighty pull, none of the other three is as effectively additive. It is as though we could aggregate any number of pennies to make up as large a capital as we desired but *not* ten- or hundred-pound notes. Under this constraint, a sufficiently large pool of pennies in the market will naturally overwhelm individual hundred-pound notes that cannot somehow be aggregated. Indeed, its victory can be so total that even the elementary particles of which atomic nuclei are made are squeezed into evanescence! It is therefore surmised that gravitation not only herds scattered materials to form stars and galaxies but that it may also crush some stars, star clusters and perhaps galaxies, into extinction and death. Gravitation is thus, to borrow a graphic phrase of Kip S. Thorne, 'both the midwife and undertaker of astrophysics'. While its former role is well understood, its function as an undertaker, that is, as a process leading to the total destruction of matter, is still a mystery. In particular, we still do not know what is the *aftermath* of the death of matter by gravitational collapse. Unfortunately, the question cannot be answered at present. The mechanism involves a quantum mechanical tunnelling process through the energy barrier against collapse which is not yet fully amenable to theoretical calculation. Nevertheless, some idealized situations can be handled.

Consider, for example, the case of a massive perfectly spherical non-rotating body like a star collapsing under its self-gravitation. As the collapse proceeds it becomes more and more difficult for the star to radiate energy. This is because, as the emitting atoms fall faster and faster towards its centre, the light radiated by them becomes increasingly red-shifted till the star 'dies', that is, disappears from view! Such disappearance can be shown to be a necessary consequence of the general theory of relativity as was first demonstrated by the German astronomer, Karl Schwarzschild, some fifty years ago.

Basing himself on Einstein's field equations for a single gravitating particle, Schwarzschild proved that the local curvature of space, which is dependent on the mass of the matter in its immediate vicinity, can ultimately close around itself so as to isolate its content from the rest of the universe and trap within it all emitted radiation provided the density of the matter is sufficiently high. The ultimate radius, the so-called Schwarzschild singularity, at which an object can disappear from view because it has trapped its outward radiation, is given by the expression*

$2\dfrac{\gamma M}{c^2}$, where γ is the gravitation constant, M the mass of the object and c the velocity of light. Since all we require to compute the Schwarzschild radius of an object is its mass, we may easily compute it for any object. For example, for an object as massive as the sun, it is barely three kilometres whereas its actual radius is 700,000 kilometres or nearly $2 \cdot 10^5$ times longer. Consequently the solar material has to be squeezed some $(2 \cdot 10^5)^3$ or $8 \cdot 10^{15}$ times tighter before it can be accommodated within its Schwarzschild sphere. Table 7 shows the value of Schwarzschild radius and density for four typical objects including the entire universe.

TABLE 7

Object	Schwarzschild radius	Schwarzschild density in gm./cm.³
100 lb. sphere	$0 \cdot 7 \times 10^{-23}$ centimetres	$3 \cdot 2 \times 10^{73}$
The sun	$2 \cdot 9$ kilometres	2×10^{16}
A quasar of 3×10^{12} solar masses	One light year	2×10^{-9}
The entire universe	10 billion light-years	$\sim 10^{-29}$

The table is only an exercise in arithmetic. For obviously it is not every object that can actually collapse within its own Schwarzschild sphere. All it is intended to show is that the more massive an object, the *less* dense it has to be to be encompassed within its Schwarzschild sphere through gravitational collapse.

* If the Schwarzschild radius of a star is r, then its gravitational red shift $z = \dfrac{\gamma M}{Rc^2} = \left(\dfrac{r}{2R}\right)$. Now R can at most be $\dfrac{3}{2}r$, if that, so that z cannot exceed 1/3.

The density drops rapidly from such unimaginably dense concentration as the Schwarzschild density of a 100 lb. ball or even a star to that of a quasar and the universe. Indeed it is remarkable that the computed Schwarzschild density (10^{-29} gm./cm.$^{-3}$) for the entire universe is quite close to the density of visible matter as actually observed ($7 \cdot 10^{-31}$ gm./cm.$^{-3}$). Consequently if the universe is finite (which is not yet established) and if its mass is entirely contained within a radius of ten billion light years, it might begin to suffer a gravitational collapse as required by oscillating relativistic cosmological models. Gravitational collapse therefore has important cosmological implications so that it is necessary to surmise the conditions under which an object, whether a star, galaxy, quasar or the entire universe, can implode gravitationally.

Current studies suggest that only the most massive stars can collapse within their Schwarzschild sphere after they have exhausted their supply of nuclear fuels. Less massive stars probably collapse to a certain point whereafter they perhaps explode as supernovae. But the rub is that gravitational collapse, even if it were occurring somewhere in the universe, would be difficult to observe. For when it happens, a curious situation develops in that no observer on the object can signal to another situated outside its Schwarzschild sphere, because all electromagnetic signals such as light rays that may be sent by him get trapped forever within the ambit of its gravitational sweep. Thus imagine an observer P tied to a particle on the surface of the star, who can communicate with another outside observer Q by means of light or radio signals. Obviously while Q remains where he is, P continues falling towards the centre of the sphere along with the particle to which he is anchored. The outside observer Q will find that the star is getting smaller and smaller, but he will never live to see it sink into its Schwarzschild singularity even if he were immortal. But for P there is only a finite lifetime before the incomprehensible and unimaginable cataclysm of complete collapse.

This curious result arises because of the increasing distortion of space-time with the continual contraction of the star under its own gravitation. Since by hypothesis the mass of the star exceeds Mcrit, the contraction of the star can never be halted by pressure or other local forces with the result that the star will go on compacting more and more. As it compacts, the gravitational field near it becomes stronger and stronger and the curvature of space-time becomes more and more. Physically the effect shows itself in the continual time dilation of signals sent by P to Q. Suppose P sends two signals at an interval of one second by his watch. These signals do not arrive at Q at an interval of one second by Q's watch. They arrive at a much longer interval – the interval becoming increasingly longer as P approaches the Schwarzschild sphere. Once P crosses into the sphere, no signal sent by him will ever reach Q even if Q waited for it infinitely long. Q therefore sees P approach his Schwarzschild sphere asymptotically, that is, more and more slowly but actually never. In particular, Q will never know what happened to P as he crossed the Schwarzschild barrier. But as seen by P himself the collapse would end rapidly in a 'singularity' of infinite density and zero volume, wherein infinitely large gravitational forces arise deforming matter and light beyond recognition and squeezing them out of existence. Although the emergence of 'singularity' in a theory is usually a sign of its breakdown, its future reformulation is not likely to affect the dynamics of collapse up to and across the Schwarzschild threshold.

The conclusion that gravitational collapse of an object implies its extinction assumes, as we mentioned at the outset, that the collapsing object is perfectly spherical and non-rotating. The mathematical analysis of more realistic situations when a massive object is non-spherical as well as rotating is so difficult that the question of its gravitational collapse cannot be definitely answered. Nevertheless, some insight has been gained by a remarkable theorem recently proved by Roger Penrose. Penrose's theorem says that in realistic non-spherical collapse too a critical stage can be reached beyond which no communication with the outside world is possible. Once this communication barrier has been crossed, one or both of two 'pathological' events may occur. First, the geometry of space-time may develop a singularity – a state of infinite density and zero volume – as in the ideal spherical case already considered. Second, and even more curiously, the collapse event may all of a sudden open up a new universe with which we had no previous contact. Stated more precisely, Penrose's theorem says that if a space-time geometric structure known as a 'trapped surface' evolves during the collapse of a star, and if our universe obeys certain rather plausible conditions, the result of the collapse must be either the emergence of a 'singularity' or a sudden link of our universe with another not previously apprehended. Of the three conditions the universe must satisfy for the validity of Penrose's theorem only one is really troublesome. It requires that at some initial moment before the collapse begins the universe must have infinite volume. Most relativity experts, however, surmise that Penrose's theorem remains valid even when the universe is finite, although the conjecture has yet to be proved.

The curious disappearance of massive objects within their Schwarzschild or Penrose 'singularity' and the still more bizarre phenomena predicted as occurring inside the Schwarzschild sphere such as the outright disappearance of matter and light have their own credibility gaps. There seems no way of bridging them by any ocular evidence. For, as already mentioned, a gravitational collapse can never be seen by any outside observer. All he can do, in the spherically symmetrical collapse for example, is to calculate the energy released as the object shrinks to its Schwarzschild radius. It is found to be $\frac{1}{2}Mc^2$ which is half the total latent energy of the mass M of the object. The yield is so great that it is about 100 times more efficient than any nuclear reaction. Thus an energy of 10^{62} ergs such as quasars seem to pour out into space could be produced if a body of only 100 million solar masses were to collapse to the Schwarzschild radius. But no one knows if this kind of gravitational collapse is actually possible, or, if it is, what mechanism could account for the exchange of energy from the gravitational field to the relativistic electrons whose spin around magnetic lines produces the observed synchrotron radiation of the quasars.

The provisional failure of gravitational collapse to account for the quasar phenomenon has prompted a large number of novel ideas. Some suggest that quasars are primeval fragments of the initial explosion with which the universe began its career from a singular state of superdense concentration some ten billion years ago. Others consider them to be protogalaxies in the making, a chain of supernovae exploding in the core of distant galaxies, a series of collisions of a large number (10^9) of ordinary stars packed in a small volume such as a sphere barely six light years across, or matter-antimatter annihilation in the depths of intergalactic space. But none of the suggested ideas has really worked so far. No wonder Howard McCord was provoked to write:

> These lights called quasi-stellar,
> Anarchic, observe not the curfew
> Of thermodynamics: they run on
> After they have run down, or
> Shine with a brighter fire
> Than they have fuel for.
>
> This is not right.
> For if I must pay for jot with tittle
> And am not to be both big and little,
> How is it they
> Can play
> According to neither Hoyle
> Nor Boyle?

CHAPTER 13

Nuclear Physics and New Cosmology

In Chapters 10 and 11 we described two cosmological theories – those of Dirac and Eddington – whose mainspring is a deep underlying connexion between the universe and the atom. That such a connexion probably exists and, when fully understood, will reveal some entirely novel aspects of the universe of which we can have no premonition now is plausible enough. But the rub is that a feeling of mystification pervades our current knowledge of the atomic or nuclear particles. The brute facts of the empirical situation are piling up thick and fast, and the theory designed to bring them all together is lagging far behind nuclear experimentation. It is therefore hazardous to rear a cosmological superstructure on the shifting quicksands of nuclear theory.

Nevertheless, the recent discovery of new types of nuclear particles known as antiprotons and antineutrons has sparked a fresh proliferation of several new cosmologies based on the Eddington-Dirac surmise of a likely link between the outer cosmos and the microcosmos within the atom. For the existence of such entities seems by analogical reasoning to suggest the possible existence of a hidden imperceptible universe of antimatter, the stuff out of which its constituents, the so-called antistars and antigalaxies, might be made. But what is antimatter? To understand its nature it is necessary to recall a few of the more salient features of ordinary matter.

That all matter, that is, the material of which everything around us is made, consists of tiny, indivisible particles called atoms is an old speculation introduced by the ancient Greek philosopher Democritus as a cohering principle to reduce the complexity of phenomena. According to Democritus, the perpetually changing façade of phenomena is merely the effect of simple movements of elementary entities, the eternally stable atoms. In many ways his view has proved truer than he knew. Among its numerous successes, one of the most spectacular is the rational explanation of the behaviour of gases by the kinetic theory. The kinetic theory envisaged a gas as a swarm of elementary particles in random motion. On this basis alone it could, for example, be proved that the temperature of a gas is merely a manifestation of the average kinetic energy of its constituent particles. If a gas is heated, the heat simply shows itself in the faster random motions of its particles so that their average kinetic energy increases.

While experiment confirmed this conclusion of the theory, it nevertheless occurred to Boltzmann, one of the creators of the kinetic theory, to inquire whether the 'elementary' particles of the kinetic theory were really elementary. For when one has postulated some indivisible entity, particle, or atom, one is inevitably intrigued by its assumed indivisibility and is led to inquire into its own internal structure. This is exactly what Boltzmann did when he posed the following problem: if the 'indivisible' entities of kinetic theory are tiny elastic spheres like miniature billiard balls as the theory assumes, then that sphere will have some internal structure of its own despite the smallness of its size. Accordingly, some part of the heat energy delivered to a gas may be expected to excite the internal motion of its particles instead of all of it going to increase their motion as a whole.

This, however, has not been observed, and the whole of heat energy is found to be kinetic, that is, due to the motion of the particles as a whole. Naturally Boltzmann was greatly puzzled by this. Unfortunately he could never fathom the reason for it as it is due to the quantum nature of the mechanical properties of the atom, a new phenomenon discovered only recently, many years after Boltzmann's death.

Prior to the discovery of the quantum properties of atomic phenomena, it was believed that any system could accept radiant energy like heat or light in any amount no matter how small. However, by the beginning of the twentieth century it was found that this rule no longer applied to small systems of the size of an atom. An atomic system could absorb (or radiate) energies only in discrete finite amounts called quanta, so that if the energy supplied happened to fall below the first quantum required to excite the internal motions of the atom, it would just fail to affect its internal structure and the atom would behave as an indivisible entity. Since at room temperature the kinetic energy of a gas atom is about one-fortieth of an electron volt against the fourteen to 100 electron volts required to excite the internal motions of an atom, it is no wonder that Boltzmann's anticipation failed to materialize.

What is true of the atom holds equally for its nucleus, except that still higher energies, of the order of 10^5 to 10^7 electron volts, are required to show their internal structure in turn. This is why the revelation of the structure of the atomic nucleus has depended on the construction of giant electronuclear machines such as cyclotrons, betatrons, synchrotrons and cosmotrons designed to produce energies of this high order.

As a result of long and patient research by means of these machines, it is now revealed that an atom consists of a number of electrons, elementary particles with negative charge, orbiting around a much heavier central nucleus very much as the planets are revolving around the sun. This analogy with the planetary system is no mere surface resemblance. It holds in several respects. First, the main mass of the atom is concentrated in the atomic nucleus exactly as that of the planetary system lies in the central sun. Secondly, the radius of the central nucleus is only a tiny fraction of the radius of the orbit of the outermost electron as is the case in our planetary system. If we choose to represent the central nucleus by a dot the size of an ordinary period, the outermost electrons revolving around it would extend on the same scale to the size of a room (about 10^3 cm.), the whole system having been magnified a hundred-billion-fold.

It therefore seems that the inside of an atom is about as empty as the interstellar void. For if we *reduced* the stellar universe to the same extent as we have *magnified* the atomic, the nearest stars would be apart to the same extent (10^3 cm.) as the outermost electron and the nucleus. This is one reason for the surmise that the microcosmos within the atom may well be a smaller edition of the outer cosmos.

Subsequent nuclear research, however, has complicated considerably the earlier picture of an atom as a miniature planetary system. The most important of these latter-day complications is that the atomic nucleus itself has been shown to have a complex structure of its own. It has been smashed and is found to consist of two kinds of elementary nuclear particles – protons and neutrons. Of these the proton is better known as it figured in the earliest atomic model of the hydrogen atom suggested by Lord Rutherford. It has the same *positive* electric charge as the *negative* charge of an electron, though it is about 1,836 times as massive.

The neutron, however, is less familiar, being a new arrival in atomic theory. But it is so like a proton in many ways – as in mass, for example – that physicists have given them both a common name – the nucleon. The only way the neutron is distinguished from its counterpart, the proton, is that unlike the proton it bears no electric charge and is electrically neutral.

This discovery of the atomic nucleus as a closely packed jumble of nucleons has in turn raised a new problem. Since opposite electric charges attract, we can understand how the revolving electrons can keep on moving in their orbits under the electrostatic attraction of the protons in the nucleus. But by the same token since like electric charges repel, it is not clear at first sight why the protons compacted within the narrow confines of the nucleus do not fly asunder.

The only way to solve the difficulty is to postulate the existence of new types of attractive forces of a non-electric nature which are able to fuse the nucleus together. From experiments we know that these new kinds of nuclear forces must be very strong since we require very high energies of the order of several million electron volts to wrench a proton out of the nucleus in which it is embedded. While they are very strong within the nucleus, they vanish to extinction even at its periphery, so that at a distance of only 10^{-11} cm. from the nucleus they are already undetectable.

It is the law governing these non-electric nuclear forces between proton and proton that Eddington tried to deduce by *a priori* reasoning in order to link the outer cosmos with the microcosmos within the atom. He argued that when a proton and an electron are close together, the lines of force emanating from one practically all end on the other so that the system is self-contained. It can therefore be inserted in any environment without disturbance. But the proximity of two particles of like sign is a very different affair. The lines of force originating from one of them are repelled by the other so that they must terminate elsewhere in the universe. Consequently, the insertion of a pair of charges of like sign, such as two protons, does cause considerable disturbance in the rest of the universe, whose action has to be superadded to the usual electric interaction of the charges according to Coulomb's law. On the basis of this argument Eddington deduced the law of nuclear forces which contained two constants of nature, namely the radius of curvature of the Einstein universe and the cosmical number N described in Chapter 11.

But it is now clear that we cannot arrive at the basic law of nuclear forces by such considerations. For in the first place, the nuclear-attraction law must hold not only for electrically charged particles like protons but also for electrically neutral neutrons whose proximity cannot provoke the reaction of the rest of the universe as that of electrically alike particles, like protons in the manner envisaged by Eddington. Further, the Japanese physicist Yukawa, by daring to inquire into the quanta of the nuclear force field on the analogy of the quanta of the electromagnetic field, was able to predict the existence of nuclear particles, called mesons, which have a mass intermediate between that of an electron and a proton and which were later found in the cosmic-ray showers. While there is a certain amount of truth in Yukawa's picture, the great proliferation of meson particles in recent nuclear and cosmic-ray research makes it difficult to assess how much truth there is in it. What is certain is that this law of nuclear forces is still a mystery that remains to be unravelled.

Another mystery of nuclear physics that until recently exercised the minds of many physicists a good deal was this: the existence of an electron with a negative charge and a proton with an equal positive charge but about 1,836 times as massive does create a bit of dissymmetry which at first sight is difficult to understand. One would have expected a proton to have the same mass as an electron, considering that their charges are equal in amount but opposite in sign, in order to provide a complete positive analogue of an electron. But since it is not so, to restore perfect symmetry one would need to postulate two new types of particles, a positive analogue of electron and a negative analogue of proton.

While the existence of the former was predicted on theoretical grounds by Dirac as early as 1928 and was actually discovered by Anderson in 1932 during the course of his studies of cosmic rays, the latter was discovered much later, in October 1955. By using a proton beam of six billion electron volts a group of physicists at the University of California in Berkeley succeeded in creating pairs of positive and negative protons (protons and anti-protons) out of the vacuum. The discovery of antiprotons in turn suggests that one may also expect to produce the antineutron, an analogue of the neutron, even though at first sight the distinction between the two is not clear since the neutron is electrically neutral and neither would possess an electric charge. However, since a neutron, even though electrically neutral, does give rise to a magnetic field, we can envisage its analogue, the antineutron, as an electrically neutral particle exactly like a neutron but with a *reversed* magnetic field.

The discovery of analogues of electrons, protons and neutrons has posed the problem of whether there could not exist antimatter, a sort of mirror image of ordinary matter built out of these analogues of elementary particles. Thus, for instance, a hydrogen atom consists of an electron orbiting around a central nucleus consisting of a proton. Now one could easily imagine a positive electron (positron) revolving around a nucleus of antiproton or negative proton. Such a structure would be an exact analogue of hydrogen, the so-called antihydrogen, a sort of mirror image of an ordinary hydrogen atom. Similarly a nucleus consisting of eight antiprotons and eight antineutrons surrounded by eight positrons orbiting around it would constitute an atom of anti-oxygen and so on for every anti-element all the way up to antiuranium. All these anti-elements could well be the raw material of possible antiworlds including, in principle, antilife. The point is whether such antiworlds of antimatter do actually occur somewhere in the universe even though we know that it is quite impossible for the two to exist cheek by jowl within the confines of a single stellar complex such as our own solar system. For in such a condition of close proximity they would mutually annihilate themselves in a massive catastrophic blaze. But this is not to say that matter and antimatter cannot coexist in splendid isolation of each other in separate galaxies or for that matter perhaps in separate stellar systems of the same galaxy. For all we know some of the galaxies or even stars we see in our Milky Way may well be made of antimatter. For, as far as is known at the moment, the light emanating from objects made of anti-matter, being electromagnetic in character, is electrically neutral and would behave no differently from that radiated by those of ordinary matter so that whole galaxies and stellar systems made of antimatter may exist without our being able to tell the dif-ference from astronomical observation.

There is, however, one possibility whereby the existence of antimatter could perhaps be revealed to us. It arises because the two kinds of matter would show different Zeeman effects in a given magnetic field. The Zeeman effect is a splitting of spectral lines resulting from the action of a magnetic field on the electrons of atoms or molecules. When a magnetic field with the same direction acts on positrons, the splitting has the opposite sense and in this way antimatter could be distinguished from matter. But the test would yield no result, if the magnetic fields associated with antimatter have the opposite direction from those associated with matter. There is, indeed, no way of telling whether any object in the heavens is composed of matter rather than anti-matter.

Nevertheless, considerations of symmetry have led some cos-mologists to suggest that the universe is composed of equal parts of matter and antimatter even though no compact system can exist as half matter, half antimatter. The most interesting matter-antimatter cosmology offered so far is that of the Swedish physicist Oskar Klein. According to Klein, the zero state of the universe was an extremely tenuous cloud of gas or rather a plasma of protons and antiprotons in equal numbers with per-haps a sprinkling of electrons and positrons. Such a primordial cloud, called ambiplasma, was initially a bloated sphere with a radius of a trillion (10^{12}) light years and a density of one particle (ordinary or anti) per 10^{12} cm.^{3}. As a density it is a mere figure of speech for a cosmic cloud of this density and of volume a million times *larger* than the sun would weigh less than a drop of water! At this dispersion the particles and antiparticles naturally have even less chance of ever making a rendezvous than two lone bedouins lost in the Sahara. But when in course of time the cloud contracts under its own gravitation to a radius of a few billion light years, the particles remain still wide apart although now and then a proton and antiproton do collide. Their mutual annihilation releases energy mainly in the form of radiation. With increasing compacting of the cloud the proton-antiproton collisions become more and more frequent and the radiation grows stronger and stronger. When the radius of the cloud has shrunk to about a billion light years, the radiation arising from particle-antiparticle annihilation is so strong that it overcomes the gravitational attraction. The cloud, including the galaxies that have condensed within it by that time, begins to expand. The upshot is the expanding universe we observe today.

Klein gives reasonably cogent arguments to show that the intensity of radiation needed to overcome gravitational attraction of the cloud might now be observed as the radiation from a black body emitter at 3°K. – the microwave radiation detected by Penzias and Wilson. He also claims that his calculations agree fairly well with certain observed features of our universe, as for example the relation of average density of matter to the rate of expansion in various regions. Nevertheless, his model faces two serious difficulties. First, how matter and antimatter came to be segregated in the first instance. Second, how the two are kept apart even if they were somehow segregated to form stable worlds. To secure the initial segregation, Klein postulates that ambiplasma contained magnetic fields either at the outset or in the course of development of the cloud. He then proceeds to show that under the combined action of gravitation and electromagnetic forces, by a process essentially akin to electrolysis, protons and electrons were siphoned off in some regions and their antis in others. The details of the process have not yet been fully worked out and it is not clear whether it will succeed. Even Klein and his supporters concede that the first difficulty has not yet been fully solved.

However, granting that the initial particle-antiparticle segregation is somehow secured, Klein has to invent yet another mechanism whereby antiworlds are enabled to exist side by side with our own kind of world. The mechanism is the cosmic variant of the Leidenfrost phenomenon concerning the behaviour of a drop of water placed on a hot plate. If a drop of water is thrown on a hot plate at a temperature moderately above boiling point (100°C.), it will evaporate very rapidly. If, however, the temperature of the plate is very much higher and the plate is concave enough to hold water without its rolling off, a teaspoon of water placed on the plate may remain there without boiling off for as long as ten minutes (see Figure 37). The reason is that rapid evaporation of the water at the interface with the hot plate forms

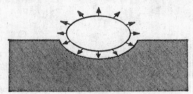

Leidenfrost Phenomenon

Figure 37. A layer of vapour at the interface insulates the drop from the hot plate itself. (From H. Alfvén, 'Antimatter and Cosmology', *Scientific American*, April 1967. Copyright © 1967 by Scientific American, Inc. All rights reserved.)

a thin layer of vapour that insulates the main body of water from the hot plate itself. Klein believes that matter and antimatter can be kept apart in the same way by a very hot layer at the interface between them. The hot layer would be produced by the annihilation reactions between particles and antiparticles meeting at the interface. We know that the vast stretches of space around the planets of the solar system, around stars and even around galaxies are occupied by thin clouds of plasma subjected to magnetic fields. Suppose such a plasma surrounding an object made up of matter borders on a magnetized antiplasma surrounding another made of antimatter (see Figure 38). Collisions between protons and antiprotons will produce as end products of their mutual annihilation high-energy electrons and positrons. These particles will spiral around the lines of magnetic fields and form a very hot layer of ambiplasma in which the electrons and positrons go on annihilating one another. Such a layer, according

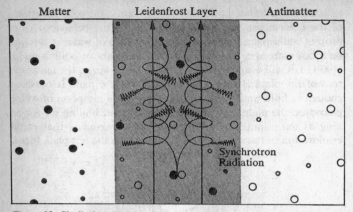

Matter   Leidenfrost Layer   Antimatter

Figure 38. Similarly a hot, insulating Leidenfrost layer might be formed at the interface between a world of matter and a world of antimatter. Such a layer would be produced by the annihilation reactions between particles and antiparticles meeting at the interface. Collisions between protons and antiprotons would produce high-energy electrons and positrons, which would spiral around the magnetic-field lines in the layer, emitting radio waves. (From H. Alfvén, 'Antimatter and Cosmology', *Scientific American,* April 1967. Copyright © 1967 by Scientific American, Inc. All rights reserved.)

to Klein, could act as a kind of curtain between matter and antimatter and thus insulate the two worlds from each other.

One may expect to detect such Leidenfrost curtains of ambiplasma by discovering certain specific emissions of energy from the borderlands of matter and antimatter. For in these borderlands proton-antiproton annihilation is continually occurring. But we know that when a proton meets an antiproton the annihilation process produces a number of mesons that rapidly decay through a complicated series of steps. The final outcome is energy in the form of gamma rays, neutrinos, one or two electrons and one or two positrons. Consequently Leidenfrost curtains of ambiplasma could be discovered by detecting gamma rays and neutrinos originating within or outside of our galaxy or by the synchrotron radiation emitted by electrons and positrons spiralling around magnetic lines. Gamma radiation could, in principle, be detected but only by devices such as scintillating counters in a space laboratory as they are scattered and absorbed by the earth's atmosphere. Neutrino radiation is virtually undetectable as neutrino astronomy is not yet a reality in the observational sense. The only detectable radiation is that emitted by the high-energy electrons and positrons spiralling around magnetic lines. Since radio galaxies and the mysterious quasars emit precisely the latter type of radiation Klein and his followers simply conclude that at least some, if not all, of the star-like radio objects including quasars arise from matter-antimatter annihilations in Leidenfrost curtains of ambiplasma.

It is true that the identification of radio galaxies and quasars with Leidenfrost curtains of ambiplasma is based on certain theoretical calculations of the spectrum of radio emission from a hypothetical magnetic ambiplasma made by the Stockholm school. But the claim has not won general acceptance. Obviously the quasar riddle is too much of a cosmic mystery for the Klein model to unravel. All that may be said in its favour is that it makes our universe of receding galaxies the outcome of a kind of *radiation* explosion issuing from particle-antiparticle annihilation rather than that of a giant primeval atom as in the big-bang cosmology of Lemaître and Gamow. It therefore postulates a more credible zero state than the big-bang cosmology. But that is about all. There is no well-established theory on which one could base the evolution of Klein's ambiplasma into the universe we know like the relativity theory on which big-bang cosmology is reared. This is why Klein's model faces serious difficulties right at the outset, namely how the initial matter-antimatter cosmic omelette came to be unscrambled in the first instance and later prevented from scrambling all over again. On the other hand, big-bang cosmology has only one really serious difficulty virtually at the end of its deductive process when its underlying equations break down at the 'creation' event. In other words, the difficulties of Klein cosmology diverge from its zero state – precisely the corresponding point to which those of big-bang cosmology converge.

However, the appeal of the argument in favour of the existence of an anti-universe of some kind based solely on the particle-antiparticle conjugate symmetry is not easy to resist. Indeed symmetries of this kind have in the past guided physicists to a number of remarkable discoveries from neutrinos to radio galaxies. They are responsible for several searches now under way like the magnetic analogue of an electric charge, the so-called magnetic monopole and the quark, whose existence is suggested by more recondite symmetries of elementary particles. There is an unwritten precept in modern physics, often facetiously referred to as Gell-Mann's 'totalitarian edict' which requires that in physics 'anything which is not prohibited is compulsory'. At least one more cosmologist besides Klein, namely Goldhaber, has succumbed to Gell-Mann's maxim sufficiently to suggest a revision of the Lemaître-Gamow theory of creation to remove its tacitly assumed asymmetry. It will be recalled that according to this theory the universe is supposed to have begun with the explosion of a 'primeval giant atom', which contained all the *nucleons* of the universe but *not* its antinucleons. Goldhaber seeks to inquire into the logical structure of a cosmology which attempts to preserve symmetry between nucleons and antinucleons in the initial conditions of creation but which leads nevertheless to an apparent present asymmetry of matter in 'our' cosmos.

According to him, we can restore the conjugate particle-anti-particle symmetry in the Lemaître-Gamow theory by postulating that in the beginning there was a single particle called 'universon' containing the mass of the whole universe. At some time presently unknown the 'universon' broke up into a particle and its anti-particle, called 'cosmon' and 'anticosmon', each possessing a large nucleonic charge but of opposite sign. This break-up of the 'universon' is similar to the spontaneous decay of a fundamental particle into a particle and antiparticle very much like the decay of a neutral meson into two mesons of equal and opposite charge. After the break-up of the universon into 'cosmon' and 'anti-cosmon' the two parts flew apart with large relative velocity. Later the 'cosmon', to which Goldhaber assigns the role of Lemaître's primeval atom and therefore positive nucleonic charge, decayed, possibly through many intermediate steps, some ten billion years ago into nucleons to form our present expanding cosmos as in the Lemaître-Gamow theory.

As for the 'anticosmon', it may or may not have decayed by now, since spontaneous decay is a process governed by chance. If it did decay to form an 'anticosmos', there would be a gigantic unseen 'twin' of the universe about us. This twin universe would be the exact counterpart of our own cosmos except that it would consist of antiplanets, antistars and antigalaxies, that is, structures analogous to what we find in our own cosmos but made up of antimatter.

Goldhaber's speculation, as he himself points out, raises more questions than can be answered at the moment, such as, for example: is the direction in space defined by the line connecting the centres of the cosmos and 'anticosmos' detectable, or for that matter is the 'anticosmos' itself detectable, and, if so, how far away is it and how fast is it receding from us? In this respect it shares with its other rival described in this chapter the qualities of a certain quaint *bizarrerie*, piquant rawness and honey-moonish extravagance inevitable in first glimpses of a vast *terra incognita* that is only just beginning to be explored.

CHAPTER 14

The Last Dusk of Reckoning

AN important aspect of natural processes which must be taken account of in classical as well as in relativity mechanics is their irreversibility. This means that once a process has occurred it cannot go back on itself. Thus no white dwarf doomed to change its celestial light into mournful gloom can climb back to its prime as a main-sequence star any more than an aged Faust can recover his youth or a faded rose blossom again. That is why our most deeply planted though ill-defined feeling is one of that inexorable irreversibility of the march of history which Omar Khayyám expresses with superb oriental fatality in one of his quatrains:

> The Moving Finger writes and having writ,
> Moves on; nor all your Piety nor Wit
> Shall lure it back to cancel half a Line,
> Nor all your Tears wash out a Word of it.

But will this moving finger continue to write forever or will there be a last dusk of reckoning after which there shall be no more writing or reading? The only science at present qualified to attempt even to guess an answer is that of thermodynamics. For its laws alone embody the indubitable irreversibility of natural phenomena, which the more fundamental laws of physics like Newton's laws of motion completely ignore. Thus if we consider the flight of a tennis ball from one player to another, we find that it follows a parabolic path. If the second player were to hit it in such a way as exactly to reverse its velocity on arrival there, it will retrace the same parabolic path in reverse.* The reason is that its equation of motion remains the same when we change the sign of the time symbol from $+t$ to $-t$. This therefore means that at each point of its path whether moving forward or backward it has the same kinematical features such as kinetic energy, potential energy and velocity (except for its direction). In short,

---

* Provided the frictional resistance of the atmosphere is neglected.

its entire kinematical history repeats itself on the return trip. Such a process is a reversible one, and the Newton-Einstein laws of motion govern processes of this type.

Indeed, so far as we know, not only are the fundamental laws of motion reversible but also those of optics, electromagnetism and for that matter of all *macroscopic* physics. Because these causal laws describe the behaviour of physical systems at different points of space and at different instants of time, they are written as partial differential equations with space and time as independent variables. The differential equations remain equally valid when the direction of time is reversed, that is, when we substitute therein $-t$ for t, because only second derivatives with respect to time appear. The situation is exactly the same as in the case of the equation of motion of the tennis ball of our earlier illustration.

Consider, for example, radiative or electromagnetic phenomena like the emission of light from stars. As actually observed they are irreversible. For we never see light flow into a star. And yet Maxwell's equations of electromagnetism describing the emission of light are unchanged when the time direction is reversed. Consequently the equations are as compatible with the outflow of light *from* stars as with its inflow *into* them. In other words, they fit equally *either* a point *source radiating* light in expanding spherical shells moving outwards *or* a point *sink sucking in* light from all parts of space in a spherical shell contracting with the velocity of light. How then do we derive the observed one-way emission of light from equations which are themselves as valid for its reverse flow? The answer is that the equations permit two symmetrical solutions, the so-called *retarded* and *advanced* potentials. In the former the oscillating electric charge *radiates* electromagnetic waves and as a result suffers damping of its motion. In the latter, the electric charge *receives* energy from all space and consequently begins to oscillate more and more energetically. Although both solutions satisfy Maxwell's equations, only the retarded potential is admitted as physically significant, the advanced potential being rejected out of hand as of no consequence. This may seem arbitrary but is apparently inevitable, because physicists have tried all along to make time measurement as space-like as possible by suppressing the basic difference between space and time.

As Whitrow has remarked,

physicists have been influenced far more profoundly by the fact that space seems to be presented to us all of a piece whereas time comes to us only bit by bit. The past must be recalled by the dubious aid of memory, the future is hidden from us and only the present is directly revealed. This striking dissimilarity between space and time has nowhere had greater influence than in physical science based on the concept of measurement.

It has led to the subordination of the concept of time in physics to that of space and, in particular, to the circumvention of the assymmetry of past and future which characterizes our temporal existence. Time-symmetric laws of macroscopic physics like those of mechanics and electromagnetism are an eloquent testimony of this tendency.

It may be that spatialization of time makes the fundamental physical laws 'over-descriptive' in that they yield additional solutions, which have to be rejected because of their suppression of the observed one-way drift of many natural processes. But attempts to purge the laws of their 'over-description' and make them time-asymmetric at any rate at the level of macroscopic physics have not succeeded. Physics has had therefore perforce to bring in other kinds of laws as if by the back door to account for the indubitable undirectional trend of happenings in the universe. Such laws are *statistical* laws, that is, laws governing *crowds* of events. It turns out that the observed *irreversibility* of natural phenomena emerges almost automatically when we begin to consider crowds of events or processes despite the fact that each one of them is *individually reversible*.

A crowd of individually *reversible* processes becomes irreversible in the bulk, because there are two ways of dealing with crowds of any kind. Either we specify the attribute or attributes under consideration of each and every individual in the crowd or we specify only the statistical averages of the individual attributes. The former is said to define the microstate of the crowd and the latter its macrostate.

Suppose, for instance, someone played a game of moving ten billiard balls on a billiard table. At any time he could move one or more stationary balls or stop some of those in motion. Suppose further that all that we could know of the state of motion of these billiard balls was the average number of balls at rest at any time as broadcast to us by an observer. Then any *macrostate* of the system of motion of the balls is given by the average number of balls at rest. Let the average number of balls at rest at any time be 0·1. This merely means that only one of the ten balls is at rest. But since any one of these ten balls could be at rest without making any difference to the *macrostate* as broadcast to us, there are clearly ten *microstates* of the motion of the balls (depending on which of the ten balls is at rest) corresponding to the single macrostate denoted by the average 0·1.

On the other hand, the macrostate of the motion of balls denoted by the average 1 has only one corresponding microstate. For in this case all the ten balls have to be at rest to yield the average 1, and there is only one way in which all of them can be at rest. In fact it is easy to verify that the macrostate of the system of motion can have only eleven possible values corresponding to the varying numbers of balls at rest as shown in Table 8.

It is obvious that if all the information about the motion of balls that we could have is the average number of balls at rest at any time, as broadcast to us by our hypothetical observer, we could never discriminate between any of the 252 different microstates in which five balls out of ten are at rest so that the macrostate of them all is denoted by the same average 0·5.

Further, if each ball has at any time an equal chance of being at rest or in motion, a macrostate with a greater number of corresponding microstates has a greater probability of occurring than another with a smaller number of corresponding microstates. For a glance at Table 8 shows that out of a total of 1,024 microstates of the system, 10 such states yield the macrostate specified by the average 0·1 and 252 that by 0·5. The latter macrostate is therefore 252/10 times as likely to materialize as the former. For this reason we may use the number of microstates corresponding to each macrostate as a measure of the probability of its actual occurrence.

TABLE 8

Serial number	Balls at rest at any time	Macrostates of motion (average number of balls at rest)	Microstates of motion*
1	0	0	1
2	1	0·1	10
3	2	0·2	45
4	3	0·3	120
5	4	0·4	210
6	5	0·5	252
7	6	0·6	210
8	7	0·7	120
9	8	0·8	45
10	9	0·9	10
11	10	1·0	1

Total 1,024

*The numbers shown in this column are the values of $10Cr$ for different values of r ranging from zero to 10. As is well known, $10Cr$ is the number of different combinations or ways in which r balls out of ten may be selected to be at rest. It is therefore the number of microstates of the motion of the balls corresponding to the macrostate denoted by the average $r/10$ shown in the third column.

Now the microstate of a body such as a gas enclosed in a chamber can only be defined by the specification of the position and speed of every single molecule in its make-up. But these are measurements of such a large number of molecules and of such ultra-microscopic precision that they are forever beyond our power to make. All that we can have is a description of its macrostate defined by a few statistical averages of some of the features of this enormous molecular population. For example, its temperature, which is merely a manifestation of the average kinetic energy (or speed) of its constituent molecules, is one such statistical attribute defining a feature of its macrostate.

Just as to each macrostate of the motion of billiard balls (that is, the average number of balls at rest) there correspond in general several microstates of the billiard balls, so also to each macrostate of the motion of the molecules of the body as defined by the average speed of its molecules (or what amounts to the same thing, by its temperature), there correspond many microstates of the motion of its molecules. In a body consisting of billions upon billions of molecules there are naturally a large number of different microstates of motion each of which has the same average kinetic energy per molecule, exactly as a large number of microstates of the motion of our billiard balls yield the same average number of balls at rest. The number of microstates corresponding to a given macrostate defined by any temperature, say T, is known as its thermodynamic probability.

This is a way of saying that the greater the number of microstates corresponding to the macrostate T, the greater is the probability that any microstate chosen at random will display the molar or large-scale features of that macrostate.

Consider, for example, a body all of whose molecules are moving with the same speed and in the same direction as the body itself. This completely ordered motion, in which a knowledge of the speed of any one of them leads to that of the speed of all of them, belongs to a macrostate which corresponds to a single microstate of its motion like all the billiard balls at rest in our analogy. This state of *highest* order of organization in the motion of its molecules has the *lowest* thermodynamic probability, there being only one microstate among the myriads possible. On the other hand, when the state of motion of the molecules in the body is highly disordered, the number of microstates corresponding to its macrostate is much more numerous, so that its thermodynamic probability becomes exceedingly great. It is this thermodynamic probability (or its logarithm called entropy) that provides us a measure of the disorder we require for a quantitative determination of the irreversibility of natural processes.

Boltzmann deduced from these considerations the famous second law of thermodynamics, which stipulates that heat passes spontaneously of itself from a hot to a cold body and not in the reverse direction thereby giving *irreversibility* the sanction of a major limit law of nature. Boltzmann's argument in proof of the second law will be clearer if we bear in mind the following analogy. Suppose we have two committees of representatives of teachers and students. The teachers predominate in one and the students in the other. Suppose further that a number of persons picked at random from the first committee are exchanged with an equal number also picked at random from the second. Since the teachers predominate in the first and the students in the second, the exchange is likely to reduce the proportion of teachers in the first and increase it in the second. Consequently the average age of the first group would decrease while that of the second would increase. This is what is most likely to happen. But it is not altogether impossible that the exchange may result in the influx of more teachers from the second committee in return for a corresponding number of students from the first. If this happens, the average age of the first committee would increase while that of the second would decrease. Boltzmann's point is that if in this universe of teachers and students we start with committees initially possessing a high degree of order or organization, as shown by the compacting together of teachers and students, we are likely to end up with committees having an increasingly smaller degree of order by greater intermingling of teachers and students.

To revert to Boltzmann's argument, consider now two chambers containing a gas at two different temperatures. Since, as mentioned before, the temperature of a gas is merely the average kinetic energy of its constituent molecules, it follows that the molecules of the gas in the hot chamber have on the average greater energy than those of the cold. In other words, there are more faster-moving molecules in the former than in the latter. Now when the chambers are connected, an approximately equal number of molecules will move from one to the other. The exchange results in the influx into the hot chamber of a larger number of slowly moving molecules (students) from the cold chamber in lieu of the faster-moving molecules (teachers) of the hot chamber. The reverse is the case with the cold chamber.

The *average* energy of the molecules of the colder chamber, and consequently its temperature, increases at the expense of the molecules in the hotter chamber, whose average energy correspondingly decreases. Heat thus flows from the hotter to the colder chamber. This means that the macrostate of the motion of the system at the outset when its constituent molecules were segregated into faster- and slower-moving molecules in the two chambers has a greater degree of order and therefore lower thermodynamic probability (or entropy) than after mixing when the faster- and slower-moving molecules have been randomly intermingled. In other words, the motion of the system has changed from a macrostate of lower thermodynamic probability or entropy to one of higher entropy or disorder.

The flow of heat from the hotter to the colder chamber is therefore as overwhelmingly inevitable as the entropy-increasing intermingling of slow and swift molecules of the two chambers. *Per contra*, the reverse flow of heat from the colder to the hotter chamber is equally overwhelmingly improbable. We may therefore reinterpret the second law of thermodynamics forbidding the flow of heat from a cold to hot body as the statement that any closed or isolated system of particles, that is, any ensemble which is shielded from either gaining energy from or losing it to anything outside automatically tends to an equilibrium state of maximum entropy, if not already in that state. Boltzmann was thus led to identify this tendency of closed systems to maximize their entropy with the direction of time's arrow.

Unfortunately, Boltzmann's statistical explanation of the irreversibility of time's arrow is not free from paradox, as was shown by J. Loschmidt, E. Zermelo, G. N. Lewis and others from time to time. Thus Lewis argued that increase or decrease of entropy in a closed system of particles is tied up with our subjective knowledge of the state of its internal motions. Increase in entropy occurs only when a *known* distribution goes over into an *unknown* distribution, and the gain of entropy which characterizes an irreversible process is *loss of information*. To prove his point, Lewis considered a simple, but typical case of three identifiable molecules in a closed cylinder with a middle wall provided with a shutter. Since each of three particles could be in either of two positions to the right or left of the partition wall, there are in all only $2 \times 2 \times 2 = 8$ distributions. Lewis showed that the entropy of the general unknown distribution of these particles is greater than that of any *known* distribution such as a particle pair on the left and the remaining one on the right as shown in Figure 39. He showed that an increase in entropy occurs when, after trapping any one *known* distribution, we open the shutter. If, however, we start with the shutter open with all eight distributions

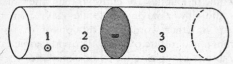

Figure 39.

occurring one after another and then close the shutter, no change of entropy occurs. He therefore concluded that gain in entropy always means loss of information and nothing more. Consequently it is a purely subjective concept.

There is no doubt that modern work on information theory reinforces Lewis's argument that entropy is a measure of our *ignorance* of the actual distribution of any system of particles. It is an idea already implicit in Maxwell's demonstration long ago that a molecular ego or sentient being, the so-called Maxwell's demon, if provided with information about the state of molecular motions of a gas, could actually make heat flow from a colder to a hotter region in violation of the second law of thermodynamics. Nevertheless, the identification of entropy with loss of information, with its implied postulation of an observer or knower, does not mean that the concept of entropy is arbitrary or subjective any more than an observation is arbitrary merely because the latter involves the participation of the observer. Because of his involvement it was once claimed that 'objective reality has evaporated' from quantum mechanics as it does not represent particles but rather our 'knowledge' or 'consciousness' of particles. The claim has been exploded with the demonstration by K. R. Popper and others that quantum mechanical concepts are fully as 'objective' as those of classical mechanics, the observer playing an identical role in both. It is the same with the newly established equivalence – gain of entropy = loss of information. There is therefore no need to purge Boltzmann's statistical reformulation of the second law of any subjective or irrelevant 'anthropomorphic elements'. Lewis's demand to do so, nevertheless, would amount to throwing the baby out with the bath water. Moreover, the temporal asymmetry of past and future whose suppression Lewis demands is an incontrovertible fact of life. It cannot be exorcized by such arguments as Lewis has advanced in support of his theory of symmetrical time, powerful and ingenious though they are.

More serious are Loschmidt's *reversibility* and Zermelo's *periodicity* objections to Boltzmann's reasoning. According to Loschmidt, the basic reversibility of *micro*-processes cannot lead to an irreversibility of *macro*-processes. For since the probability of a particle having a given velocity is independent of the sign of the velocity, depending as it does on its square, the principle of dynamic reversibility requires that to each *micro*state of motion of a given (isolated) system there will correspond another *micro*state of motion when it passes through the *same* configuration but with all velocities reversed in direction. Hence in the course of time separation or segregation processes should occur as frequently as mixing or shuffling processes. Consequently the entropy of the system should decrease as often as it tends to increase so that it cannot account for the observed uni-directional drift of time.

The *periodicity* objection of Zermelo is based on a famous dynamical theorem of Poincaré which states that under certain conditions the initial state of a system of particles in motion will recur infinitely often. Zermelo therefore claimed that molecular processes should be cyclical instead of unidirectional as postulated by Boltzmann.

Both the objections may be countered by the fact that Boltzmann's statistical proof of the second law of thermodynamics hinges only on the *average* variation of the entropy of an isolated system and therefore does not rule out the possibility of decreases in its value. Indeed fluctuations from the average must occur. But the frequency of occurrence depends on the size of the system. Thus, as was shown by M. V. Smouluchowski over fifty years ago, even if we consider a very small sphere of radius, say, 5×10^{-5} centimetres, containing air at room temperature and pressure, the mean recurrence time of a 1 per cent fluctuation from the average is of the order of 3×10^{60} years, exceeding the putative age of the universe by a factor of 10^{50}. This is only another way of saying that any such fluctuation from the average is all but impossible. While enlarging the sphere will make the chance of such a fluctuation even more remote, diminishing it alters the situation in a radical manner. For if we reduce the radius of the sphere just considered by no more than a factor of about 10^{-5} centimetres, the mean recurrence time of a 1 per cent fluctuation is reduced to only 10^{-11} seconds. It therefore appears that microscopic phenomena can have no intrinsic time direction, if this is to be defined in relation to internal entropy increase.

The aforementioned order-of-magnitude calculation of Smouluchowski would seem to make more plausible the idea of time reversal first mooted by R. P. Feynman in particle physics. In trying to explain the curious phenomenon of pair-creation and pair-annihilation, that is, the spontaneous transformation of a γ-ray photon into a pair of electron and its antiparticle, positron, or conversely the collision of the latter two to produce a γ-ray photon, Feynman argued that a positron could be regarded as an ordinary electron 'travelling backwards in time' somewhat like an old Rip Van Winkle 'growing back to his babyhood'. To understand Feynman's idea, imagine an idealized situation in which elementary particles are shot from a point A to another point B across a barrier or a field of force, PQ, which slows down their motion (see Figure 40(a)). Originally the particles travel towards P with a certain velocity v which the barrier reduces to u, the decrease for different velocities v being defined by postulating a constant difference of kinetic energy. On emergence from the barrier at Q, the particle recovers its original velocity v to ensure conservation of energy. If now we represent the situation by drawing a space-time graph of the motion with the vertical line as the time axis as shown in Figure 40(b), the path of the particle is seen as the broken line A P' Q' B'. The main feature of the path is that the two terminal segments AP' and $Q'B'$ are parallel since the velocities of the particle outside the barrier are the same, namely v. But the segment $P'Q'$ is refracted towards the vertical or time axis as the velocity on

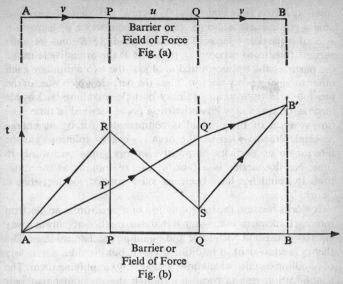

Figure 40. (a) shows particle moving from *A* to *B* across the barrier or field of force *PQ*; (b) is the space/time graph of the motion of the particle moving with velocity *v* outside the barrier but with velocity *u* < *v* within the barrier.

entering the barrier is reduced to *u* < *v*. In other words, the path of the particle is obtained by drawing two parallel lines from *A* and *B'* and then joining the points *P'* and *Q'* where they cut the vertical time axis through *P* and *Q*, the termini of the barrier.

However, we could in theory repeat the construction by drawing two parallel lines *AR* and *B'S* cutting the verticals through *P* and *Q* at *R* and *S* as shown in Figure 40(b). The path *ARSB'* also meets the requirements of the construction in that *AR* and *SB'* are indeed parallel though the path *RS* inside the barrier goes 'backwards in time'. At first sight this solution of the problem is as bizarre as Dirac's solution of the fishermen problem recounted in Chapter 9. For it makes the particle at *R* travel 'backwards' in time before its emergence out of the barrier at *S*. But Feynman nevertheless used it successfully to explain pair creation in the following way: We could say that at a certain time a particle left A. Some time later there was a pair creation at S of a particle (electron) which proceeds to B' and its antiparticle (positron) proceeding to R. At R the antiparticle meets the particle which was emitted at A and the two annihilate each other to yield a γ-ray photon. Thus the particle which was, in the previous description, an ordinary particle travelling backwards in time, becomes now an antiparticle going forward in time. The time-reversal of the original is compensated for by its charge reversal. When we do so, the negative energy solutions can be interpreted as particles of positive energy going backwards in time. No doubt this is a fanciful way of looking at the situation. It would not have been accepted but for the fact that it 'works'.

It works because the interpretation of a positron as an electron moving backwards in time can be handled mathematically in a way entirely consistent with logic and quantum mechanical laws. The theory is equivalent to traditional views, but provides a new way of handling certain calculations and greatly simplifying them. The simplification results because although the phenomena of pair creation and annihilation seem to involve three separate particles – one positron and two electrons – in Feynman's theory there is only one particle that goes sometimes forward in time at others backward. Consequently what we observe as a positron is simply the electron moving momentarily backwards in time. Because our time in which we observe the event runs uniformly forward, the time-reversed electron appears to us as a positron. We think the positron vanishes when it hits another electron. But it is just the original electron resuming its forward time direction. The electron executes a miniscule zigzag dance in space-time hopping into the past just long enough for us to see its path in a bubble chamber as the path of a positron moving forward in time. Feynman's theory thus shows that if we look at things from the worm's-eye-view of the particle itself there are no anomalies of pair creation and pair annihilation since in reality there is only one particle. But the anomaly is removed by introducing another, namely, that of time-reversal even though on the microscopic scale.

While some like Reichenbach consider the irruption of time reversal in the microscopic field 'the most serious blow that the concept of time has ever received in physics', an even bigger blow has been delivered it in the past decade.

The blow is the revelation now being vouchsafed in increasing measure that some of the symmetries of space and time hitherto assumed to underlie the fundamental laws of nature do not seem to hold perfectly on the *subatomic* scale. The first to break down in 1956 was what may be called space symmetry (P), the symmetry involved in the law of conservation of 'parity' in nuclear physics. This is the law which states that in every physical process a certain characteristic of the system called its 'parity' remains unchanged. The law belongs to the class of conservation laws like that of energy which are the most powerful principles of physics. Unfortunately, unlike energy, parity, having no analogue in classical physics, is not readily translated into everyday language. The closest one could come to it is to say that 'parity' is the formal expression of the behaviour of the wave function of a quantum mechanical system when the space coordinates describing the system are inverted through the point at the origin. If, for the sake of simplicity, we consider such systems as can move only along the x-axis, the inversion merely means changing the x-coordinate of every particle of the system into $-x$ or, what amount to the same thing, replacing any point P of the system with its coordinate x by its mirror image P' as reflected in a plane mirror OL at the origin (see Figure 41). Such an analytical transformation may therefore be visualized by viewing the system not directly but its optical image as reflected in an

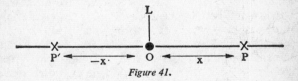

Figure 41.

ordinary plane mirror. A mirroring procedure of this sort will thus produce a counterpart of any given physical system. The full context of the law of conservation of parity is that the 'mirrored' system so produced is also a possible system. That is, it can be as fully actual as its original. This may seem odd as the 'image' is only an optical illusion of the real. But the point is *not* that the image itself is real, but that the image *can* be made real. That is, it is possible to have another system conforming to the image specifications fully as real as the original. Consequently the parity principle is only an embodiment of the reflection symmetry of nature. It merely asserts that the world of the looking-glass is a possible world even though like the left-hander it may look odd. Thus a looking-glass book may seem written in an odd kind of language; but it is a possible language that *could* be read. It is the same for all artifacts like clocks and gloves. There can be no real distinction between an object and its mirror image such as a left-handed and right-handed glove in so far as their right to exist goes.

However, it has recently been realized that there are certain rare processes whose mirror images are *not* realizable in our

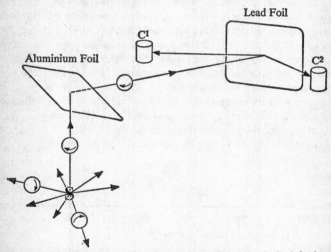

Figure 42. Electrons from a β-active source tend to spin to the left about their direction of flight as can be shown by scattering them twice, first from an aluminium foil and then from a lead foil. The two symmetrically placed counters C1 and C2 then count at slightly different rates.

world. Consider, for example, the experimental set-up exhibited in Figure 42 wherein a piece of radioactive phosphorus at S is emitting electrons in all directions. The piece is enclosed in a suitable box of half-inch perspex (not shown in the figure) so that all electrons except those travelling upwards and passing through the top of the box are stopped. On emerging from the top they strike a sheet of aluminium where some of them are deflected by a right angle and escape through a horizontal extension of the box, only to strike a lead plate placed near it. Since the whole arrangement is entirely symmetrical about the line of flight of the electrons approaching the lead plate, we should expect that of the electrons that are deflected by the lead plate equal numbers should go to the right and to the left. But the experiment reveals a pronounced bias towards the right rather than to the left – contrary to expectation. About 10 per cent more electrons are recorded on the right counter than on the left even though the arrangement is quite symmetrical, the two positions of the counter, namely $C1$ and $C2$, being the mirror image of each other.

The asymmetry observed arises from the fact that in such radioactive decays emerging electrons behave (speaking analogically) like spinning bullets shot out of a rifle rather than cannon balls fired without spin from a cannon. A non-spinning cannon ball shot out of a cannon looks essentially the same in a reflected system. But a rifle bullet, spinning in a definite sense, say, right-handed about its line of flight, will appear left-handed in a mirror – an essentially distinguishable situation. In a precisely analogous fashion in certain radioactive processes such as in the afore-mentioned experiment, the neutron whose decay results in the emission of an electron behaves like a rifle not a cannon. It happens that whenever a particle that has spin gets sent out as a result of such radioactive decays it tends to spin in a left-handed way about its direction of flight. It is this spin that causes an asymmetry in the deflection pattern the experiment reveals. Deeper examination of the experiment, as also of certain others like the celebrated radioactive cobalt 60 experiment of Madame Wu's group performed in 1956 at the instance of T. D. Lee and C. N. Yang, shows that there are certain processes whose mirror reflections are not possible so that their parity is not conserved. A typical case is the β-decay of a neutron*

$$n^\circ \;\rightarrow\; p^+ \;+\; e^- \;+\; \sqrt{}^-$$

neutron   proton   electron   antineutrino

whose 'parity' is *not* conserved. That is, such decays look essentially different when viewed in an inverted coordinate system or as reflected in a mirror. No stronger violation of the parity conservation principle or P symmetry could be produced even though such decay processes are intrinsically very rare.

The observed breakdown of P symmetry has led to interesting speculations about possible violations of other kinds of symmetry – those of charge (C) and of time (T). Prior to 1956, physicists believed that each of these three basic symmetries held separately throughout nature in that 'reflection' of any process in any of the three P, C or T 'mirrors' was as fully realizable as the original. Reflection in C and T 'mirrors' really means transformation of the process in a specified way to produce its symmetrical counterpart, *not* with respect to its spatial configuration – that is achieved by ordinary reflection in a plane or P mirror – but with respect to the other two attributes, namely its charge content and temporal flow. To illustrate, the 'reflection' of any process in the T 'mirror' is its time-reversal obtained by changing the sign of the time coordinate t into $-t$ in the equations governing the process, just as P reflection of a process is its inversion by the change of the spatial coordinate x into $-x$. T reflection is therefore the outcome of a metaphorical mirroring procedure (transformation) whose mathematical description is analogous to that of ordinary reflection in a plane mirror. Similarly, when we 'reflect' a process in another metaphorical 'mirror', called C 'mirror', we obtain a 'mirror image', where every particle in the original complex is changed into its conjugate antiparticle with reverse charge. Thus 'reflection' of any object like an atom in the C mirror is its antilogue, the antiatom, which is as fully qualified to exist as the original.

Till 1956 all the three P, C and T symmetries were believed to hold good independently. The 'reflection' of any given process in

*For further explanation of such decay processes see Chapter 15.

any of these three 'mirrors' taken singly or two or all three at a time was also a possible process. There were thus in all seven possible 'reflections' of any given process in the C, P, T, CP, CT, PT and CPT 'mirrors' invented by physicists to describe the inherent symmetries of the laws of nature. Of these the P mirror is already shattered with the breakdown of parity principle in certain β-decay processes. There is reason to believe that some of the remaining six 'mirrors' are also *slightly* defective. Thus a mathematical argument based on the observed spin direction of electrons and positrons that are respectively emitted by negative and positive muons (elementary particles) seems to suggest that C reflection is no more an *exact* symmetry than P reflection is. If the P and C mirrors are known to be slightly defective, the two defects were believed, until quite recently, to cancel themselves out. Consequently the CP image, the outcome of two successive reflections, namely the P image reflected in C mirror or vice versa, was thought to be a possible outcome in our world. It certainly seemed to be the case, at least for a number of phenomena that occur in radioactive decays and that violate C and P symmetries separately. But a recent experiment (1964) by the Princeton physicists James H. Christenson, James W. Cronin, Val L. Fitch and René Turlay, has yielded some indirect though controversial evidence in support of a violation of even CP symmetry. If CP symmetry is, indeed, violated, T symmetry cannot remain inviolate. For if so, the very last mirror we have, namely the CPT mirror, *cannot* be perfectly true. And yet physicists cannot afford to believe that it is even slightly defective. The belief is not because of faith in nature's innate preference for perfect symmetry. Such a belief in nature's perfection is as unreasonable as the old idea of the Greeks that planetary orbits can only be such perfectly symmetrical figures as circles. The physicists' belief in the CPT symmetry springs from the stubborn fact that the whole framework of modern physics and, in particular, of quantum electrodynamics and special relativity would collapse if the CPT mirror were found to be even slightly untrue. Consequently if the CPT mirror is to be salvaged, one has to concede that the T mirror is defective so that the fundamental laws of the microcosm are *not* strictly time symmetric.

Conceivably the recently suspected slight asymmetry of time on the subatomic scale, if confirmed, may provide a basis for the one-way orientation of time's arrow. But no one has yet been able to show how. Nor do we know whether this arrow has any connexion with other time arrows such as those provided by the increasing trend of entropy, one-way drift of radiative processes or the cosmological expansion of the universe. Indeed for that matter we have still to establish a link between the one-way drifts of each one of the three latter processes.

Consider, for instance, the radiative and cosmological chronologies which Wheeler and Feynman tried to connect without much success. In the Wheeler-Feynman theory electromagnetic radiation is described in terms of direct particle action. That is, any two charges such as P and Q interact with each other by an action which travels at the speed of light. If their distance apart is r, the action that emanates from P at any time t will begin to exert its influence at Q at time $t + \frac{r}{c}$, because $\frac{r}{c}$ is the time taken by the influence or 'action' emanating from P to reach Q travelling at the speed of light c. But this action must have an equal and opposite reaction implying that Q's reaction, though starting at time $t + \frac{r}{c}$, reaches P at t! If we call P's effect on Q a 'retarded' effect because it occurs after a time lag of $\frac{r}{c}$, Q's reaction is 'advanced' because it occurs in advance of the sponsoring action itself. Thus 'retarded' and 'advanced' effects go hand in hand. This may seem to be a defect of the Feynman-Wheeler theory. But they managed to turn it to good account in a remarkable way. They argued that the universe does not consist of just two particles P and Q so that we are not justified in calculating the reaction from Q alone. Indeed we must include the advanced effects of all other particles R, S, T, U . . . in the universe. Wheeler and Feynman were able to show that in a static infinite universe with a homogeneous distribution of charges, the combined reaction on P from all the charges including P itself is just such as to provide a purely retarded effect as actually observed. They could thus show that the choice of the retarded solution of Maxwell's electromagnetic equations is not wholly arbitrary. It is dictated by the universe.

While we may thus manage to secure the radiative arrow of time a foothold in the structure of the universe as a whole, it provides no coupling whatever with the cosmological arrow implied by the Hubble expansion of the universe. Indeed, Wheeler and Feynman assumed the universe to be static, which is time-symmetric. If therefore we repeat their calculations by reversing the sign of the time coordinate throughout, we shall get a consistent result but with pure advanced effects where there were retarded effects previously. Since both the retarded and advanced effects are equally consistent with Maxwell's equations of electro-magnetism, Wheeler and Feynman had to resort to statistical-cum-thermodynamic considerations to choose between the two. Hoyle and Narlikar claim that this is unnecessary. In a very recent communication they try to show that the choice of the correct (retarded) solution can be made independently of extraneous thermodynamical-cum-statistical considerations pro-vided the basic postulate of Hoyle's steady-state theory in its original version is conceded. But recourse to a much beleaguered theory is indeed too slender a basis for establishing the irreversi-bility of time via that of radiative processes and Hubble expan-sion of the universe. It seems there is as yet no way of doing so except by direct resort to statistical-cum-thermodynamical con-siderations in the Wheeler-Feynman way. Perhaps Eddington was right in excluding both radiative processes as well as Hubble expansion of the universe as insufficiently fundamental to provide a basis for time's direction. Given the initial and boundary con-ditions for starting the reverse of a radiative process or an ex-panding universe, the reverse event is certain to occur. But the irreversibility arising from the switch of an ordered arrangement to a disordered one like shuffling a new deck of cards seems al-most irrevocable. For even if we continue shuffling it at random for billions upon billions of years, the original order of the pack is not likely to re-emerge. Consequently the shuffling or mixing processes from which Boltzmann deduced the second law of thermodynamics seem to be irreversible in a stronger sense than the radiative events or the cosmological expansion. Indeed the latter is deemed to reverse *itself* in oscillating models, and the equations governing them allow it too. It may be that a new time arrow that recent evidence suggests may be built into some of the most elementary particle interactions will be as fundamental as that of the thermodynamical arrow. But no one knows yet exactly how to build it. We have therefore for the present no alternative but to equate time's arrow with the tendency of the entropy of every (closed) system of particles to increase to an upper limit.

The inexorable trend of entropy, or degree of disorder of statistical systems, towards a maximum has led to interesting and varied speculations about the ultimate fate of our universe. For if, instead of considering isolated systems like the gas in the hot and cold chambers, we now envisage the entire universe as one vast ensemble of irregularly moving particles, each one of its macrostates by the same token is likely to be followed by another of higher thermodynamic probability of greater disorder and disorganization. It therefore follows that the universe as a whole, too, is continually tending towards a state of maximum disorder when all temperature differences shall have been wiped out, putting an end to heat flow from any place to any other in the universe. The most conspicuous manifestation of this inevitable march of the universe towards such a motionless and dead level sea of heat is the enormous and continuous outpour of stellar radiation into empty space. Such an ultimately quiescent state of the universe when all the embers of its activity will have faded out of existence is the heat death of the universe – the last dusk of reckoning after which the moving finger shall write no more.

Many attempts have been made to evade the cosmological consequence of the second law predicting a final *fin du monde*. Thus it has been objected that the second law, which has been found by actual experience to apply only to closed systems, may not hold for the universe as whole, as the universe may not be a closed system. Such, for instance, would be the case if it were unlimited in extent. But, as Weizsäcker has pointed out, this theoretical objection does not really invalidate the argument leading to the heat-death conclusion.

The course of events in a finite part of the world could continue

300

forever only if energy were to flow into it continually from some-where. But in the absence of such an influx of energy and in the face of its continual dissipation by stellar radiation into empty space, any observable finite part of the universe must inevitably march towards its heat-death doom. There is only one escape from this conclusion, suggested by Tolman on the basis of his synthesis of relativity and thermodynamics. Tolman has shown the possibility of a certain class of oscillating model universe which even though expanding and contracting irreversibly, seems able to evade attaining an unsurpassable state of maximum entropy. Without positively denying the possibility of our actual universe's reaching a state of maximum entropy, his finding thus does allow some relaxation in the 'rigour of our thermodynamical thinking'.

In sum, the validity or invalidity of the heat-death conclusion is one of those many inferences of cosmology in which the basic inadequacy of physical observations on the cosmic scale makes cosmologists select those solutions out of the many that appeal to them emotionally, aesthetically, metaphysically, morally, or otherwise. Thus to an Eddington the conception of an oscillating universe continually running down and continually rejuvenating itself seems from a moral standpoint wholly retrograde. On the other hand, to a Tolman the idea of the beginning and end of the universe is so repugnant as to drive him to hunt for oscillating models of the universe so that there need be no limit in time either forward or backward. However, while it may seem hazardous at present to make any final statements about the application of the second law to such enormous ensembles as the whole universe, it does apply rigorously to each local ensemble such as the sun, earth or star. But if all the stars by their incessant outpourings of radiation into space are exhausting themselves inexorably into a state of degeneracy known as the white-dwarf stage, there is little to warrant the assumption that the universe can rejuvenate itself. If so, the stars are setting and the caravan draws a step nearer to the dawn of nothing.

PART III

Origins

CHAPTER 15

The Origin of the Elements

IF, as shown on page 201, matter in the universe cannot be infinitely old, it must have originated at some finite time in the past. We can roughly date its origin because some of the ninety or so naturally occurring elements are radioactive. That is to say, they decay spontaneously into other elements which accumulate. By measuring the accumulation of the decay products and knowing the rate at which the parent elements decay it is possible to calculate the 'age' of the parent elements. Actual calculation shows that radioactive elements are seven to fifteen billion years old depending on whether the parents were made all at once cosmologically and have been decaying ever since, or have been synthesized gradually by astrophysical processes like nuclear reactions inside stars. A possible clue as to which of the two processes whether cosmological or astrophysical actually originated the elements could perhaps be provided by the prevailing element abundances in the universe, if only we could determine it. For obviously the universe is much too large for us to obtain a truly representative sample of its material content. The closest we can come to it is to examine the materials in our own immediate neighbourhood. We do so in two collateral ways. First, we observe from afar *spectroscopically* the element abundances prevailing in distant objects like the sun, near-by stars, gaseous nebulae and interstellar gas. Second, we directly analyse samples of material of the earth, meteorites and cosmic rays. Naturally the compositions obtained for different objects by different methods do not agree in detail. But in many cases the differences can be explained plausibly enough to warrant the assumption that they are basically the same. For example, the precise proportions of elements are not exactly the same in the sun as on the earth, because some gases that exist in the sun were largely lost from our planet when it was formed. If we allow for such explicable differences in the stellar, solar, meteoritic and terrestrial abundances, we find that they are all essentially the estimates of primeval solar system material. We can therefore make an overall abundance compilation by selecting the abundance determination which seems most secure. The best known such compilation is by H. E. Suess and H. C. Urey. Though really parochial, it is, *faute de mieux*, presumed to be universal in the hope that conditions in the still more distant galaxies are similar to those in our own. As we saw in Chapter 6, it is a very risky extrapolation which Gamow has skilfully exploited to rescue his cosmology from a serious difficulty.

The results of the great Suess-Urey compilation are shown in Figure 43. It is a plot of the abundances of nuclides on the logarithmic scale against their corresponding mass number, that is, the number of protons and neutrons composing the nuclide. It will be observed that the most abundant nucleus in the universe is hydrogen (about 72 per cent by mass) followed by helium (about 26 per cent by mass). The next few elements of increasing mass numbers – lithium, beryllium, boron – are far scarcer; while carbon, nitrogen and oxygen are very abundant. The still more massive elements – neon, magnesium, silicon, sulphur, argon and calcium – are nearly as common. Thereafter the curve plunges down to rise again to another peak around mass number 56 of the iron nucleus. Beyond mass number 90 or 100 the level is more nearly constant with a few sharp peaks superimposed.

The crucial question at issue is whether the observed pattern of element abundances is the outcome of an all-at-once cosmological event or of gradually occurring astrophysical processes. An instance of the former type is the initial explosion of the primeval atom postulated in the big-bang cosmology of Gamow. As it has already been described in some detail in Chapter 7, we will not dwell on it any more except to recall that the absence of stable nuclei of mass numbers 5 and 8 obliges it to ascribe the origin of hydrogen and helium to the initial cosmological big bang but that of the bulk of heavier elements to subsequent astrophysical processes like stellar nucleogenesis. Unfortunately primeval helium production of about 27 to 30 per cent by mass under the postulated conditions of the big bang is still a serious embarrassment on two counts. First, as noted earlier, it is uncomfortably

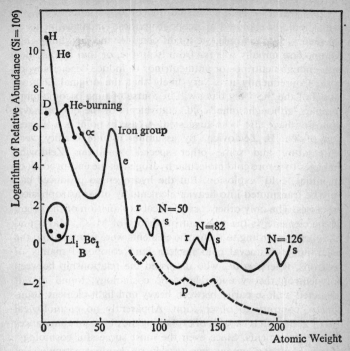

Figure 43. Schematic curve of atomic substances as a function of atomic weight. (Based on data of Suess and Urey.)

large. If the universe started with such a cosmic abundance of helium, the present helium abundance (around 26 per cent by mass) should have been somewhat larger considering that much more helium is likely to be produced than destroyed in subsequent astrophysical processes. Second, the very oldest population II stars which are supposed to have condensed out of the helium-rich primeval material should show much higher helium abundance than they actually do. Very recent studies, tentative though they are, seem to show that the helium content of some older population II halo stars in the galaxy at their surface is less than 1 or 2 per cent by mass.

307

It has therefore been suggested that other variants of the big bang should be explored to escape from these embarrassments of the theory. For example, we might consider the possibility that the universe initially was far from isotropic, or that it was filled with enough neutrinos or antineutrinos to induce degeneracy, or both. Consequently it is very likely that the original Gamow model of the 'hot' big bang will in course of time become more complex, although some 'cold' equivalents of the hot primeval explosion have also been suggested. A case in point is a suggestion of Ya. B. Zeldovich. By assuming infinite density, zero temperature and some other special conditions Zeldovich manages to secure cold molecular hydrogen as the end-product of an initial 'cold' explosion. But the hydrogen so produced can only be transmuted into heavier elements by other astrophysical processes. The only other cosmological explanation of the origin of the elements is the polyneutron theory of M. G. Mayer and E. Teller. According to them, the elements were formed from the break-up of primeval cold nuclear fluid composed mainly of neutrons. Peierls *et al.*, who discussed the relationship between polyneutron theory and relativistic cosmology, found that it predicted a close parity between heavy and light element abundances, contrary to observation. Apparently no cosmological theory at present is capable of explaining the observed abundances of *all* the elements. Since even the most successful cosmology, namely that of Gamow, has to rely on gradual astrophysical processes like nuclear reactions inside stars to account for the origin of elements heavier than helium, we now take up the study of stellar nucleogenesis.

As we have already seen, some synthesis of heavier elements has been occurring in stars ever since their formation from primeval hydrogen. Indeed normal stars like the sun maintain their radiation output by fusing hydrogen into the next heavier element, helium. As, growing older, they run short of hydrogen, they sustain themselves for a while by fusing helium into still heavier elements. The problem of element formation is therefore closely related to that of stellar evolution with the further complication that the heavy elements observed in a star have not always been synthesized in the same star. This is because material

THE ORIGIN OF THE ELEMENTS

synthesized in one generation of stars is continually blown into interstellar space by nova outbursts, supernova explosions and more slowly by stellar winds, to reappear in stars of a later generation condensing out of interstellar gas and dust. Consequently any astrophysical theory of element synthesis has two aspects. In the first place, it must show that physical conditions do occur in which element synthesis takes place to produce the observed abundances. Secondly, it must also provide that synthesized material can be removed from its place of birth and incorporated in other objects. At the moment the two aspects of the dual problem have not progressed equally. While the aspect of element synthesis is fairly well understood, that of ejection and recycling of the processed material is quite underdeveloped. Indeed the former is an achievement in physics of singular brilliance inspired by modern cosmology, in particular by the earlier version of the steady-state cosmology of Hoyle, Bondi and Gold. Although the original version has now been discarded, its offshoot, the theory of stellar nucleogenesis, is a permanent asset. Since it is a neatly reasoned blend of nuclear physics and astronomy, we shall begin our exposition with a brief account of nuclear transformations on which it is based.

As we know, atoms consist of nuclei surrounded by electrons and nuclei consist of a mixture of protons and neutrons, collectively called nucleons. Since the mass (A) of every atom is very nearly an integral multiple of the mass of the lightest element, hydrogen, whose nucleus is a single proton, it is natural to think of a stepwise process of element formation by adding one more nucleon to its predecessor. The addition of one more proton to the nucleus of an atom turns it into the nuclide of the next higher element in the periodic table. But the addition of one or more neutrons alone will turn it into a different species or isotope of the same element. Accordingly if we are to understand the building of atomic nuclei it will be necessary to invoke some processes whereby protons can turn into neutrons and vice versa. Such processes are now well known. They are radioactive decays whereby some nuclei spontaneously transform themselves into those of other kinds with the emission of radiation. The *observed* radiation is of three types: α-particles, which are nuclei of helium of mass number 4 (2 protons and 2 neutrons), β-particles which are electrons, and γ-rays which are short wavelength and hence very energetic photons. To the three *observed* radiations we must also add a fourth *inferred* radiation, the emission of neutrinos, a species of particles already encountered in Chapter 3. As noted there, Pauli had to expressly postulate them to balance the energy books of β-decay. Their existence has since been more securely established even though neutrinos being particles of negligible rest mass and nil charge interact so rarely with other particles that they are all but unobservable.

The reason for spontaneous radioactive decays which give rise to these emissions will be obvious from Figure 44 which is a

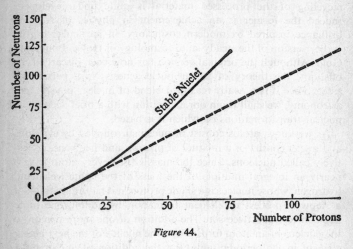

Figure 44.

plot of neutrons versus protons in stable nuclei. As will be observed, for lighter stable nuclei the numbers of protons and neutrons are approximately equal. But for heavier nuclei there is a slight excess of neutrons over protons to ensure stability. This is a consequence of the strong electrostatic repulsion exerted by protons on one another which must be balanced by attractive nuclear forces between nucleons. As more and more protons pile up in the nucleus its stability requires a comparatively greater number of neutrons to counterbalance the excessive electrostatic repulsion of many protons. The plot of Figure 44 shows that an atom of a given atomic number Z, that is, having Z protons, has only a very narrow range of possible numbers (N) of neutrons, if it is to remain stable. But this is not to say that atomic nuclei with either too many or too few neutrons relative to the number of protons present for stability cannot exist at all. They do, but they transmute themselves into more stable ones either by the initial protons switching into neutrons or vice versa by what are called β-processes. If the number of neutrons is excessive, the transformation of a single neutron into a proton will reduce the number of neutrons while increasing the number of protons. To conserve electric charge, such a transformation requires the emission of a negative electron, and, to conserve energy and angular momentum, of an antineutrino, the antiparticle of neutrino. Accordingly the β-process of neutron decay may be symbolically represented in equation form as

$$n^{\circ} \quad \rightarrow \quad p^{+} \; + \; e^{-} \qquad \sqrt{}^{-} \qquad (1)$$

neutron	proton	electron	antineutrino
electrically	positive	negative	electrically
neutral	charge	charge	neutral

The electron leaves the nucleus and is observed as β-particle but the departing antineutrino is hard to detect. In case the nucleus has too few neutrons, the inverse reaction

$$p^{+} \rightarrow n^{\circ} \; + \; e^{+} \; + \; \sqrt{} \qquad (2)$$

proton	neutron	positron	neutrino

occurs whereby a proton becomes a neutron with the emission of a *positive* electron or a positron and a neutrino. But an unstable nucleus if it contains too many nucleons exceeding, say, 210, may also stabilize itself by α-decay, that is, by emission of packaged nucleons. Such a package is the α-particle, the helium nucleus consisting of four nucleons, that is, two protons and two neutrons.

In sum, there are three processes available to an unstable nucleus to stabilize itself. The negative β-decay process (1) decreases the number of neutrons by one thereby increasing nuclear protons by one. The positive β-decay process (2) does just the reverse as is naturally to be expected. But α-decay decreases both neutrons and protons by two involving as it does the loss of a complete nucleonic package. These three processes are schematically depicted in Figure 45. Very often a succession of α- and β-decays with accompanying γ-decays to carry off excess energy is required before a nucleus acquires stability. These decay

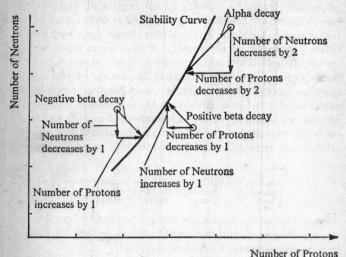

Figure 45. The circles denote the initial position of the unstable nucleus. The arrows indicate how alpha and beta decays tend to bring it to a stable configuration shown by crosses on the stability curve.

processes are of fundamental importance in understanding the synthesis of elements in stellar interiors, where the energy-producing nuclear reactions give rise to nuclei which beta-decay and emit neutrinos. In some stars the *reverse* α-decay occurs whereby the nuclei capture α-particles instead of ejecting them. There is indeed a wide diversity of nuclear reactions going on within stars depending on their mass, chemical composition, density, age, stellar population type, etc. Burbidges, Fowler and Hoyle enumerated in their fundamental paper of 1957 some eight different types of such nuclear reactions. Later developments have added one more. These nine reactions may be divided into three main groups:

(a) Exothermic or heat-producing nuclear reactions,
(b) neutron capture processes,
(c) miscellaneous processes.

(a) *Exothermic reactions* occur successively as a star's internal temperature rises with age. We mentioned in Chapter 2 two such reactions, namely proton-proton chain and carbon-hydrogen cycle whereby hydrogen is burnt to yield helium 'ash'. Later, with the exhaustion of hydrogen, the star switches to burning helium and carbon. Besides the three hydrogen, helium and carbon burning reactions there are two more called a- and e-processes. They occur at still higher temperatures, about 100 to 200 times higher as will be seen from the reaction temperatures of each process shown in Table 9 below:

TABLE 9

Nuclear process	Temperature
Hydrogen burning	2×10^7 degrees K.
helium burning	2×10^8 degrees K.
carbon burning	5×10^8 degrees K.
a-process	10^9 degrees K.
e-process	4×10^9 degrees K.

(b) Neutron capture processes are of two kinds – slow and fast. If the conditions in the star are such that the nuclei can capture them only slowly, it is called slow neutron capture or more compactly the s-process. Its rapid counterpart is the r-process.

(c) Finally there are two miscellaneous processes called the p-process (proton capture) and the x-process whose nature is as yet so ill understood that it has been given the mystery prefix x.

We have already referred to some of the aforementioned reactions in the context of stellar evolution in earlier chapters. Our present concern is to describe the part they play in the synthesis

313

of elements. As a condensation either of primeval hydrogen or of interstellar gas and dust becomes a luminous star, it begins to burn hydrogen. In the small main-sequence stars like our own sun, protons (hydrogen 1) combine in a steady chain through deuterium (hydrogen 2) and light helium (hydrogen 3) to ordinary helium (helium 4) which remains as the 'ash' of the hydrogen burning process. This is the proton-proton chain. In more massive main-sequence stars, where central temperatures are higher, the hydrogen burning is catalysed by the presence of carbon nuclei but the end-product is helium as before. This is the carbon-nitrogen cycle wherein protons combine successively with carbon 12 to make carbon 13, nitrogen 14, nitrogen 15, which then combines with a proton to produce oxygen 16, which immediately α-decays into carbon 12 and helium 4 (α-particle). The original carbon 12 after all these transformations is thus regenerated at the end of the cycle like a catalyst in a chemical reaction.

In the course of time as stocks of hydrogen are depleted, the star begins to depart from the main sequence to become a red giant resorting to helium burning which starts off at the very centre of the star and spreads outwards. The process leads to the formation of carbon 12 and oxygen 16 through the intermediary of the unstable beryllium 8 nucleus. This is why we can observe 'carbon' stars in whose light the absorption bands of carbon molecules and compounds are abnormally strong. In some of these carbon stars there is evidence of carbon 13 present in appreciable amounts compared with carbon 12. Apparently the carbon 13 being brought up from the interior by convection currents is the outcome of hydrogen burning at some places.

With the exhaustion of helium, the core again contracts until carbon begins to burn. It forms neon and sodium together with more hydrogen and helium. The latter are high energy α-particles resulting from the α-decay of neon nuclei. These α-particles are captured to form those abundant nuclei whose atomic masses are multiples of four: neon 20, magnesium 24, silicon 28, sulphur 32, aragon 36, calcium 40. Extreme effects of the α-process are shown by some very faint white dwarfs which show only magnesium and calcium in their spectra. Other elements in this mass range are produced by subsidiary reactions such as proton captures.

We have seen that as a star ages and exhausts its original stock of nuclear fuel, it begins to burn the 'ash' of the preceding burning process. The process obviously cannot go on indefinitely. It reaches its dead end when the star passes the stage of burning magnesium, silicon, sulphur, etc., to produce still heavier nuclei of iron group in the vicinity of atomic mass 56. The reason is that the balance of nuclear and electrostatic forces in the atomic nucleus reaches its peak strength around mass number 56 so that the nuclei of the iron group of elements are the most stable. Indeed they are so stable that any nuclear reaction involving them *absorbs* energy instead of releasing it. The star cannot therefore balance its energy budget by burning the iron group of elements to produce still heavier ones. Consequently a qualitatively altogether new state of affairs emerges. A chaotic profusion of nuclear reactions synthesizing the various elements as well as breaking them down into lighter ones begin to occur simultaneously. If the reactions are fast enough, the disruption and synthesis rates become equal and equilibrium results. At temperatures around 4×10^9 degrees K. and at densities expected in stellar interiors, the upshot of such an equilibrium or *e*-process can be calculated by statistical mechanics. Burbidges, Fowler and Hoyle who made such a calculation found close concordance between computed and observed nuclidic abundances as shown in Figure 46.

Although we have carried the process of nuclear burning inside stars as far as it can go, we have succeeded in building elements no further than the iron peak – a mere quarter of the way through the table of atomic weights. The theory of element formation provides the answer. But it requires that material once cooked in stellar interiors is re-cooked in other stars. To explain the build-up of nuclides more massive than iron, it is assumed that a small proportion of the iron group elements is subjected to prolonged neutron bombardment in stars of the second and later generations. It will be recalled that the nuclei of carbon 13 are produced during the initial stage of hydrogen burning by the carbon-nitrogen cycle as the star switches to helium burning with the depletion of hydrogen. Carbon 13 combines with an α-particle to produce oxygen 16 plus a neutron. The neutrons so

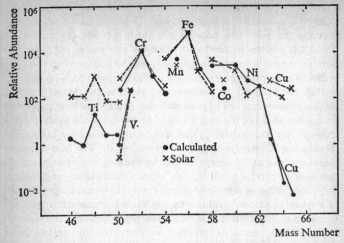

Figure 46. (From Bernard Pagel, 'The Origin of the Elements', *New Scientist*, 8 April 1965).

produced do not have to overcome electrical repulsion to react with the nuclei. When each neutron is absorbed, it adds one mass unit to the nucleus. If the nucleus is unstable, it beta-decays into the next higher element provided another neutron is not absorbed

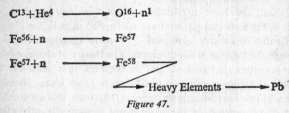

$$C^{13} + He^4 \longrightarrow O^{16} + n^1$$

$$Fe^{56} + n \longrightarrow Fe^{57}$$

$$Fe^{57} + n \longrightarrow Fe^{58}$$

$$\longrightarrow \text{Heavy Elements} \longrightarrow Pb$$

Figure 47.

before the decay. In this way the slow process or '*s*-process' chains up step by step till it terminates at bismuth 209, because neutron capture beyond it leads to elements which α-decay much faster than the neutron capture time scale of 10^5 years. This is why neutron capture products accumulate as lead and bismuth as shown in Figure 47.

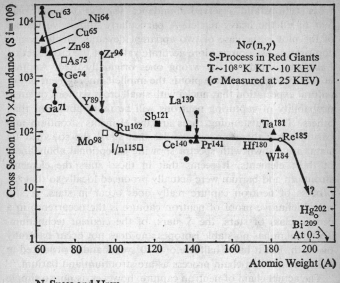

Figure 48. (From Fred Hoyle, *Galaxies, Nuclei and Quasars*, Heinemann, 1966.)

We have definite evidence to prove that heavy elements do get built up by neutron captures in the manner envisaged in Figure 47. For we expect the product of the observed relative abundance and the neutron capture cross section (essentially the probability of a particular nucleus picking up a neutron) to vary smoothly from one atomic weight to the next. This is because a nucleus with a high capture probability will more readily be moved on to the next member in the series than will a nucleus with a low capture probability. Figure 48 is a plot of the product of measured cross sections and observed abundances for approximately one-half of all the heavy nuclei. A particularly impressive feature of the plot is the case of two apparent deviants which became conformists as soon as more accurate values of the cross section were substituted for the wrong ones originally used. Another noteworthy feature of the plot is the implicit embodiment of our natural expectation that nuclei with small cross sections or small probability of capturing neutrons will be more abundant than others. The outstanding cases are the elements strontium and barium. It is therefore no wonder that stars, the so-called Ba stars, are known with spectra showing exceptional abundances of these elements. It seems that in these stars the elements strontium and barium were actually produced locally so that the process of neutron capture really does occur in stars. A still more definitive proof of neutron capture is the occurrence in a certain class of stars, the S stars, of the element technetium. Technetium has no stable isotopes and does not occur naturally on earth. But it has a half life of $2 \cdot 10^5$ years and is produced in the same neutron-chain process as are strontium and barium.

The actual chain of neutron capture, however, is not so simple as the diagram in Figure 47 might seem to suggest. For the chain must necessarily pass through only such nuclei as remain stable. Since the neutrons are captured rather slowly in periods of the order of 10^5 years, there is plenty of time for the nucleus to β-decay, if it happens to be unstable. The chain thus proceeds upwards through a 'main line' of stable nuclei only. A portion of the main line produced by the chain of neutron addition passing through the elements silver, cadmium, iridium and tin to tellurium is shown in Figure 49. It will be seen that not all the observed stable isotopes of cadmium and tin are generated in this way. The track does not pass through cadmium 116 or through tin 122 and tin 124, because these elements decay much faster in periods of the order of a day. To produce the missing nuclei it is necessary to invoke another process of neutron capture – the so-called rapid or r-process. Such a process occurs when an ultra-rapid fusillade of neutrons leads to the capture of several neutrons by the irradiated nucleus on a short time-scale of a few seconds. The nucleus thus acquires many neutrons before there

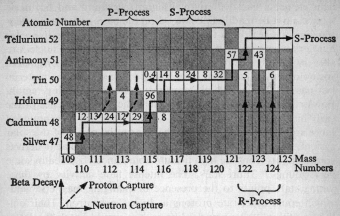

Figure 49. (From Bernard Pagel, 'The Origin of the Elements', *New Scientist*, 8 April 1965.)

is time for it to beta-decay. That such heavy elements can indeed be built up in this way was demonstrated by the first explosion of a hydrogen bomb at Bikini. The neutrons released during the explosion were captured by nuclei in the metallic shell of the bomb thereby transforming them to transuranium elements such as californium 254 and the like. Consequently very heavy radioactive elements can be built up by the *r*-process until the chain is stopped well beyond uranium at californium 254. While the *r*-process builds up the very heavy neutron-rich nuclei, the *s*-process builds up most of the isotopes with fewer neutrons. Although these processes are only necessary for elements heavier than iron, the processes will also work if lighter nuclei are irradiated with neutrons. It is therefore likely that some fraction of the material below the iron peak has also been formed in this manner.

Besides the products of *r*- and *s*-processes, there is a third group of nuclear species beyond nickel which cannot be produced by either *s*- or *r*-process. These by-passed nuclei are ascribed to yet another process (the *p*-process) involving a small amount of proton capture by neutron-rich nuclei. Finally, there is the mysterious *x* process required to explain the local abundances of light elements like deuterium, lithium, beryllium and boron as these cannot be readily understood on the basis of stellar nucleogenesis. The reason is that at the ruling temperatures in stellar interiors where nuclear reactions are occurring, these nuclei are all very rapidly destroyed. Consequently unless they originated cosmologically, they must have been produced since the material last passed through a high temperature nuclear burning region. Fowler and Hoyle have suggested that they may be relics of the early history of the solar system – a view that will be expounded more appropriately in the next chapter on the origin of the solar system. It will suffice for the present to mention that it derives some support from the high lithium abundance observed in some very young *T* Tauri stars. The intense flare activity by these young stars points to the presence of changing magnetic fields which could accelerate protons as in a synchrotron. Their collisions with abundant nuclei like carbon would break the latter down into lighter elements by what are called 'stripping reactions'. Subsequent irradiation with neutrons is needed to explain the predominance of lithium 7 over lithium 6 and of boron 11 over boron 10. Similar electromagnetic effects may also be the cause of the bizarre composition of the so-called peculiar stars. They are often observed to have strong variable magnetic fields and show a variety of overabundances and shortages of many different elements. But we do not know yet whether surface or interior reactions are responsible.

The theory of stellar nucleogenesis we have outlined is not, however, free from some serious difficulties. For instance, it has recently been criticized by Suess on the ground that it makes the solar-system material a chance superposition of matter processed in a variety of different objects. Such a chance superposition is not likely to show the regularities it actually does. In any case they are not readily understood on the theory of Burbidges *et al*. As we saw, their theory supposes that an early generation of stars built up to the iron peak from hydrogen and spewed out the products in supernova explosions. The latter generation stars built heavier elements on the iron peak by neutron captures and blew them off from their surfaces. That all the products of the diverse processes we have described came to be mixed in just the right proportions to exhibit certain regularities noticed by Suess seems implausible. The most important of them is the fact that abundances of nuclei having a given excess (N–Z) of neutrons (N) over protons (Z) lie on a smooth curve when plotted against mass number irrespective of whether the nuclide is ascribed to the s-, r-, or p-process. It does seem an improbably strange coincidence that products of the s-, p- and r- processes, which presumably took place in entirely different objects, happen to fit such a smooth curve. Furthermore Suess finds additional abundance peaks in the 'wrong' places and certain 'minor' effects that might arise not from synthesis but from fission of very heavy nuclei.

Nor is this all. There is an awkward discrepancy in helium content. If no helium has been produced cosmologically and all of it originated astrophysically as Burbidges *et al.* assume, it would account for a mere 7 per cent of the observed abundance. For the mass of the galaxy is estimated to be around 100 million suns. If its present luminosity is ascribed to conversion of hydrogen into helium (an overestimate but not a gross one) and that luminosity has been constant for its putative lifetime of ten billion years, about two billion solar masses of helium would have been produced. This is barely 2 per cent of the mass of the galaxy against the observed 26 per cent. The difficulty persists even if it is assumed that there were relatively more massive stars in the early life of the galaxy. Although most of the luminosity comes from hydrogen burning, some conversion of helium to heavier elements also takes place and recent models of medium-mass stars suggest that at any time the amount of unburnt helium is much less than would be naïvely estimated. T. W. Truran *et al.*, who have recently studied the helium problem, conclude that if the helium was produced astrophysically it must have been made in low-mass stars, which in turn implies an age of twenty to thirty billion years for the galaxy. It is uncomfortably large for any big-bang cosmology and even on the steady-state theory it would make our galaxy much older than the typically expected age of three to four billion years on the basis of the currently accepted value of Hubble's constant H. The helium problem may be solved by cosmological production but to obtain the correct mixture of all the other elements it may be necessary to assign a major role to nucleogenesis in objects of an unknown type such as the 'massive objects' in a state of gravitational collapse which have been much discussed recently. It is therefore clear that we cannot yet come to any definite conclusion about the origin of the elements.

Although thanks to the work of Hoyle, Burbidges, Fowler and Cameron we now have a good understanding of a series of nuclear processes capable of producing all of the elements, there are certain difficulties in assuming that all nucleogenesis has occurred in ordinary stars. The seat of a major part of nucleogenesis may perhaps be 'massive' objects. Alternatively it may well be the outcome of a single process as suggested by the Suess regularities to which element abundances seem to conform. But if this single process is cosmological, it is not likely to be simple. A simple cosmology producing the correct amount of helium might well fail to satisfy some other critical tests. Consequently a more complicated cosmology, which is neither homogeneous nor isotropic, might have to be considered. Even so it is not likely that it could afford to ignore the role of stellar nucleogenesis, an indubitable fact of cosmological evolution. It therefore seems that a new amalgam of an anisotropic cosmology and stellar nucleogenesis with 'massive objects' thrown in for good measure will have to be synthesized to provide the answer to the problem of element origin.

CHAPTER 16

The Origin of the Planetary Worlds

IN our survey of the current cosmological theories of the origin, evolution, and end of the universe as a whole we have had no occasion to refer to our own little world – the solar system – except for two brief remarks. First, we mentioned that the sun after swallowing its inner planets would be gradually eroded into one of those stellar deserts, the white and dark dwarfs with which our sky is being populated in increasing numbers. Secondly, we stated that the process of nebular contraction whereby Kant tried to explain the origin of the solar system was untenable. And yet how the solar system with all its planets and the planets of planets, the so-called satellites, came into being is not only the starting point of all scientific cosmologies in the past but is also one of the high points of the new.

If we leave aside the pre-scientific cosmogonies that tried to make the universe on the model of the tabernacle of Moses, we find that the first attempt to construct a cosmogony designed to explain observed cosmological data (then practically limited to the confines of the solar system) on the basis of some simple hypothesis was made during the seventeenth century. This was no mere accident. For Western Europe had entered into what Lewis Mumford has called the eotechnic phase of its history when with the dawn of the age of great navigations and the resultant overseas trade there emerged a new world view which considered the universe as potentially explicable by rational inquiry and ordinary observation.

At first, in a timid, faltering spirit the thinkers of the new age began to search for rational explanations of the universe around them. Descartes, for example, suggested how the solar system *might* have arisen *naturally* as a result of the formation of immense vortices in the plenum while still acknowledging the 'truth' of the Genesis story of creation. But later, towards the second half of the eighteenth century, when the full implications of Newton's work had been measured out, scientific theories of the origin of the solar system began to be formulated.

Almost from the outset these theories were based on two conflicting ideas. First, there was Kant's view, to which we have already alluded, that the solar system was the product of a slow development by forces working within itself over a long period of time. Secondly, there was the alternative suggested by Buffon about the same time, that the solar system came into being as an aftermath of a cosmic catastrophe caused by the interaction of another celestial body with the sun, such as an actual collision or a close encounter. The debris splashed in space by this upheaval collected later into the planets and their satellites that we observe today. It is true that the theories of the origin of the solar system have had many changes since these two hypotheses were first stated, particularly during the past fifty years. Nevertheless, paradoxical as it may seem, the more they change, the more they stay the same. For despite all their vicissitudes, their hard core remains one or the other of these two rival ideas which have dominated all cosmogonies from Kant and Buffon in the eighteenth century down to Weizsäcker and Hoyle in our own day.

Naturally we cannot decide between them unless we first enumerate the basic characteristics of the behaviour pattern of the solar system the cosmogonic theory is intended to reproduce. Only then can we infer the appropriate mechanism of its birth that best accounts for what we actually observe. Briefly, we find that our solar system consists of the sun, nine large planets, thirty-odd satellites belonging to six of these planets,* over 1,600 asteroids, the flying mountains of planetary space ranging from a mile to 400 miles in diameter, and an indefinite number of comets and meteors all voyaging together as one dynamical system in interstellar space. The whole system displays a number of features which may be broadly classified under four main groups:

a. First, there are the orbital regularities, which means that all the planets revolve around the sun in more or less the same plane in nearly circular orbits and in the same direction. The rotation of the planets around their axes (as well as the sun's own axial spin)

---

* Twelve belong to Jupiter, nine to Saturn, five to Uranus, two each to Neptune and Mars, and one (our own moon) to the earth.

also follows the same direction as their orbital revolutions. Moreover, the planetary orbits lie almost in the equatorial plane of the sun if we ignore the small inclination between the two — only about 6°.

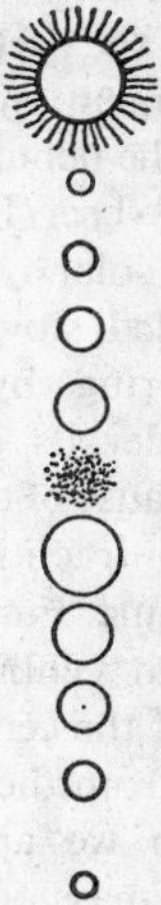

Density (in grams per cm³)	Mass (in Earth's mass as unit)	Bode's Law	Distance (in astronomical units) Actual		
				☀	Sun
4.5	.05	.4	.39	○	Mercury
4.8	.8	.7	.72	○	Venus
5.5	1.0	1.0	1.0	○	Earth
3.9	.1	1.6	1.52	○	Mars
		2.8	2.805	⁂	Asteroids
1.3	318.0	5.2	5.2	○	Jupiter
.7	95.0	10.0	9.5	○	Saturn
1.5	14.5	19.6	19.1	⊙	Uranus
2.4	17.2	38.8	30.0	○	Neptune
53 (?)	.9	77.2	39.4	○	Pluto

Not to Scale

Figure 50.

b. Secondly, all the planets are situated at regular distances from the sun. If we adopt the earth's distance from the sun as our unit of reckoning, the distance r_n of the nth planet from the sun conforms quite closely to the so-called Titius-Bode law except in the case of the last two planets Neptune and Pluto:

$$r_n = 0\cdot 4 + 0\cdot 3\cdot (2)^{n-1}.$$

Figure 50 is a schematic representation of the nine planets and the asteroids of the solar system showing their actual distances alongside the corresponding distances according to Bode's law. It is perhaps no mere accident that planetary distances follow Bode's law. For there is some evidence to show that it holds to some extent even in the case of distances of the satellites from their respective primaries also.

c. Thirdly, while the masses of the four inner planets are low and their densities high, those of the four outer or Jovian planets are high and densities low (see Figure 50).

d. Fourthly, almost the entire rotation of the system as measured by its angular momentum is packed into the planets and their satellites, whereas the entire mass of the system is concentrated in the central sun. The sun possesses over 99 per cent of the mass of the system but barely 2 per cent of its total spin or angular momentum. For this reason, slow evolutionary cosmogonies like the nebular hypothesis of Kant and Laplace have until very recently been found to be untenable.

For if the solar system did originate by the gradual contraction of an extended, slowly rotating gaseous nebula, it could only cast off gaseous rings by appreciably increasing its rotation by contraction under its gravitation. Now contraction does increase rotation because of the law of conservation of angular momentum, for after contraction the same rotation is now spread over a smaller volume. Further, a sufficiently fast spin can also generate large enough centrifugal force to overcome the gravitational attraction of the central nucleus, so that a ring of matter may be thrown off from the main body.

But when we apply these ideas quantitatively, we find that the nebular mass could never have spun so fast as to break up into separate rings. For if so, we should even today be able to observe the vestigial remains of such a fast rotation in the main solar mass, whereas we actually find it rotating very, very slowly indeed – once in twenty-five days. In fact, even if the total rotation of all the planets (about 98 per cent of the whole solar system) were to be condensed into the present sun, it would rotate about fifty times faster, that is, complete one revolution in about twelve hours. This rate of rotation is still too slow to cause any rotational breakup. For the centrifugal force at its equatorial belt as a result would increase only by 5 per cent of its local gravity, so that it would be in no danger of casting off any ring from its equator. Nor would the supposed original primeval sun distended to the orbits of the outermost planets Neptune and Pluto be in any such condition of rotational instability. For in that state the rotation would be much slower, and though gravity at its extended equatorial belt would also be smaller, actual calculation shows that the centrifugal force would still be too weak to overcome gravity. For this reason the lopsided distribution of angular momentum of the solar system between the central sun and its planets has hitherto proved an insurmountable difficulty for the nebular hypothesis of Kant and Laplace. The failure of the nebular hypothesis led to the revival of the rival idea of Buffon.

In Buffon's days it was not known that comets are all cry and no wool. What they contain most is vacuity so that they are mere luminous vacua. Deceived by their appearance, Buffon, who estimated that comets must be several thousand times more massive than the earth, postulated that the solar system originated by an actual collision of a comet with the sun. Later, when the hollowness of the comets was revealed, the comet was replaced by a more massive celestial object, another star like the sun.

At first it was believed that a close encounter of the intruder star with the sun would suffice to pull enough material out of the sun's surface by its tidal force to give birth to the planets. The initial heavier eruptions would provide material for the outer massive planets, and the secondary eruptions as the intruder receded back into outer space would give rise to the smaller inner planets. The gaseous matter splashed out of the sun by the tidal action of the intruding star soon cooled into liquid drops – planetesimals – and later into solid cores. These cores during the course of their orbital rotations around the sun captured a good deal of the lighter gases, whereby many of the planets built up the extensive gaseous atmospheres that still surround them. Further, their originally elliptical orbits were smoothed out by the resisting gaseous medium through which they swept. In this way it proved possible for tidal theories to account for two of the four basic features, namely (a) and (d), though they did not attempt to explain the remaining two, (b) and (c).

However, even the explanation of features (a) and (d) by means of the tidal theory runs into serious difficulties the moment one endeavours to submit it to a quantitative test. F. Nölke, for instance, has shown that the motion of the cores through a resisting medium is not likely to turn initially elongated elliptical orbits into nearly circular ones, so that feature (a) is not properly accounted for. Nor is the explanation of the peculiar distribution of the angular momentum, feature (d), any more satisfactory, though at first sight it does look as though the gravitational force of the intruder star could make the planetary spins several times as great as that of the sun.

In order to make the tidal idea explain this feature, the inter-action theory of the origin of the solar system has been obliged to shift its ground continually. But each shift has created a fresh tangle for every one it actually resolved. Thus when actual calculation showed that the tidal force of the intruder star would be too weak to fling the solar material from its surface so far out in space to the present positions of the planets and at the same time impart their observed rotations, Jeffreys substituted an actual grazing collision (but not a headlong one) in lieu of a close encounter originally proposed. Taking into account the viscosity of the chunk torn out of the sun, he then showed that the rotations of the required order could ensue. But even so, the angular momentum of the planets could not be more than a tenth of their actual value.

While the original difficulty remained, Jeffreys's suggestion conjured a new one. For a tangential collision of the kind en-visaged would tear material not merely out of its outer envelope but would scoop it out of deeper layers underneath. Such material excavated out of the solar interior would be highly compressed at a temperature of the order of some ten million degrees. It would therefore be too hot to stay together long enough to cool and condense into even a gaseous cloud after its release from the terrific grip of solar gravity in its interior. All the erupted material would thus be scattered into galactic space by its own radiation pressure within a few minutes after its extraction from the sun and could never condense into planets as required by the theory.

A calculation by Spitzer shows that the factor favouring the escape of the ejected material by its own radiation pressure from the sun's sphere of gravitational influence is about 100. It is therefore all but certain that such encounters could not originate a system like our own solar system even if due allowance were made for the approximations assumed in Spitzer's calculation.

Last-ditch attempts to save catastrophic cosmogonies like the tidal theory described above have been made, but with little success. For example, following a suggestion of Russell, Lyttleton put forward the view that the planets were formed not by materials scooped out of the sun by the interaction of the intruder star but by the latter's disruption of the sun's erstwhile companion, assuming that the sun originally formed part of a twin-star system. Although it is not unlikely that the sun might have been a binary originally, as the vast majority of the stars appear to have been formed as binary or multiple systems, Lyttleton's theory creates another difficulty for itself. For it has to explain how the tidal action of the intruder star during its close passage to the sun's companion managed to pull out of the latter enough material possessed with the requisite rotation to provide for the formation of the planets while at the same time giving it enough recoil or kick to remove it altogether from the gravitational field of the sun. It must explain, in other words, how the tidal filament pulled out of the sun's companion was left behind whereas the companion itself was lost.

It is true that Lyttleton was able to show that under certain initial conditions such a contingency – the removal of the solar companion from the sun's gravitational sphere with the simultaneous retention of a tidal filament pulled out of it – could come to pass. But this again leads to a fresh difficulty. For now the planets could form only inside that narrow range of distance within which the part of the tidal filament retained by the sun was spilled. This means the planets could only condense within a narrow strip of distance close to the original location of the companion before its encounter with the intruder. But actually the planets are spread far and wide. The outermost planet is nearly 100 times as distant as the innermost one. For this reason Hoyle* amended Lyttleton's theory by dropping the intruder and

* He has now abandoned it in favour of a modernized version of Laplace's nebular hypothesis.

invoking the phenomenon of supernova explosion as the cause of the solar companion's disruption.

While such an explosion would give the solar companion's remaining nucleus enough of a recoil to be shot out of the sun's gravitational ambit, the sun is assumed to retain a tiny wisp of residual gas which could condense into the planets and their satellites. But it is difficult to see how any of the material erupted by a supernova explosion, which would be at an even higher temperature than that of the material dug out of the sun's outer shell by tidal action or actual collision with an intruder, could fail to escape into galactic space. Spitzer's objection to the condensation of a tidal filament mentioned earlier would apply here with even greater force.

This is not all. If nature did originate our solar system by tidal action or collision of an intruder star with the sun, she seems to have departed from her normal practice of bountiful provision. Ordinarily, to provide for a single galaxy she produces a cluster of them, for a solitary star a whole galaxy, for a lone life a teeming population, for one oak a myriad acorns, and for a bush a lush forest. But a collision origin of planetary systems could hardly produce one in aeons. For the stars in our Milky Way are so far apart that two solitary Bedouins in their random wanderings through the African Sahara are more likely to meet than two stars in their courses through galactic space. For a scale model of stars in a galaxy is like a rarefied cloud of dust particles wherein each particle (about 0·01 inch in diameter) chosen to simulate a star is a few miles apart from its nearest neighbour. The chance that two particles in such a vacuous cloud will ever collide is an infinitesimal of about the same order of nothingness as that of a bridge player getting a complete suit of cards. If then we have to depend on a collision or even a close encounter to give birth to a system like our solar system, it can only be an extremely rare cosmic freak.

It may soon be possible to confront this conclusion about the rarity of planetary systems with observation even though any possible planetary companion of our nearest neighbour as big as our biggest planet, Jupiter, would be too faint to be visible through the 200-inch Mount Palomar telescope, the world's largest, at stellar distance. Nevertheless, the existence of unseen planetary companions of stars may be inferred from the minute perturbations that they cause in the positions of the stars to which they are attached. The method is a hypersophistication of an older technique whereby Leverrier and Adams independently discovered (on paper!) over a hundred years ago the then unknown planet Neptune by observing the perturbations in the positions of the planet Uranus caused by the former. It is, however, very difficult to apply in practice. For in searching for unseen planetary companions of other stars in this way, we have no certain way of telling a genuine perturbation from a false one, both being hopelessly intertwined in a medley of ghostly errors of observation.

In spite of great practical difficulties, Peter Van de Kamp has been able to use it with some success. His work does yield tentative indications of unseen companion objects whose masses are about one-hundredth of a solar mass. It is likely that they are planetary bodies like those in our solar system. If so, it would seem that planetary systems are too numerous to arise by such a minutely chancy affair as a stellar encounter or collision.

Nor is it possible to improve materially the chance of planet formation by substituting a gas cloud for a star as the interacting external body, in accordance with a suggestion made by the Russian academician Otto Schmidt. Certainly a star in its voyage through galactic space is more likely to meet an extended gas cloud than to meet its own compeer. Our own sun, for example, may have had a few hundred such encounters during the five billion years of its existence. But any such encounter by itself cannot originate a planetary system, any more than a seed scattered by the winds can sprout into a tree. For a planetary system to arise, the sun must wend its weary way through the cloud – wearily indeed if it is to drag the cloud material in its gravitational ambit. Calculation shows that its relative speed must not exceed half a mile per second if it is to do so.

How serious a handicap this limitation is on Schmidt's mechanism of planet formation may be appreciated from the fact that the sun's present velocity relative to the stars is twelve miles per second. If it has moved on as fast in the past as now, the chances of its ever dragging the material of a gas cloud in its wake are even smaller than those of a collision with a star. Besides, how a star behaves when it ploughs through a gas cloud is a calculation bristling with uncertainties. One such calculation by Hoyle and Lyttleton seems to show that it would gobble up vast quantities of cloud material to grow into a supergiant rather than collect in the form of rings in its equatorial plane in preparation for planet formation as Schmidt suggests. The difficulty of cloud capture is so severe that even Schmidt has been obliged to postulate the simultaneous passage of a second star in close proximity, though the approach of the second star need by no means be as close as in the tidal hypothesis of Jeans. This relaxation in the condition of approach no doubt vastly increases the probability of a sun-star encounter. But as the capture of a cloud requires in addition certain velocity restrictions, it would still be enough of a rarity to make the origin of the solar system a cosmic freak. Schmidt's alternative suggestion, therefore, is no less vulnerable on this ground than those it seeks to replace.

Nevertheless, Schmidt and his followers believe that the initial difficulty of the theory is no serious handicap to its acceptance. For if, rejecting catastrophic theories of planetary origin, we envisage it as the product of slow development, the planets could have arisen in one and only one of the following two ways: either they originated at the same time and from the same single mass of so-called solar nebula as the sun itself or they originated from pre-existing interstellar matter after the sun had been fully formed. But Schmidt and his followers claim that the former possibility is ruled out by its failure to explain feature (*d*) – the peculiar distribution of angular momentum – so that one has perforce to fall back on Schmidt's hypothesis.* It must be owned that if the initial difficulty of postulating a suitable mechanism for the capture of gas cloud can be overcome as Schmidt's followers hope it will be in the future, more recent research by Urey and others seems to favour the basic assumptions of Schmidt's

* As we shall see later on, Alfvén, Weizsäcker, and Kuiper have suggested new variants of Laplace's nebular hypothesis to explain feature (*d*). But they have to invoke cosmic processes which Schmidt considers highly improbable.

theory. It is therefore worthwhile going a little way with him.

According to Schmidt, the origin of the planets came about because the sun on its journey around the centre of the Milky Way passed through a cloud of gas and solid dust particles, a part of which it somehow managed to drag in its wake. Since the cloud, like the sun itself, was initially rotating around the galactic centre, its capture turned part of its rotation around the sun, thus flattening the cloud. At the same time, the dust particles of the cloud, separated from the gas by solar radiation, began to precipitate towards the central or equatorial plane of the flattened cloud (see Figure 51). In this collection of dust specks into a flat disc the mutual gravitational attractions between the particles increased because of the smaller separation. The result was the agglomeration of the small primordial particles of the dispersed protoplanetary material into a multitude of asteroidal bodies of different size and mass, that is, bodies of size intermediate between the primordial particles and the present planets. Some of the larger asteroidal bodies which grew faster than others became the 'embryos' of the planets, budding in time into full-fledged planets by gradual accretion of the remaining asteroidal bodies and their fragments (see Figure 51).

Schmidt's collaborators – Gurevich, Lebedinsky, Levin, and others – have employed the methods of statistical physics to show that a system of solid particles with great angular momentum and sufficient total mass must inevitably follow the evolutionary sequence briefly outlined above. In this way Schmidt has no difficulty in explaining feature (a) – the orbital regularities of the solar system. Nor is feature (d) – the peculiar distribution of angular momentum – much of a hurdle. For in Schmidt's theory, the angular momentum of the cloud and hence of the planets is not directly connected with that of the sun. It is derived from the angular momentum pertaining to the rotation of the stars and the interstellar gas and dust clouds around the centre of the galaxy. By thus relying on the virtually inexhaustible total store of the angular momentum of the entire galaxy as a source of supply of the planetary angular momentum, Schmidt has no need to stretch a point to endow the planets with the lion's share of the angular momentum of the solar system which they actually possess.

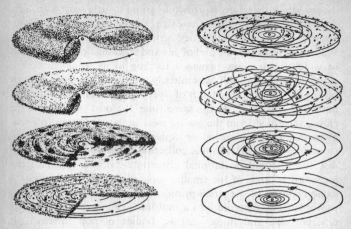

Figure 51. Two stages in the evolution of the protoplanetary cloud according to Schmidt. Left, the cloud flattens into a disc on account of its rotation and the solid primordial particles separated from gas agglutinate into a swarm of asteroidal bodies. Right, the swarm of asteroidal bodies thickens through mutual collisions and leads to planet formation by their gradual accretion. (From Boris Levin, *The Origin of the Earth and the Planets*, Central Books, 1963.)

For his explanation of feature (*b*) Schmidt considers that when planets are being formed those particles have the greatest chance of joining the 'embryo' planet whose specific angular momentum (i.e., angular momentum per unit mass) is closest to that of the embryo. Schmidt concedes that some particles, of course, may join an embryo planet other than their 'own', but such deviations would mutually cancel themselves out so that the particles may be assumed to be distributed along the sections of the axis of specific angular momentum allotted to each embryo. Since a particle's specific angular momentum is proportional to the square root of its orbital radius, Schmidt is able to prove that on the assumption of a smooth distribution of matter in the protoplanetary cloud the angular momenta of the planets and in consequence the square roots of their orbital radii would increase approximately in arithmetical progression. In other words, the orbital radius R_n of the nth planet would obey the following relation:

$$\sqrt{R_n} = p + qn,$$

wherein p and q are constants.

But to secure agreement between the formula so derived and the actual values of the planetary orbital radii, he is obliged to assign one set of values to the constants p and q for the group of four inner planets and quite another for the five outer planets. This may look arbitrary, but the division of planets into two groups of four inner or terrestrial and five outer or giant planets reflects the difference in the properties of the inner and outer zones of the protoplanetary material. For it can be shown that up to a distance coinciding with the asteroid belt the sun's heat is appreciable, but quite close to absolute zero beyond it. As a result, in the small inner zone of the terrestrial planets warmed by solar heat, only particles of non-fusible stony matter and metals with high density could survive the solar heat stroke. In the huge outer zone of the giant planets sheltered from the sun's radiation by its greater distance, the temperature of the particles was so low that volatile substances froze onto them – water vapour, carbon dioxide, methane, ammonia, and related compounds. In this way Schmidt not only justifies his choice of two different values of the parameters p and q in the distance law given above but also provides a natural explanation of feature (c) of the solar system – that of high density and low mass of the inner terrestrial planets and low density and high mass of the outer giants.

Schmidt's cosmogony thus provides a fairly plausible explanation of all the four main regularities of the solar system. An additional merit is its emphasis on the accumulation of solids from smaller-size primordial particles after their separation from gas by solar radiation, a feature that is in better accord with present-day geochemical research. Such research provides valuable complementary clues to the solution of the cosmogonic riddle in that by considering the present composition of the earth's crust – its mantle of water and atmosphere – we can visualize some of the earlier stages of our earth's evolution on a basis independent of remote cosmological origins.

Thus in 1953 the celebrated geochemist, Harold C. Urey, showed by a study of the present abundance of relatively volatile elements such as mercury and boron on the earth's surface that it was unlikely to have ever been exposed locally to temperatures of more than a few hundred degrees centigrade and may not have been hotter than the boiling point of water. Further, a hot molten* earth in its early accumulative stage could never have produced the earth we know today but only an arid, oceanless planet. For molten materials could never hold any interior water to be released later on cooling. Urey reinforced this conclusion by his study of other planets like Mars, whose equatorial bulge was shown to be inconsistent with a core-mantle structure. It is best accounted for by assuming that about 30 per cent of the planet is iron distributed uniformly throughout the rocky globe. Such a condition could never have come to pass if Mars had ever been molten. These and other recent observations† of Urey seem to favour Schmidt's fundamental assumption that planets were formed by the accumulation of primordial dust at relatively low temperatures. Indeed, Struve's interesting recent discovery of a region about Sco B where there is no hydrogen but iron and presumably other non-volatile elements is in agreement with this. Here at any rate there is evidence that a separation of the type Schmidt envisages is a cosmic possibility, having occurred at least once somewhere. For these reasons Schmidt's theory is the only one among *interaction* cosmogonies that is not yet completely *hors de combat*, even though its inability to specify an acceptable mechanism whereby the sun could capture a dust-gas cloud is still an unresolved difficulty.

* The high temperatures of a few thousand degrees centigrade that are known to prevail now in the earth's interior and the existence of a core of molten iron beneath the mantle of rock at one time led to the belief that the earth had passed an earlier molten stage. We now know that this internal heating is the outcome of the slow accumulation of heat released through the breakdown of radioactive elements, a small admixture of which forms part of earth's substance. Since the rate of heat generation is known, precise calculation confirms that thermal energy generated by such radio-active disintegration accumulated over long periods of a few billion years of the earth's existence is sufficient to raise its interior temperature to the extent actually observed today.

† See page 358.

Since interaction cosmogonies have produced a difficulty for every solution, they have obliged some cosmogonists to essay new resurrections of Laplace's nebular hypothesis. It seems possible now to amend it to overcome the main stumbling block to its acceptance – its inability to account for the peculiar distribution of the angular momentum of the solar system, feature (*d*). For the newly discovered laws of cosmic electrodynamics and turbulence can be invoked to determine the behaviour of a contracting gaseous nebula instead of only gravity.

We saw in Chapter 4 how stars condense out of interstellar gas clouds of cosmic composition, that is, consisting of the same stuff as stars are made of with hydrogen as their main constituent. It is from some such primeval clouds that our sun originated. As the sponsoring cosmic cloud, the solar nebula, shrank under its own gravitation, a stage was reached when the main mass condensed into an embryo sun at the centre. This embryo sun soon became dense enough to spark the nuclear reactions in its interior and enable it to shine like a star while still surrounded by a thick and extended cloud of residual gas and dust. It is this residual gas and dust of the primeval solar nebula that provided the source material for the formation of planets. Now the process of planet formation may be envisaged as taking place in four alternative ways proposed by Alfvén, Whipple, Weizsäcker and Hoyle.

In Alfvén's scheme the story of planet formation begins when the embryo sun, condensing out of the interstellar gas cloud, has accumulated about half the present solar mass within a sphere distended to the outermost planet, that is, between ten and 100 times the radius of the earth's orbit. The remaining half of the solar material is still uncondensed, spread in the form of a gaseous envelope around the central solar embryo extending to a distance of about 0·1 light year. Alfvén calls this gaseous envelope the 'initial cloud'.

Since half the material of the primeval cloud has already condensed into the embryo sun at the centre, the atoms of the 'initial' cloud now start falling rapidly towards the sun, pulled in by its gravitational attraction like a cosmic Niagara cascading into the deep from all directions. However, midway through their fall, before the cascade plunges into the solar deep, it is halted

337

by the intervention of electromagnetic forces that now come into play and begin to dominate the situation. The reason is that the falling atoms of the gas cloud are stripped of their satellite electrons, so that the cloud, instead of consisting of electrically neutral atoms, becomes a swarm of ionized particles, that is, electrically charged protons and electrons.

Ionized particles, however, are very sensitive to the sun's magnetic field. They cease to fall any more towards the sun, as their fall is now impeded by the electromagnetic forces which are far more powerful than the sun's gravitational attraction. A calculation by Alfvén shows that the electromagnetic force exercised by the sun's magnetic field on a proton moving in the earth's orbit exceeds the sun's gravitational attraction by a factor of 60,000. At larger distances this excess factor no doubt diminishes greatly, but even so, the electromagnetic forces maintain their lead by a comfortable margin. Thus at a distance as remote as that of the outermost planet Pluto it is still 250. Consequently once the cloud atoms are ionized, their motions are governed almost solely by electromagnetic forces of the sun's magnetic field.

The sun's magnetic field, however, not only prevents the initial cloud from collapsing into the sun but also sieves the cloud material according to its chemical constituents. For atoms of different elements require different conditions for their ionization. Thus helium is the hardest to ionize. Consequently it is the last to stop falling into the sun. The result is that helium continues to fall long after the rest of the elements in the cloud have been stopped. This is Alfvén's *A*-cloud.

The next lot of elements in point of difficulty of ionization consists of hydrogen, oxygen and nitrogen. They thus fall out of the cloud after helium has already been segregated. Their segregation out of the rest of the material is quite rapid. Only a small portion of the cloud now remains. It contains two groups of elements, one consisting of carbon and sulphur and the other of iron, magnesium and silicon. The former group of elements, being harder to ionize, gets separated out as *C*-cloud first, leaving the latter as the main constituent of the *D*-cloud.

In other words, the different chemical constituents of the

'initial cloud' in their fall towards the sun are segregated into four main clouds on account of different voltages and temperatures required to ionize atoms of different elements. Of course, Alfvén is very far from claiming that the segregation of the elements of the initial cloud is chemically pure. Our description of the process is only an idealization of a very complex process. In actual practice what happens is that elements of each group form the most abundant constituent of their corresponding cloud mixed with elements of all other groups as 'impurities'.

Alfvén then proceeds to examine quantitatively at what distances from the sun the four segregated clouds are likely to be stopped. But here he faces a difficulty. For we have no direct experience of what happens when a gas falls in towards a magnetized central body. Nevertheless, basing his argument on a theory of cosmical electrodynamics built up by blending deductive theory with empirical observation, he shows that B-cloud is stopped in the region of the terrestrial planets, Mercury, Venus, earth and Mars; C-cloud in that of the giant planets Jupiter, Saturn and Neptune; and D-cloud beyond Neptune. As for A-cloud, Alfvén himself doubts its existence. We may therefore ignore it in our exposition, treating it as a part of the B-cloud.

In this way Alfvén offers a natural explanation of the difference in the chemical composition of the planets, instead of having to import into the theory later *ad hoc* assumptions like that of evaporation for the purpose. Thus the D-cloud, consisting of elements easiest to ionize and the first to be halted beyond Neptune, provides material for the formation of only relatively small bodies like Pluto and Neptune's satellite Triton because of the failure of the weak solar magnetic field to produce a complete group of planets at this distance. The C- and B-clouds, which are stopped later, closer to the sun, are halted at appropriate distances for the formation of regular sets of planets as well as their satellites. Alfvén's analysis shows that with the exception of only two small Martian satellites, all the regular bodies of the solar system (i.e., bodies moving in coplanar and nearly circular orbits) can be accounted for if the mass is accumulated in those regions where theory shows that the B-, C- and D-clouds ought to be formed.

Alfvén has now to account for the peculiar distribution of the angular momentum, feature (d). To do so he invokes a process called 'acceleration', which imparts angular momentum from the central sun to a falling cloud stopped by ionization. What happens is this: Owing to turbulence the initial cloud is in rotation, so that the cloud material has the same average rotation or angular momentum per unit mass as the central sun, which after all has precipitated out of the same primeval cloud as Alfvén's initial cloud. But the solar material, having condensed a good deal, will be spinning much faster than the more distant cloud material, even though both materials have the same average angular momentum per unit mass.

However, as soon as the cloud material is ionized (a process which, as we saw, takes place at a considerable distance from the central sun), the electromagnetic forces tend to accelerate the cloud material to the same *angular velocity* or spin as that of the central sun. This is Alfvén's 'acceleration' process. It is a consequence of Ferraro's theorem, according to which *all* parts of a magnetic line of force in a conducting medium will tend to move with the same angular velocity. If there is a difference in the angular velocity of materials situated along a magnetic line of force, magneto-hydrodynamic effects will be produced and the result will be a transfer of rotation or angular momentum.

A case in point is the transfer of angular momentum produced in one of the laboratory experiments of Lundquist, wherein the angular momentum of a rotating disc at the bottom of a cylinder is transferred to a mirror floating in mercury contained in it by means of a magneto-hydrodynamic wave (see Figure 52). If the rotating disc is rotated without the mercury's being in a magnetic field, the floating mirror on top remains stationary. But if the cylinder is placed in a vertical magnetic field, rotation of the vibrating disc below is transmitted to the floating mirror on top and can be detected by the angle through which an incident beam of light is reflected by the mirror.

But in cosmic space there seems at first sight no conducting material like mercury in our experiment to effect the transfer of angular momentum from the central sun to the gas cloud. And in the absence of some sort of conducting material between the

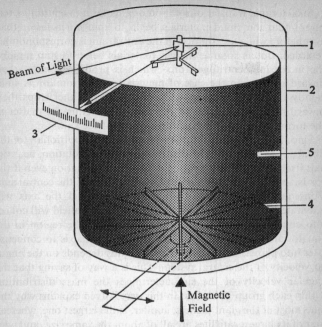

Figure 52. Arrangement of the experiment on magneto-hydrodynamic waves in mercury: (1) floating mirror, (2) stainless-steel cylinder, (3) scale, (4) vibrating disc, (5) mercury. (Lundquist, 1949; from H. Alfvén, *On the Origin of the Solar System,* Oxford, 1954).

two, no transfer of momentum is possible. Without being dogmatic, Alfvén suggests that some ionized matter does leak all along the lines of force from the regions where the gas clouds are stopped by ionization down to the central body. With the whole space between the central body and the regions where the gas clouds are halted permeated by leaked ions, the difficulty concerning a conducting medium is resolved, or so it seems. Assuming the existence of such a conducting medium, it can then be shown that the required transfer of angular momentum will take only a few years in the case of planets and a few days in the case of transfer from the planets to their satellites.

Such a rapid transfer of rotation or angular momentum merely means that the cloud materials begin to spin much faster than ever before while the rotation of the central sun is correspondingly retarded. But as soon as the gas materials begin to spin rapidly, the increasing centrifugal force which is produced brakes their inward motion. In the case of materials whose rotation is sufficiently rapid, the increase in centrifugal force either stops their inward motion altogether or reverses it, collecting the matter in the form of rings in the equatorial plane of the sun.

Since the centrifugal force* is directly proportional to the distance of the material from the central axis of rotation, near the axis itself it will be too weak to counteract gravitation even if the gas is accelerated to the same angular velocity as the central sun. Consequently, all matter from a region around the axis will gravitate into the central sun, while that further afield will collect into concentric rings in the equatorial plane. The region in the equatorial plane between which the material that is to condense later into planets and satellites collects thus depends on the angular velocity of the central system. This is a way of saying that the angular velocity of the sun determines the mass distribution within each group of bodies. In this way Alfvén explains why the innermost of the giant planets, Jupiter, is the largest one, whereas four of its largest satellites are all of about the same size, and why in the Saturnian system the satellites are smaller the closer they are to their primary.

When the gas accumulates in the equatorial plane at certain distances from the sun, it is still mainly ionized. The next process that occurs is its condensation, which is supposed to occur in two stages. First, the gas condenses into small grains. Next, these grains agglomerate into larger bodies. There thus results a swarm of small solid or possibly liquid bodies moving in circular orbits around the sun. These small bodies condensing out of the minute grains finally accumulate into a series of larger bodies of the size of the planets and their satellites, although this further consolidation of small bodies may under certain conditions be interrupted.

Alfvén believes that these anti-consolidation conditions actually

* The centrifugal force on a particle of mass m spinning with angular velocity w at a distance r from the axis of rotation is mw^2r.

materialized in the regions where the material of the asteroids and Saturn's rings originally collected. Consequently, the further process of consolidation of the accumulated material into a planet or a satellite was inhibited for good. This is a reversal of the common view that the asteroids and the Saturnian rings are the fragments of a broken-up planet and a satellite.

Alfvén's view is that the original fragments never agglomerated into a whole body at all. The reason in the case of Saturnian rings is that its matter accumulated within the Roche limit of its primary planet, Saturn. Inside this limit the tidal forces of the more massive primary planet on a satellite exceed the self-gravitation of the less massive satellite. The satellite therefore cannot stay whole and is smashed to pieces by the tidal forces of its primary. According to the older theory, at some time in the remote past one of Saturn's satellites ventured to stray into an orbit within the Roche limit and got broken up. Alfvén, on the other hand, considers that grains of matter that make the ring originally collected within Roche's limit so that they could never consolidate into a single satellite.

Likewise, the reason the asteroids failed to agglutinate into a planet is the low density of matter in the region where they are situated. It is a region of comparative gap between the penetration limits of B- and C-clouds, where mass density is very small. There is no room in Alfvén's theory for the existence of another planet with which the parent planet of the asteroids could have collided to produce them according to the more usual view.

Alfvén's cosmogony is indeed a marvel of ingenuity. Although it is still considered to be highly speculative in many ways, its basic ideas, such as the transfer of angular momentum, may nevertheless contribute to the solution of the problem of the origin of our solar system. It has, however, an insufficiently guarded Achilles' heel. The main motive force for working Alfvén's mechanism of planetary births is the system of electromagnetic forces which arise by the action of a solar magnetic field of enormous strength on ionized atoms of the surrounding 'initial' gas cloud. Unfortunately for the theory, the existence of both the solar magnetic field and ionization in the gas cloud has been denied.

Take first the solar magnetic field. If the sun were assumed to have the same radius at the time the planets were formed as now, Alfvén's mechanism would require a surface field of 3×10^5 gauss to make it work at all. But it is doubtful if the present sun has any magnetic field at all. The most accurate measurements have at best revealed a very weak magnetic field of a few gauss only or even less. It is true that at the time of the formation of the planets the youthful sun might have been very different from what it is now. But from what we know of stellar evolution, the sun seems to have altered but little during the past five billion years of its life. It is unlikely that such a powerful magnetic field could have decayed in its entire lifetime since, according to Cowling, the decay time of a solar magnetic field is ten billion years.

Although Alfvén does not hold this calculation as being in conflict with his own hypothesis, on the ground that it is based on the unlikely assumption that there are no internal motions in the sun, he has not given any amended calculation on more realistic assumptions. In its absence it is difficult to appreciate how a solar field of such gargantuan magnitude could have evaporated almost completely in what after all is barely 5 per cent of the sun's life, unless it is assumed that the sun at the time was distended to the dimensions of the orbit of Mercury.

Nor is the evidence in favour of the postulated ionization of the atoms in the initial cloud any more conclusive. There are two possible causes of ionization – solar radiation and collisions between the atoms. A calculation by Ter Haar seems to show that both would be unable to ionize the cloud material to any appreciable extent. Thus solar radiation is powerless to ionize even magnesium atoms (which, as we saw, are the easiest to ionize) anywhere outside the orbit of the innermost planet, Mercury. The atomic collisions are still less effective. They could ionize barely one atom out of ten billion even under the conjunction of most favourable conditions.

In the earlier version of his theory Alfvén simply assumed that the atoms in the gas cloud continue to increase their kinetic energy by falling freely in the gravitational field of the sun until it becomes so large that ionization by collision takes place. In his revised version Alfvén has, in view of Ter Haar's criticism,

344

examined the process of ionization in the cloud in some detail. Basing his position on analogies with known electromagnetic effects like that of electric discharge, Alfvén concludes that the postulated ionization of the cloud atoms is not invalidated by Ter Haar's calculation.

In view of doubts about the very existence of the cosmic processes Alfvén invokes, some cosmogonists have been on the lookout for other ways of modernizing Kant's cosmogony. As mentioned before, three additional ways of doing so have been suggested, by Whipple, Weizsäcker and Hoyle. In Whipple's modernized version the planets begin to form when the bulk of primeval nebula of cosmic dust and gas has shrunk from its original radius of about one light year to that of a few light hours, so that it extends to the limit of the outermost planet, Pluto. All this while, the cloud has been condensing in comparative calm for millions of years. But this cosmic calm is only a rather long prelude to the coming storm that now breaks loose with cataclysmic fury. The reason is that once the cloud has compacted within the comparatively narrow confines of the solar system, its central core grows into a veritable gravitational giant.

It now begins to tear tornadoes of dust and gas from all its outlying parts. They – these peripheral whirlwinds of dust and gas – sweep in long spirals clanking around its girdle like manacles around a solar Atlas. But before long, on account of its ever-mounting self-attraction, the girth of our gravitational Atlas – the central core of the cloud – collapses suddenly into what by comparison is a mere Pleiades waist. When this happens the central core condenses into the sun. But during the last stage of its ultra-rapid collapse, a number of separate whirlwinds of dust and gas spiralling inwards are left stranded in their orbits, broken loose from the collapsed core. Some of these whirlwinds or eddies which are rather close to the sun are sucked in or are blown away by its radiation. But those that are far enough away to retain their identity later agglomerate into planets in more or less the same fashion as in other theories. The satellites of planets are formed by an appropriate number of repeat performances of the same process on a smaller scale.

In his quantification of these ideas Whipple assumes the original cloud to possess negligible rotation in order to account for the low angular momentum or rotation of the sun. He further assumes that the separate eddies of dust and gas that are to condense later into planets have from the very moment of their formation the angular momentum or rotation that they are observed to possess. In this way the explanation of the peculiar distribution of the angular momentum feature (d) is thrown gratuitously into the theory at the very outset, instead of emerging from it as its consequence. Nor is its elucidation of other features any better. It does not attempt to explain (b) and (c), while its explanation of (a) is half-hearted and weak. Moreover, on the basis of the theory we should expect the larger planets to form nearer the sun whereas we actually find them to be farther away.

We therefore proceed to examine the next alternative on our list, namely Weizsäcker's version of the story of planetary origin. The leitmotif of his cosmogony is the behaviour of an extended gas cloud under its own gravitation and turbulence. As we saw in Chapter 6, Weizsäcker envisaged the universe of galaxies as a collection of discrete eddies raised by gravitation and turbulence in the primeval gas cloud of cosmic composition soon after the initial explosion of Lemaître's giant atom that 'created' the universe and set it on its career of expansion. We also saw that such a cosmic eddy or galaxy, initially an irregular gas cloud, would develop into a rotating disc with the main mass concentrated in the centre but different parts of the system moving with different velocities. As a result, there are set up powerful viscous stresses which tend to accelerate the slowly moving outer parts and retard the faster-moving inner ones in an endeavour to make the whole system rotate uniformly like a rigid body.

But such an equalization of the differential rotation throughout the system entails a loss of energy accompanied by a transfer of angular momentum or rotation from the inside to the outside of the system. This twin process – of dissipation of energy and transfer of rotation – is possible because mass with more than average rotation disappears into outer space while the rest of the mass with low rotation becomes concentrated at the centre, thus providing the necessary energy. This is why in the course of time the central part of the main mass changes into a uniformly rotating core with its peripheral parts condensing into a rotating disc – the spiral arms – which subsequently escape into cosmic space. In other words, turbulence and gravitation guide cosmic eddies of galactic dimensions from irregular forms to ellipticals via the spirals.

The evolution of planetary systems of stars is merely a repetition of the same cosmic theme in a minor key. For, as we mentioned earlier, within each large eddy – the galaxy – there grow afresh smaller eddies, the protostars. As before, these smaller eddies develop from irregular gas clouds into a uniformly rotating central core containing the main mass of the system, with a small fraction of its material condensing in the form of a thin rotating disc around the inner core. While the bulk of the material in the rotating disc disappears (like that of spiral arms in the larger edition of this cosmic drama) into interstellar space, a small residue is retained in the form of planets orbiting around the central star.

In Weizsäcker's theory the story of planet formation therefore begins when the irregular gas cloud – the primeval solar nebula – has condensed into a central core surrounded by a flat disc of gas rotating around it in its equatorial plane. While the central sun is at this time nearly in its present condition in respect to its principal attributes such as mass, radius and luminosity, the rotating disc of cosmic gas and dust has about 100 times the present mass of all the planets, or about 10 per cent of the solar mass. Now such a disc of rotating gas would be turbulent for the reason that its Reynolds number would be of the order of 10^{14} and thus greatly in excess of the critical value of about 2,000. But since gravitation predominates greatly over other forces such as gas pressure and turbulent viscosity, the turbulent motion of the gas would no longer be random. In the absence of turbulence, each gas particle in the disc would move in an elliptical orbit around the sun under the influence of the sun's gravitation. But when turbulence and gravitation take over joint control of their movement, with gravitation as the senior partner, the result is that all particles with the same period of rotation and therefore the same average distance from the sun are channelled into a number of turbulent eddies situated at the same average distance from the sun.

Without further justification, Weizsäcker assumes that these turbulent eddies form a regular pattern with the same number of principal eddies (actually five) in each successive area between concentric circles, as shown in Figure 53. Since along those circles where the rings of vortices meet, large viscous stresses are produced, there arise secondary eddies on the circles separating the main vortices. These are Weizsäcker's well-known 'roller-bearing' eddies. It is in these secondary roller-bearing eddies that

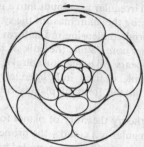

Figure 53. Weizsäcker's system of primary vortices. The outer arrow indicates the direction of rotation of the whole disc, while the inner arrow indicates the direction of rotation in the vortices. The sun is at the centre of the whole system.

conditions for favourable condensation of the disc material into planets arise. Consequently we may expect the planets to form at distances from the sun corresponding to the radii of the circles separating the main eddies. This enables Weizsäcker to deduce the Titius-Bode law of planetary distances, feature (*b*). But since he actually assumes that the number of large eddies in each concentric ring is the same, the 'proof' really begs the question.

Weizsäcker has, however, less difficulty in accounting for other features. Feature (*a*) is a simple consequence of the condensation of planets in the 'roller-bearing' eddies. The planets will naturally begin to rotate in a counter-clockwise direction, if the whole system is rotating in a counter-clockwise direction in agreement with observation. The rotation in the large eddies is in the opposite, clockwise direction. During their formation and immediately thereafter, the planets will be surrounded by extended atmospheres which later develop into satellite systems.

According to Weizsäcker, the entire evolution of the system from a disc to discrete planets takes about ten to 100 million years, which is also the time the disc takes to disappear because of the dissipation of energy. Weizsäcker also gives an explanation of the peculiar distribution of the angular momentum in the solar system, feature (*d*). We noted earlier how the evolution of a non-uniformly rotating gas cloud proceeds by a continual tendency on the part of viscous stresses set up inside it to equalize the rotation throughout the system by dissipation of energy and transfer of rotation from inside to outside. It therefore follows that although initially the recently formed sun at the centre of the solar nebula is rotating much faster than the material at the outskirts of the disc, the aforementioned process results in due course in a slowly rotating central sun surrounded by a faster-rotating gaseous disc. Now Weizsäcker assumes that in course of time the light elements evaporate into interplanetary space, carrying with them their high rotation or angular momentum, while the matter falling into the sun does not possess any angular momentum. In this way he tries to explain at the same time both features (*c*) and (*d*).

Not that Weizsäcker's theory has been able to withstand all the critical scrutiny to which it has been subjected since its formulation; quite the contrary. Its explanation of feature (*d*) has been shown by Ter Haar to be largely untenable. Nor is there any reason in view of the erratic character of turbulent gases to believe that such a symmetrical pattern of principal eddies as postulated by Weizsäcker could ever arise under the conditions envisaged.

Nevertheless, the theory has the great merit of putting forward a new fertile idea of extraordinary power: that we and our planetary worlds are the debris of cosmic turbulence tempered by gravitation. If indeed these worlds are the outcome of the solar system's internal evolution without interference, further development of Weizsäcker's idea is the most hopeful line of advance at present. Basing his position on it, Kuiper has greatly extended Weizsäcker's work.

In an effort to replace the hypothetical and, as we found, untenable system of eddies shown in Figure 53, Kuiper employs Kolomogorov's spectral law of turbulence to determine the actual pattern of eddies that are likely to arise in the solar nebula. He shows that they would be unequal in size but that on the average the larger eddies would arise farther away from the sun. A diagrammatic sketch of the turbulence pattern according to Kuiper is shown in Figure 54.

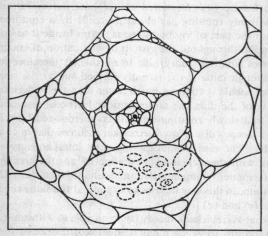

Figure 54. Schematic distribution of eddies in solar nebula according to Kuiper. Sun is at centre. 'Interior' eddies are shown in only one instance.

Further, Kuiper gives reasons for believing why planets are unlikely to condense in the secondary or 'roller-bearing' eddies as envisaged by Weizsäcker. First, the lifetime of such roller-bearing eddies is found on computation not to exceed 1 per cent of the period of revolution of the main eddy, which works out to below one year, or much less. Secondly, because of violent turbulence in these secondary eddies, there would certainly be a great increase in temperature. Both on account of their ephemeral character and their comparatively high temperature, the secondary eddies appear to Kuiper less suitable than primary vortices for fostering planetary condensation. This means in effect throwing overboard Weizsäcker's explanation of feature (*b*), Bode's law. But there need be no disappointment at the closure of one possible avenue of explanation of feature (*b*). For Bode's law is shown to be no 'true' law in the sense that it is an inevitable consequence of turbulence controlled by gravitation. It is rather found to be a result of the initial density distribution of the solar disc.

To arrive at the density distribution of the disc, Kuiper examines in detail the break-up mechanism of the solar disc which leads to the formation of planets. He finds that the question of whether the disc material at any point stays or disappears depends to a large extent on its gravitational stability. It will be gravitationally unstable if the difference in the attractive force on any two neighbouring elements exceeds the attractive force between them. For in that case the tidal forces of the sun will overcome the self-attraction of the elements resisting their break-up. It can easily be calculated that if the density of the material at any distance r from the sun falls short of the critical value $6M/\pi r^3$, where M is the solar mass, it will be gravitationally unstable and otherwise stable. Kuiper calls this critical density $6M/\pi r^3$ at any distance r the local Roche density at that place. As will be observed, it is solely a function of the distance from the sun of the place to which it pertains.

Whether the actual density at any place exceeds its local Roche density or not determines whether the material there will stay together or separate. Consequently if a condensation somehow forms at any place with a density well in excess of its local Roche density, then it will extend its gravitational domain in such a way that the mean density within the domain will be the Roche density of the region while beyond it the tidal forces will break up the material. This is why the entire disc fragments itself radially into separate spheres of action each having an average density equal to the local Roche density. On the basis of these ideas, it is possible to calculate the mass of the separate condensations into which the disc breaks up, if the distances to the sun at the time of their formation were known. Assuming that they formed at the present distances of the planets, Kuiper computes that, save for minor deviations, all the primeval condensations had about the same mass. This identity of the mass of each condensation provides a new interpretation of Bode's law.

The equality of the mass of condensations may, however, seem to raise a difficulty for Kuiper's theory. For the masses of the planets are actually found to vary widely. For instance, the mass of the largest planet, Jupiter, exceeds that of the smallest, Mercury, by a factor of 6,360. But the present mass differences of the planets are attributed to the subsequent evolution of the condensations – protoplanets – into planets whereby, in consequence of the physical and chemical processes associated with condensation and sedimentation, the different protoplanets condense into planets a varying proportion of their original material – from about 10 per cent in the case of proto-Jupiter to a mere 0·01 per cent in the case of proto-Mercury.

The complementary chemical studies of the origin and development of planets undertaken by Urey confirm these ideas. According to Urey, the protoplanets in the location of the four inner planets developed considerable heat after their initial condensation from the solar disc. As a result, the lighter gases such as hydrogen and helium rose to the surface and escaped and were irretrievably lost to them, leaving the heavier elements. The protoplanets in the location of the major planets, formed as they were at much remoter distances from the sun, escaped the heat bath of their closer counterparts and were thus able to retain the lighter elements. In this way Kuiper and Urey are able to explain feature (c).

Another consequence of the evolution of the planets from protoplanets is the formation of satellite systems in an analogous way. Kuiper has examined the process of satellite formation to account for the origin of not only the regular satellites, that is, those bodies which move in nearly circular orbits with small inclinations, but also of irregular ones having elongated orbits and large inclinations. But it will involve us in too many and too recondite details to go into them here. One aspect of this process may, however, be described as it has a bearing on the explanation of the peculiar distribution of the angular momentum of the solar system, feature (d).

During the course of their contraction leading to the formation of planets, all of the protoplanets are strongly affected by solar tides, which impose a forced rotation on them with the same period as their orbital rotation. This is a consequence of the fact that condensations occur at the critical local Roche density and for this reason affect equally all protoplanets both near and far. For at the local Roche density the solar tidal forces and internal attraction balance, so that if the tidal forces at the larger distance of, say, Neptune weaken, they have only to overcome a correspondingly weaker self-attraction owing to the lower initial density of the material there, which is the same as the local Roche density.

But as the protoplanets condense, density increases continually so that solar tidal forces become progressively less effective. With the attenuation of the braking effect of solar tides, the rotation of the planets tends to increase rapidly with contraction because of the conservation of angular momentum. The planets thus begin to rotate with periods smaller than their orbital periods, the rotation being direct for all distances from the sun.

However, the crux of the explanation of feature (d) is that the present angular momentum per unit mass of the planets is only a fraction of the original rotation momentum when the protoplanets were still distended. Solar tides first build it up by transfer of momentum from the sun to the protoplanets but reduce it later in the subsequent contraction period, wherein they are greatly assisted by the escape of the gases carrying their angular momentum with them from the periphery of the contracting protoplanets.

All in all, while many sub-problems connected with the origin of the solar system still await their solution, it does seem that the Weizsäcker-Kuiper-Urey theory of planet origins explains the main characteristics of the solar system better than any other with the possible exception of Schmidt's. One particularly satisfactory feature of the theory is that it invokes no freak cosmic process for the formation of planets. In fact, it shows that planetary formation is only a special case of the almost universal process of binary-star formation. It appears that all but ten per cent of the stars are born as twins, triplets, quintuplets or ntuplets. Even some of those that seem single now may owe their solitude to their breaking loose from their erstwhile companions. Kuiper's calculations show that whether a nebular disc condenses into planets or into a multiple-star system is mainly dependent on its density. It is only a narrow range of density that permits planet formation. Above this range the disc would condense into one or more companion stars, and below it into innumerable small splinter condensations like the asteroids or even less. It would therefore seem that the very existence of our earthly abode has been a matter of touch and go. A little more density in the region of Jupiter and our solar system would have blazed into a binary-star system instead of a series of dark planets subsisting on borrowed light. While our own emergence has no doubt been a matter of chance, it is a chance that is by no means negligible. According to Kuiper's reckoning, one in every 100 or 1,000 stars produces a nebular disc of density appropriate for planetary formation.

Nonetheless, Kuiper's theory, too, is not immune from criticism. For instance, Ter Haar blames it for its inability to account for the requisite flattening of the gaseous disc of the original solar nebula towards its equatorial plane for two reasons. In the first place, the gravitational field of the sun cannot be neglected in comparison with the gravitational force due to the disc itself, as Kuiper seems to have done. Secondly, Kuiper's use of thermal velocity instead of the turbulent velocity in his derivation of the height of the disc is said to be inadmissible. Ter Haar's estimate of the disc after making allowance for these two factors is forty times greater than Kuiper's, thus making impossible the attainment of densities required by his theory.

This leaves the field to Hoyle's recent revival of Laplace's nebular hypothesis in an endeavour to find the mysterious x-process responsible for the local abundances of light elements. We have already alluded in the last chapter to the difficulty their genesis presents. After trying various possibilities Hoyle came to the conclusion that the origin of these elements is intimately tied up with that of the solar system itself. Fortified by the high lithium abundance observed in some very young T Tauri stars, he suggests that terrestrial meteoritic abundances of light elements all the way from deuterium through lithium to beryllium and boron probably arose when our sun was itself a T Tauri star just about to emerge on the main sequence out of the parent interstellar cloud. There are two very peculiar features of these light element abundances which enable him to postulate the conditions that must have prevailed at that early phase of the origin of the solar system. First, the abundance of deuterium (about one part in 10^4) that we find in terrestrial water is far in excess of what may be expected (about one part in 10^{17}) from the complex of nuclear reactions now going on in the sun. Second, lithium 7 is about twelve times as abundant as lithium 6, whereas they should be produced in about equal proportions if they originated from spallation reactions, that is, reactions in which an energetic incident particle hits a nucleus and produces a spray of particles. As we saw in the last chapter, these energetic particles could arise from the intense flare activity of the early solar T Tauri. Their collisions with other nuclei like carbon would break them down into lighter elements by stripping reactions. However, it is necessary to invoke their subsequent irradiation with neutrons in order to obtain the observed predominance of lithium 7 over lithium 6. But all available neutrons produced by spallation would be captured by hydrogen and nothing would be left over to react on a constituent like lithium 6 because of the excessive hydrogen abundance in the primeval solar nebula. After all, hydrogen forms the vast bulk of ordinary solar material. Hoyle is therefore of the opinion that any theory of planetary origin that envisages primitive protoplanets forming out of solar-type material with its big hydrogen excess cannot be right.

To purge the planetary material of its hydrogen excess in the immediate vicinity of the sun – the region occupied by the inner planets – he requires that the original hydrogen of the solar nebula was pushed by an outward motion away from the sun leaving behind *solid* particles of magnesium, silicon and iron. These dropped out of the gas simply because they have low saturation vapour pressures at the prevailing temperatures (about a few hundred degrees C.). It is from these solid drop-outs that the inner planets later aggregated. The hydrogen, the gases and other materials that did not condense at a temperature of two to three hundred degrees C., continued to move outwards and provided the source material of the outer plants. Hoyle thus attempts at one stroke to account for the observed abundances of light elements as well as the high density and low mass of the inner planets and the low density and high mass of the outer ones – feature (c).

However, the real hurdle in the way of resuscitating Laplace's old idea is feature (d). To overcome it, Hoyle harks back to considerations of star formation, which seem to show that the original angular momentum of the primitive sun must have been about a thousand times the angular momentum of the present-day sun. In other words, according to Hoyle, the angular momentum of the original solar condensation was somehow stored in the planetary material and not in the sun. This, however, raises the problem of showing how the sun's rotation was slowed down to its present level. In order to brake the solar spin, Hoyle requisitions a magnetic coupling between the contracting solar condensation and planetary material in the outer disc of the fast rotating solar nebula whereby the former's angular momentum is transferred to the latter – an idea borrowed from Alfvén. But Hoyle uses the magnetic field only to carry Maxwell-Faraday stress through space and thereby transmit angular momentum from the sun to the planetary material and not to secure any chemical segregation of the solar cloud as Alfvén does. As a result, he does not require the gigantic magnetic fields we encountered in Alfvén's theory. An order-of-magnitude calculation shows that the magnetic intensity at the surface of the primitive sun need be only about one gauss to effectively transmit the angular momentum from the sun to the planetary material. In short, Hoyle envisages that the protosun when it reached its T Tauri stage began to spin fast enough to develop an outer disc of planetary material as in the old nebular hypothesis of Laplace. The latter's main stumbling block, its lop-sided angular momentum distribution between the core and the disc, is then resolved by recourse to magnetic torque coupling.

Unfortunately, however, the conditions needed to make the process work are not likely to materialize. Consider, for example, the temperature requirement. To prevent the solid particles of magnesium, silicon and iron from being blown away by the out-ward streaming gas so that they do get dropped at the sites of the inner planets, it is necessary that the surface temperature of the collapsing protostar remains only a few hundred degrees at least till the time the solar nebula condenses to about 200 times its present radius. But if the behaviour of a gas cloud of one solar mass contracting under its own gravitation outlined on page 87 is any guide, its surface temperature is likely to be around 4,000°K. and luminosity about 100 times that of the present sun by the time the solar nebula shrinks to a ball with the prescribed radius (see Figures 10 and 11). Consequently the segregation of the heavier elements from the hydrogen at the locations of the inner planets cannot occur in the simple manner envisaged by Hoyle. In fact, as we showed in Chapter 4, this is the stage when the energy released by the gravitational contraction begins to break the molecules and ionize the resultant atoms. Accordingly the ambient magnetic field, which judging from the behaviour of T Tauri stars like NX Monocerotis in the final stages of condensa-tion from the interstellar medium may well be much stronger than Hoyle postulates, will act on the ionized particles leading to a good deal of chemical separation as well. It therefore follows that the magnetic field will not merely transmit angular momentum from the inner solar core to the outer planetary disc as Hoyle demands. It will also govern the transport of the ionized materials in the disc itself – a feature entirely ignored in Hoyle's calcula-tions. Hoyle's resurrection of the old nebular hypothesis of Laplace therefore raises almost as many difficulties as it removes.

From the foregoing review of the latest theories it would seem that none of them provides a satisfactory and conclusive answer to the cosmogonic riddle. The main difficulty in formulating a viable theory of the origin of the solar system is the lack of concrete information on which to base it. Most extant theories attempt to account for the purely dynamical regularities of the solar system embodied in features (a) to (d) with varying degrees of reasonableness and plausibility. But obviously we need additional clues to the system's early history from the detailed behaviour of gas clouds on the threshold of stardom as well as study of meteorites. Unfortunately both studies are extremely complicated and do not readily yield helpful clues.

Consider first the condensation process. We still do not know the complex initial conditions such as the intensity of prevailing interstellar magnetic fields and the precise state of turbulence, which undoubtedly play a paramount part in determining its subsequent evolution. In fact, because of our ignorance we cannot even decide whether it will condense as a binary or as a single star with a planetary system. While we do know that during its early contraction a protostar may be rotating more or less as a rigid body, we do not yet know how the magnetic field trapped within it brakes its spin nor how it governs its subsequent break-up. That this is a major difficulty is clear from the fact that even purely accidental configurations of the magnetic field are likely to affect the subsequent evolution of the protostar in a very significant way. Nor have any stellar models been constructed to simulate the behaviour of such broken fragments so that we cannot tell how fast it contracts or whether it contains large convection zones that would tend to scramble the internal magnetic field and to isolate it from that of the surrounding medium.

According to all present indications a protosun contracting under its own gravitation will develop a surface temperature between 1,000° to 4,000°K. during the state of its rapid collapse from the orbit of even the outermost planet Pluto to that of Mercury in less than a hundred years. It is then necessary to envisage a rapid cooling process because, as Urey has pointed out, the absence of any observable fractionation of elements less volatile than mercury and its compounds in the case of meteorites and of the earth indicates generally low temperature accumulation. Consequently in his 'Boundary Conditions for Theories of the Origin of the Solar System' surveying the relevant physical and chemical evidence, Urey greatly emphasizes the importance of solid bodies and the physical and chemical processes to which they were subjected rather than that of gases and their behaviour in gravitational fields. If so, a theory of the origin of the solar system has to meet two rather contradictory conditions. For while the rapid collapse of the condensing solar nebula leaves in its wake wisps of hot materials at temperatures in the range 1,000° to 4,000°K. it has somehow to secure its instant cooling to enable its aftermath to be dominated by solids and their separations from large amounts of gas. Alternatively one may revert to Schmidt's idea whereby a fully formed sun passing through a cloud of gas and solid dust particles somehow drags a part of it in its wake as planets. It is here that observation could provide some guidelines. Unfortunately, it is difficult to obtain them because we cannot observe at stellar distances all the nuances of either Schmidt's accretion process in the case of stars voyaging through thick interstellar gas clouds, or of the condensation process, leading to planet formation in the case of young T Tauri stars, some of which may now be in the very act of originating their planetary systems if they do arise in this way.

As for the clues that the study of the chemical and physical properties of meteorities may provide, we find that the resulting interpretations are still largely controversial. Take, for example, J. H. Reynold's recent discovery that the Richardson meteorite contained an anomalously large abundance of xenon 129 relative to other xenon isotopes. While Hoyle claims it to be a confirmation of his theory of spallation reactions in the primitive solar nebula, Cameron, on the other hand, views it as evidence that Hoyle's process is unlikely. The reason is that Reynold's result allows a dual interpretation. It could arise either from the decay of xenon 129 formed from xenon 128 by addition of a neutron as a consequence of spallation reactions, or it could be a survival from a period of nucleosynthesis in the galaxy, which enriched the interstellar medium with fresh radioactivities a relatively short time before the formation of the solar system. It happens that Hoyle adopts the former alternative and Cameron the latter. It is not easy to decide between them, as it is necessary to consider a very large number of physical and chemical processes in many fields of astrophysics.

We may expect to find more definite clues when we land on the moon and begin its systematic exploration. Because of the absence of erosion and the minimal volcanic activity there, our satellite has preserved a record of events that has long since been wiped out from the earth. Accordingly on-the-spot moon probes during the next few years will perhaps enable us to extrapolate back-wards in time with a great deal more confidence than at present. But even so these probes, no matter how complete, will not give us complete understanding of the processes involved in the formation of the solar system. It seems we shall have to await the invention of new observational techniques enabling us to scan the finer details of planetary formations now under way either around young *T* Tauri stars condensing under their self-gravitation or fully mature stars voyaging through thick inter-stellar clouds of gas and dust.

PART IV

Life in the Universe

CHAPTER 17

Life in the Universe

WE have seen how matter in its constant development emerges in ever new forms from cosmic gas clouds and eddies to stars and their satellites. But the more we split, powder, and pulverize these kaleidoscopic evolving forms, the more clearly they reveal their fundamental unity. At these deeper atomic and subatomic levels all their endless diversity dissolves and appears as a mere shadow-play of the come-and-go of a relatively few types of elementary particles manifesting a perfect identity of behaviour everywhere. For their constituent atoms, electrons and protons – whether they remain buried in the bowels of our own good earth or waltz wildly on the face of remote stars or lie quiescent in the lonely depths of interstellar space – obey the same basic laws. The great variety of forms in which they nevertheless manifest themselves is merely the outcome of their widely different environments. The same atoms which reveal themselves in one situation as a clod of saturated earth appear in another as the source of a radiant speck of light or a faintly audible rustle of cosmic whiff. This essential uniformity of behaviour of atomic particles throughout the wide universe long ago posed an intriguing question: if some terrestrial atoms somehow managed to cast themselves into the human mould, could they not wake to a similar ecstasy of life and consciousness in other celestial habitats? Many man-made myths from Micromégas, Selenite and Bel Abon to mooncalf pastures, Martian canals and flying saucers testify to the continual persistence of a naïve belief in the power of life to irrupt in extra-terrestrial abodes as unlike one another as Sirius, our moon and Mars.

We now know better. But the earlier illusion about life's ubi-quity was a consequence of our total ignorance of the real com-plexity of the problem. As usual, the greater the ignorance, the more pompous the verbiage devised to mask it. *Generatio aequi-*

voca, generatio primaria expressed the belief that fully formed living organisms arise miraculously out of non-living matter. Since one miracle deserved another, life's miracle on earth could be repeated with equal facility and abundance everywhere. This is why some sober pioneers of stellar astronomy like Sir William Herschel felt fully authorized 'on astronomical principles' to populate even the sun with innumerable inhabitants. If there is now greater sobriety in the appraisal of the possible existence of life in other worlds, it is because of our greater understanding of life and its processes.

We no longer define a living being, as Woodger did some time ago, as 'an X in addition to carbon, hydrogen, oxygen, nitrogen, etc. plus organizing relations'. Perhaps the mysterious X-component was thrown in to provide a foothold for some sort of mysterious spirit which could either be equated to a soul after the forthright fashion of medieval theologians like St Thomas Aquinas, or to a more devious if less tangible *élan vital* of the vitalists and creative evolutionists like Bergson and Lloyd Morgan. Modern biochemistry has put an end to all this mystical rodomontade of the creative evolutionists and their allies the vitalists and telefinalists. For living substances have now been shown to be built up of the same kinds of atoms as are found in non-living matter with no evidence of any vital spirit entering when it wakes to life, or leaving it when it dies. If therefore we prune away the gratuitous X-element in Woodger's definition of life, life's essence would seem to reside solely in the organizing relations. In other words, life merely becomes, to use Haldane's apt phrase, 'a self-perpetuating pattern of chemical reactions'.

How such patterns of self-subsisting chemical reactions came into being initially is still a mystery, though it is no longer fashionable to conceal our ignorance of it under cover of Latin phrases of ancient vintage. Earlier we noted one form of it which became quite untenable when the exploration of microscopic life showed all cases of apparently spontaneous generation of life to be illusory as they are really instances of emergence from still minuter living organisms. Life, it seemed, could subsist in ever-minuter forms like the fleas of fleas in the doggerel:

Big fleas have little fleas
Upon their backs to bite 'em.
The little ones have lesser ones
And so *ad infinitum*.

The discovery that little bacteria always arise from still lesser ones now swung the pendulum to the other extreme. The old formula of spontaneous generation was reversed into *omne vivium ex vivo*, life always emanates from life. That is to say, each living organism, however primitive, requires a parent organism from which it springs by some biological process such as budding, fission, spore formation, or sexual reproduction.

Unfortunately, even this volte-face, though resting on an ultracritical scrutiny of a wide series of observations, has run into serious trouble to account for the origin of life. For if life on earth was sparked by pre-existing living organisms a few billion years ago, these sparking organisms could only be some sort of seeds of life – the so-called cosmozoa – that came into our world from interstellar space wafted about by light pressure or riding on Noah's arks of meteorites. But no cosmic Noah's cargo of life, however rudimentary, could possibly survive the hazards that are now known to beset an interstellar voyage. If it is not killed outright by exposure to the ultra-violet radiation implied by the excitation of aurorae, it is certain to be snuffed out of existence by the continual bombardment of cosmic rays from both of which dangers all our terrestrial life is shielded by the blanket of atmosphere. Besides, no viable bacteria or germs or microbes have ever been discovered in meteorites despite numerous attempts to find them. The recent claim of G. Claus and B. Nagy about the presence in certain meteorites of microscopic objects said to be 'fossilized remains' of primitive cells has not been substantiated. There seem to be insuperable difficulties in accounting for any way in which such presumed organisms could have come into existence or even maintained themselves on the dry surface of the putative place of their birth, presumably asteroids. In any case, it merely compounds the problem by adding to the complexity of life's emergence that of its transport from its extraterrestrial place of origin to the earth.

With the rejection of both brands of ancient verbiage concerning the origin of life, many biologists seemed for a time to have come to the end of their tether. They considered the phenomenon of life quite inexplicable without recourse to a *deus ex machina* of some sort – a God, spirit, purpose, entelechy, nisus, or some similar mystical principle. We may well appreciate the despair that drove them into such vitalistic mysticism. For until about sixty years ago the complexities of the self-organization of a single cell, not to speak of even the most primitive organism alive, were so enormous as to make any analysis of intercellular events all but hopeless. Since then the discovery of radioactive but stable isotopes of elements like carbon, phosphorus, nitrogen – the so-called tracer elements – whereby chemical compounds passing into a cell may actually be labelled and recognized in the subsequent metabolic products of the individual dose administered, has provided a new biological instrument of immense power. This, along with the invention of electrically driven ultra-centrifuges as well as new techniques much more delicate than those of light-microscopy, such as X-ray diffraction and electron-microscopy, has enabled biochemists to reveal the detailed structure of the ultimate building-blocks of cells, namely atoms, inorganic ions, molecules of water, amino acids, fats, sugars, proteins, nucleic acids and so on. As a result there has been a steady though slow veering away from the mystical approach of the vitalists.

These new techniques have revealed that life is a vast chain of chemical reactions catalysed by minute substances called enzymes essentially similar to any ordinary chemical reaction in the laboratory which needs to be sparked by a catalyst. The complexity of the life process that lured the vitalists into mystical by-ways lies in the vastness of the chain of interrelated chemical reactions but not in the nature of the individual reactions per se. Thus to account for the whole metabolic process of a single cell we require a thousand species of enzymes.

The isolation of all the enzymes involved in the functioning of any particular type of cell is therefore even now an enormously complex task though by no means as hopeless as the earlier biologists imagined. But all the complexity springs from having to dovetail into one consistent pattern a vast number of inter-related chemical reactions that are otherwise quite ordinary. They take place according to the usual physiochemical and thermo-dynamical laws and like all laboratory reactions are accom-panied by certain transformations of matter and energy. Nor is there anything peculiar about this energy or matter. It is plain ordinary energy of everyday physics having no affinity with Bergsonian *élan vital* or Shavian life force, even as matter is ordinary commonplace matter.

Consequently, a naturalistic explanation of vital phenomena in terms of physics and chemistry is in principle possible now even though the present state of these sciences is unequal to the task of unravelling all the knots in the tangle of life. In the first place, we still do not know exactly how life took its first great leap forward when non-living matter built itself into a simple organization like a bacillus or an alga. Life's break-through of the great chasm of inanimate matter has left no trace of its daring irruption. We get a fair glimpse of the course of evolution of unicellular organisms into various types of living beings existing today from the fossil remains of extinct species. But incipient life did better than Brutus's lowly climber. It did not merely turn its back on the ladder by which it attained its upmost round; it actually burned it and consigned its ashes to the limbo of oblivion. As a result we have no clues whatever to guide our speculations into the origins of life.

Secondly, the evolution of unicellular organisms from inani-mate matter was certainly a much more complex and intricate affair than their further evolution into the animals and plants of today.* This is why even the manufacture of a synthetic cell able to mimic the ordinary vital activities will still provide only a very partial answer to the riddle of life in that it will leave uncovered the mystery of reproduction. An ersatz cell, for all its powers of self-regulation, will perhaps be as sterile as a mule or a castrated bull. Recent work by Perutz and his team

---

* It has taken biologists the best part of a century to complete even this relatively simpler task.

would seem to show that to fathom the secret of reproduction we have to delve deeper still into the stereo-chemical three-dimensional arrangements of atoms within molecules of life which by virtue of their curious shapes enable life to proceed and reproduce itself.

But the synthesis of cells that imitate the normal activities of life and the manufacture of those seeds of life we call genes and chromosomes are merely preliminary steps in unravelling the mystery of life's origin on earth. To follow life's *actual* genesis from non-living matter to its initial incipient forms, we need to know in addition the conditions prevailing on our planet at the time these forms sought to secure their first faltering foothold.

We can date life's emergence from inanimate beginnings rather precisely. The Cambrian rocks, the oldest group in which fossils have been found, are evidence of an abundant and varied life as long as 500 million years ago. Fossils of many thousands of different kinds of Cambrian animals unearthed by palaeontologists show that the shallow seas of the times must have teemed with life. Certainly many kinds of trilobites, brachiopods and worms flourished on the floors of the Cambrian oceans and burrowed in the muds that underlay them, for they have left a fossil record of their existence. Considering that all the great phyla and many classes of animals represented in the Cambrian fauna had advanced far along the evolutionary road despite their primitiveness, it seems that at least three-quarters of animal evolution must have occurred before the onset of Cambrian times some 500 million years ago. Consequently plants and animals must have in all probability lived in the earth's oceans for at least 1,500 million years before the Cambrian period, even though all the pre-Cambrian animals and plants except the calcareous algae, lacking hard parts, have left no satisfactory fossil record of their former existence.

To divine the genesis of these pre-Cambrian forms of life, we must therefore first construct the conditions that prevailed on earth some two billion years ago as also the geochemical and biochemical processes that could come to pass in the hurly-burly of our earth's infancy. If this commotion was a consequence of its formation by Schmidt's process* of slow accumulation of solid particles, as present evidence seems to suggest, the slowly growing globe must have remained cold except for slight warming from the heat generated by the impact of the infalling cosmic material. This lukewarm heat bath during the accumulative stage of earth's career caused the emission of gas and water vapour of which there were small quantities in the earth's mantle of rock. In fact, such an emission of gas and water from the earth's interior continues to this day though on a greatly reduced scale as, for instance, is the case during volcanic eruptions when the earth belches out vast quantities of water vapour, carbon dioxide and other combustible gases into the atmosphere.

It was by means of some such process on a much vaster scale that the interior gases and water vapour escaped to the earth's surface. The condensation of water vapour led to the formation of rivers, lakes, seas and oceans that constitute the earth's hydrosphere, and the gases formed its atmosphere.

While the aqueous content of the hydrosphere has remained materially the same during the aeons that have elapsed since the formation of our planet, it has been suggested that not more than one-millionth of the original atmosphere has survived the subsequent vicissitudes of its life. The underlying rationale of this shrewd surmise of Harrison Brown and Hans Suess is that of the tell-tale vestiges of inert gases like helium, neon, krypton, xeon and so on, remaining in our atmosphere. Their present residue has been estimated at less than one-millionth of what could be anticipated on the basis of their original concentration in the earth's atmosphere. Since these gases are chemically inert, they could not have disappeared by chemical combination with other materials of the earth. It therefore follows that they, along with other gases that were present in the earth's primeval atmosphere, flew off almost completely into space during the initial warming phase of its existence, lukewarm though it was.

Later, when the surface cooled with the cessation of the infalling cosmic material and the impacting together of asteroidal bodies, a secondary atmosphere was formed from gases com-

* See page 333.

bined or occluded in the solid body of the earth. What precisely was the composition of this secondary atmosphere is uncertain, so that the nature of the photochemical reactions that preceded the evolution of living matter cannot be correctly surmised. If it contained oxygen in any quantities as our present atmosphere does, an ozone screen absorbing all ultra-violet radiation would have formed in its upper reaches. There would thus be a lesser diversity of possible photochemical reactions than in the absence of oxygen. For this reason, as well as others into which we need not enter here, Haldane is of the view that the earth's early atmosphere had little free oxygen and consisted largely of hydrogen, ammonia and methane like those of the other planets, and perhaps carbon dioxide like that of Venus.

Under the influence of ultraviolet solar radiation, metastable organic molecules would be formed out of such atmospheric material. For instance, organic molecules can be formed by photochemical production of formaldehyde from carbon dioxide as has been actually demonstrated by S. L. Miller in Urey's laboratory where amino acids among other compounds were formed by passing an electric discharge through a mixture of gases. Since Miller's classic experiment many workers like C. Ponnamperuma, C. Sagan and R. Mariner, to name only a few, have shown that even more complex but specific life-forming molecules like the base adenine, the pentose sugar, ribose, the lipids, etc., can be formed synthetically by merely subjecting a mixture of simple gases to various types of excitations such as are likely to occur on primitive earth. It is therefore natural to expect that such molecules would be formed under conditions then prevailing on the earth. Once formed, these organic substances could have dissolved in the primitive seas, lakes and pools of water. It was earlier believed that in the absence of any bacteria to destroy them, the waters of the seas and lakes must have held in solution an unimaginably complicated mixture of mineral salts, amino acids, alkalis, colloidal particles of clay, and other minerals. This vast emulsion of carbon compounds, or Haldane's 'hot dilute soup' was supposed to be the medium in which catalytically active molecules or enzymes multiplied, thus providing a material basis for life to originate and grow. After all, living matter in its most fundamental aspect is really nothing but a highly organized system of such enzymes dispersed in aqueous solution.

It is now realized that sufficient quantities of organic materials like amino acids could not have accumulated in the seas to allow their condensation into proteins to occur. If the experimental simulations are any guide, the amount of biological molecules likely to have been produced from the gases in the primeval atmosphere would be too small to yield Haldane's hot thick soup in the seas and oceans. Such a thick concentrated 'soup' is more likely to arise at several places in small, probably fresh-water, pools continually subject to drying by evaporation rather than in a continuously wet environment. As the pools dried out, the concentration of organic material in them might well have become sufficient to allow them to react to form polymers.* Indeed it may be that life began in micro-pools on the surfaces and in the crevices of rocks or the soil where minerals may have played a part in catalysing the polymerizations. Once the living systems had arisen they could then have evolved in the sea into which they would certainly have been washed. It is only some such scheme that can meet the apparently contradictory requirements of high concentrations of biological molecules to allow polymerization and of a continuously aqueous environment to allow the evolution of living systems once they had arisen. Whatever the site of life's first irruption – seas, pools, micropools or rocky crevices – the experimental simulation of conditions prevailing on primitive earth leaves no doubt about the natural origin of life.

Much has been made of the statistical improbability that a random mixture of substances could happen to form enzyme proteins or other complex and specialized structures by a purely 'fortuitous concourse of atoms'. Lecomte de Noüy, for instance, has argued that if the universe is merely a medley of randomly moving molecules and if the ultimate particles like electrons build up atoms, and atoms build molecules, and molecules colloidal particles, and so on, then it is practically impossible that the ultimate particles could have organized themselves by pure chance in such highly complex structures as human beings, or even living cells and protein molecules for that matter.

* See page 376.

Thus the odds that a moderate-sized molecule consisting of, say, 2,000 atoms and possessing the same degree of asymmetry as a protein molecule should arise by random processes are roughly one in 10^{318} against it. These odds are so large that we might as well concede that the occurrence of life is a miracle requiring an act of special creation and thus hand over the problem to theology or metaphysics. But the argument in principle is of a piece with that of a lucky bridge player who, having been dealt a succession of particularly bumper hands, cites the event as proof of providential intervention. The fact is that given the cards, a dealing mechanism, and sufficient time, all distributions, even the most remarkable one of each player getting a complete suit, must sometime or other take place.

The answer to de Noüy's argument therefore lies in showing that the basic laws governing the behaviour of matter are such as could have led to the genesis of life by a natural process rather than by the fiat of an Almighty God. To do so we have to search for and formulate the organizing principles that led to the emergence of life, such as Newton's law of gravitation that made the solar system what it is, or Maxwell's equations governing electromagnetic phenomena. In the case of life's genesis from inanimate beginnings, the basic organizing principle is the chemical law of valence which itself is a consequence of the inherent structure of the atoms.

In Chapter 13 we pictured the atom as a miniature solar system* of a central nucleus of protons and neutrons sticking together in a tiny droplet with a number of satellite electrons orbiting around it. The number of orbiting electrons – its atomic number – and the pattern of the electronic dance they perform within the atom is the key to its chemical behaviour, because it is in consequence of the interactions between their satellite electrons that atoms exert on one another forces which bind them in various combinations called molecules. To delve into the nature of the chemical forces that bind the various atoms together we have therefore to discover how the electrons within the atoms arrange themselves as we pile more and more of them around the nucleus in going to higher and higher atomic numbers.

* See pages 269 and 309.

Modern quantum theory has shown that they arrange themselves in 'shells' or groups of electrons in such a way that the members of a group are approximated at the same radial distance from the nucleus, differing rather widely from the radial distances of other groups. Thus the hydrogen atom with atomic number 1 has only one electron revolving in the first shell. The helium atom with atomic number 2 has both its electrons orbiting in the first shell while lithium, with atomic number 3, has two electrons in the first shell and a third in the second. The neon atom, with atomic number 10, has two electrons in the first shell and eight in the second.

How many electrons revolve in each shell depends in the first place on the number of satellite electrons (or the atomic number of the atom) and the *saturation* number of each of the successive shells in which the electrons orbit. The saturation number of a shell, by the way, is the maximum number of electrons that can possibly flock together in that shell. Pauli discovered a principle of great power known as the exclusion principle from which he deduced the saturation number of each shell by a purely enumerative process. He showed that the saturation number of electrons orbiting in the nth shell of any atom is $2n^2$ so that the saturation number of electrons in the first, second, third, fourth . . . successive shells of an atom are 2, 8, 18, 32 . . . respectively.

Atoms whose shells contain the saturation number of electrons behave like satiated beings, having no tendency to appropriate any more electrons from other atoms and thus attempt to combine with them. This is why helium (both of whose electrons revolve in the first shell), neon (in which two electrons revolve in the first and the remaining eight in the second shell), argon (in which two electrons revolve in the first, eight in the second, and the remaining eighteen in the third shell), etc. form a remarkable series – the so-called noble gases. They are inert gases which do not enter into any ordinary type of chemical reaction because their atoms, with full electron shells, have almost no tendency to combine with any other.

On the other hand, those atoms whose electron shells are only partially full have a strong tendency to combine whereby they are able to share each other's outer electrons and thus form what is called a covalent bond. This type of bond is of great importance in the chemistry of living organisms because it explains the great power of carbon to form a multitudinous variety of compounds on which depends the capacity of carbon to function as a veritable backbone of living matter.

The reason for this is that a carbon atom has four electrons in its outer (second) shell and needs eight for saturation. Since hydrogen has only one electron in its outer (first) shell and needs two for saturation, a carbon atom can, like a Turkish pasha, herd four hydrogen atoms into its chemical harem – the covalent bond filling its incomplete outer shell with four electrons from four hydrogen atoms, each one of which is in its turn assured a stable shell of two electrons by appropriating an extra one from the carbon. This is why a single carbon atom is able to marry itself to a quartet of hydrogen atoms producing the gas methane which figures so prominently in the evolution of planetary atmospheres.

But methane (CH_4) is only the first of a series of carbon compounds – the hydrocarbons – which is the outcome of the great chemical affinity of the carbon atom whereby it can form a succession of related substances each of which differs from adjacent members only slightly in molecular structure and chemical properties. Thus the series of hydrocarbons represented by the formula C_nH_{2n+2}, where n is the number of the compound in the series, begins with methane (CH_4), the principal ingredient of natural gas, and progresses through ethane (C_2H_6), propane (C_3H_8), butane (C_4H_{10}), and pentane (C_5H_{12}), etc., to octane (C_8H_{18}) and beyond. These are the principal ingredients of gasoline, kerosene and diesel fuel.

Not only is a great variety of compounds possible because of the increasing value of n, but once n is fairly large a great variety of *structures* is possible with the *same* number n. The reason is that the structure of a molecule is the particular way in which each atom is linked by valence bonds to each of the other atoms. With increasing n, the number of such ways of patterning the valence bonds of atoms multiplies rapidly. Thus, while only one such way or structure is possible in the case of methane (CH_4), ethane (C_2H_6), and propane (C_3H_8), two different structures can be drawn for the next hydrocarbon in the series, (C_4H_{10}), namely,

```
    H   H   H   H                    H   H   H
    |   |   |   |                    |   |   |
H—C—C—C—C—H    and    H—C—C—C—H
    |   |   |   |                    |   |   |
    H   H   H   H                    H   H   H
                                         |
                                     H—C—H
                                         |
                                         H
```

Accordingly two related or *isomeric* substances corresponding to the two aforementioned structures actually exist. The former is normal butane and the latter its cousin isobutane. Clearly, as n increases, the number of possible isomeric variations increases rapidly. As we saw, butane ($n = 4$) has only two. The next member pentane $n = 5$ has three, but it can be shown that n need only be 20 to yield 366,319, and at $n = 40$ the number of different structures theoretically possible is over sixty-two billion!

The structure possibilities of carbon compounds that we have hitherto explored are merely the outcome of different ways of drawing its chemical diagram. The purpose of such a diagram is not to represent the true geometrical configuration of the atoms or molecules of the substance but rather to specify the valences of various atoms. In other words, it gives only the topological relations of the atoms within the molecule. It can therefore happen that two substances may have the same composition, molecular weight, and structure, but different configurations of atoms within the molecule. Such, for instance, is actually the case with tartaric and racemic acids. Although both have to be assigned the same structure, they differ in a number of their characteristic properties such as densities, melting points, solubilities and crystalline forms. Such substances are known as stereoisomers. Isomerism and stereoisomerism are thus important attributes of carbon compounds which have been made to add grist to the mill of life because they help provide incomparably finer gradations in properties than does the development of the methane series.

The methane series and its isomers are only one example of

many series of substances based on long chains of carbon atoms. The flexibility of the carbon atom enables it to keep one of its four chemical bonds in abeyance and thus to form another series beginning with benzene, whose formula C_6H_6 is represented by

Benzene

The benzene ring itself can act as a unit in chains of rings, two such rings producing naphthalene of moth balls and three anthracene:

Naphthalene   Anthracene

The variations in these ring themes are as endless as those of the structures in the higher members of the saturated hydrocarbons considered earlier.

All this by no means exhausts the versatility of the carbon atom in forming a wide diversity of carbon compounds. It employs yet another device to produce variety, namely polymerization, a term that has now become a household word with the development of artificial plastic materials like nylon. As is well known, polymerization is a chemical process whereby two or more molecules of the same substance (monomer) unite to yield a molecule (polymer) with the same percentage composition but a higher molecular weight which is therefore an integral multiple of the original molecular weight. Thus, for instance, styrene (C_8H_8) is a colourless liquid with an aromatic odour. On standing for weeks at room temperature or for a few days at 100°C. the pure liquid polymerizes, becoming at first an increasingly viscous solution of polymer in monomer and finally a clear, odourless, glassy solid. The product is a mixture of polymer molecules of the formula $(C_8H_8)_n$, where n has a wide range of values and averages several thousand.

Thus compounds of carbon by chemical bonding, isomerization (both structural and stereo), multiplication of benzene ring structure and polymerization provide an infinite variety of complex materials that form the 'solid' core of living substances. They are mostly networks of carbon atoms whose prolific power to associate with others by the aforementioned means we have already noted. One such network is the class of polymers of high molecular weight which are composed of long chains of amino acids linked together by peptide bonds, that is, bonds formed between the carboxyl group COOH of one amino acid and the amino group H_2N of the next with the elimination of water. Thus, for instance, the amino acid, glycine, $H_2N \cdot CH_2 \cdot COOH$, unites with another, alanine, $H_2N \cdot CH(CH_3) \cdot COOH$, whereby hydrogen of the (H_2N) amino group of the former combines with (OH) of the carboxyl group (COOH) of the latter to form water and another amino acid, alanylglycine:

$$H_2N \cdot CH(CH_3) \cdot COOH + HHN \cdot CH_2 \cdot COOH$$
Alanine $\qquad\qquad\qquad$ Glycine

$$\rightarrow H_2N \cdot CH(CH_3) \cdot CO \cdot NH \cdot CH_2 \cdot COOH + H_2O$$
Alanyl-glycine $\qquad\qquad\qquad\qquad$ Water

Such a chain or network of amino acids is called a polypeptide.

The number of different types of similar chains, that is, protein molecules, which one may build up from even twenty known amino-acid residues, is practically unlimited, just as one may construct a virtually infinite vocabulary (mere chains of letters) with a limited alphabet. Thus a fairly light protein like haemoglobin contains about 600 amino-acid residues. But the possible diversity of molecules of this size with twenty amino acids is theoretically $20^{600}(=10^{781})$, being the number of different ways of filling 600 places when each place may be filled in twenty alternative ways.

When we recall that the total number of particles in the universe according to Eddington* is only 10^{79} (if one may use 'only' in juxtaposition with such a gargantuan number), we may appreciate the vastness of 10^{781} and thus obtain an inkling of the quasi-infinite mutability of protein structure and its properties. No wonder there are 100,000 different protein species in the human body itself and that the manifold substances belonging to the protein family are as diverse as hair, nails, skin, bone, cartilage, muscle, connective tissue, nerve fibre, blood haemoglobin, hormones, insulin, egg albumin, feathers, wings and shells of insects, and so on.

Nevertheless, the fact that all this immense variety of protein materials is merely the outcome of permuting a score of amino acids seems to suggest that the whole diversity of living organisms has an extremely simple chemical basis. The basis is simply the quantum mechanical theory of electronic shells we have already outlined, so that emergence of life is merely the self-realization of the potentialities of atomic electron states. Noüy's objection to a naturalist origin of life on account of its statistical improbability disregards the chemical basis of life. It tacitly assumes that *all* permutations of atoms are *equally* likely in total disregard of the fact that most permutations are excluded by the electronic structure of atoms while some others are heavily favoured. His argument therefore is of a piece with the statistical 'deduction' that since the chance of even a single meaningful sentence in Shakespeare's works arising by a random strumming on a typewriter is infinitesimal no such work could ever arise! The fallacy lies in ignoring the mechanism responsible for stringing letters of the alphabet in one case and atoms in the other. If we pay due regard to the chemical proclivities of atoms built into their electronic structure and treat the question of origin of life as a problem of chemical kinetics and thermodynamics, we find that emergence of life, far from being a highly improbable event, is all

* See page 247.

but inevitable given the right conditions. Thus A. B. Zahlank has very recently shown how life could arise as a result of random and purely chemical events. Taking catalysis as the main agent and using information theory and topology he arrives at a set of systems with about six catalysts per system which could lead to the emergence of a sequence of chemical reactions involving replication, performed on metabolites supplied by the environment. Although such schemes are as yet purely formal, they do expose the error in denying a natural origin of life on the grounds of its statistical improbability. In fact, as we mentioned before, given the cards, a distribution process and sufficient time, almost any distribution, however remarkable and however low its prior probability, must occur. Likewise, given Haldane's 'hot dilute soup' of protein precursors like amino acids and other organic compounds spread over an ocean covering a large fraction of our planet, and given a time span of some two billion years, it is easy to imagine the emergence of some molecules with the self-duplicating property of a gene – a molecule of life.

Once formed, such molecules would multiply until the accumulated supply of one of the immediate components was exhausted – a foretaste of Malthusian pressure of population on the means of subsistence. A premium would then be put, as pointed out by N. H. Horowitz, on a variant able to catalyse the formation of this component from a protein precursor still present in abundance – foreshadowing evolution by mutation and natural selection. Selection would also favour molecules capable of appropriating others and breaking them down to reorganize their constitution according to the devourer's own pattern – prototypes of future predators.

But all this is nothing more than an outline of a basis for merely beginning a probe into the origins of life from inanimate roots. To complete it and thus close the great gap preceding the establishment of cellular organisms, we still need to know far more of intracellular histology, microphysiology, geochemistry and cosmogony. When we know enough of these subjects to be able to specify the magnitude and complexity of the smallest possible self-reproducing molecule system on the one hand, and of the cosmogonic processes of planetary formation on the grandest scale on the other, we shall be able to suggest an environment that is both a cosmogonic possibility as well as capable of breeding the microscopic molecules of life – the genes and chromosomes.

It will then be possible to demonstrate that under the conditions that actually prevailed on earth, the emergence of life was inherent in the basic laws governing the behaviour of matter everywhere. To be sure, the evolution of inanimate molecule into man via the missing link of a self-catalysing protein molecule may seem a miracle. But modern science does give us some glimmer of the Ariadne's thread running through the entire gamut of life from the 'subvital' autocatalytic particle of protein all the way to man as an increasingly complex crescendo of self-sustained patterns of chemical reactions with their constant *va-et-vient* of ordinary atoms. This is why we need only depersonalize Omar Khayyám's 'they' into the basic laws of the behaviour of matter to be able to conclude with him:

> With Earth's first Clay They did the Last Man's knead,
> And then of the Last Harvest sow'd the Seed.

But would not the same laws that kneaded man out of earth's first clay do the same for stars, planets, and their satellites both in our own Milky Way as well as in other galaxies? There is a school of thought prevailing even now that has a sort of mystic faith in the fitness of almost every celestial environment to develop and evolve its own species of living beings whose bodies and organs are supposed to be adapted to the peculiar circumstances of their particular abode. Even the strictly disciplined and controlled imagination of that great imaginer, Olaf Stapledon, could not resist the temptation of toying with the idea that life and intelligence could perhaps emerge and persist in the incandescent environment of stars in the forms of myriad self-conscious flames of life.*

But it is clear that he did so merely to incorporate into the myth he set out to create a belief that is a part of our cultural matrix even though it is beyond the limits of contemporary scientific credibility. For life, as we have seen, is an attribute of

* See Olaf Stapledon's *Last and First Men*. Penguin, 1963, p. 269.

matter that appears only in highly complex and therefore correspondingly fragile structures. Even a protein molecule, not to speak of other more elaborate structures like cells and animal tissues, is a complex and delicate entity, extremely sensitive to heat.

The molecular hurly-burly that heating inevitably lets loose proves too much for the delicately balanced constituents of proteins as well as living tissues. It coagulates proteins, decomposes amino acids and deactivates enzymes. No living tissue, cell or bacterium seems therefore able to withstand even boiling water. The hottest niche that living organisms have so far been known to occupy is in Yellowstone Park, where bacteria have learnt to survive in hot springs at 76°C., but none has been found in any of the hotter natural springs in spite of life's incredible resourcefulness to colonize any possible ecological corner. Apparently 80° to 90°C. is a fundamental upper limit to which life can accommodate itself.

The possibility that living organisms made of sterner stuff like silicon in lieu of carbon would be able to survive in a very much hotter environment has no doubt been suggested because silicon ranks next to carbon in its power to combine with other atoms to form a multitude of compounds. But unlike carbon, silicon atoms show little tendency to combine with one another in long chains and rings. This is why the silicon counterpart of the hydrocarbon series does not stretch very far and why even those hydrosilicons which do exist are far more fragile than their carbon prototypes. Consequently silicon is no match for the prolific power of carbon to act as a base for the build-up of complex structures that are indispensable precursors of life.

Since living matter, wherever it may occur in the universe, presumably requires a vast variety of complex compounds from which to synthesize itself, carbon alone among the elements seems qualified to serve as its backbone. A silicon cell therefore is not even a plausible biological 'perhaps' so far as we can imagine at present.

But even if it were, it would not materially extend the temperature range up to which life could ascend. For every chemical compound, whether of silicon, carbon or anything else, can be broken down with sufficient heat. Indeed the more complex its structure, the more easily it disintegrates. We may therefore rule out all possibility of life in any of its variety and adaptations in the stars, as even the coolest of them have surface temperatures of 3,000°C. – much too high for any except the simplest compounds to exist.

But if excessive heat inhibits life of all kinds, too much cold is equally fatal. For vital processes – as our experience here on earth shows – depend ultimately on the energy received from the sun. If this supply had been much smaller, it is doubtful if life could have ever been sparked out of its inanimate slumber. It therefore follows that only planets located at just the correct distances from their central star to give them the right amount of starshine and warmth can be suitable abodes of life in any possible form. If they are too near their central star, no life can originate in the midst of the hell fires to which their close proximity exposes them. On the other hand, if they happen to be too far away, life may remain congealed forever in the frosty cold of outer space.

But it is not merely their distance from the central star that has to be right. So must be their other features such as mass, axial rotation, atmosphere, hydrosphere, ellipticity of orbit and so on. Thus if a planet or satellite like our own moon fails to combine the right blend of these other conditions, its location at the right distance from the sun will be of no avail. For the moon, having too small a mass to retain any atmosphere, is both airless and waterless. Further, since it spins around itself slowly in about a month, a lunar day is as long as a terrestrial fortnight. In consequence, the surface rocks of its equatorial regions are alternately grilled by a fortnight's solar radiation undiluted by atmospheric absorption and chilled by an equally long Stygian night. The noon temperatures at the subsolar point thus soar to 120°C. and drop to −150°C. at midnight.

Nor is there any wind or water to temper the effect of such violent oscillations of surface temperature. In such a lunar chiaroscuro of light and shade, no live organism, even if it could get there somehow, could escape being alternately roasted and put into a deep freeze.

Then again, if a planet's orbit is highly elliptical, as in the case of Mercury, the seasonal variations in the intensity of solar radiation may prove too great for life to develop and take root. Thus Mercury at its closest approach to the sun receives two and a half times as much sunshine as at its farthest. This alone gives rise to too violent fluctuations of temperatures but the violence is fantastically amplified because of the coincidence of the period of its axial rotation with that of its revolution round the sun. As a result it presents approximately the same face to the sun as the moon does to the earth, so that it is a fearful furnace of molten lead, lava and tin on the sunny side and an equally terrifying Cimmerian nightmare of frozen gases on the other eternally dark, unlit side. These conditions, whose rigour is in no way mitigated because of the virtual absence of wind and water, prevent it from harbouring life of any kind whatever.

It is therefore small wonder that barely two (excluding our own earth) out of the nine planets and their numerous satellites of our solar system have managed to acquire the right combination of conditions that might provide a possible home for life in some form or other. They are Venus and Mars. Barring these two, the other planets beyond Mars are so far off that the punch of the sunrays is attenuated to a mere fraction (4 per cent) of their terrestrial intensity by the time they reach even the closest of them, Jupiter. Consequently, Jovian surface temperature is 138°C. *below* the freezing point of water, which is cold enough to liquefy many gases. On Jupiter, therefore, and still more on other outer planets beyond, the solar energy, the sole ultimate motive power behind life in any form, is much too feeble to lure life out of the dismal frost of these hostile worlds.

But even Venus and Mars, in spite of the conjunction of several favourable features, may happen to draw a blank and turn out to be lifeless leaving only the earth as the sole stage in our solar system for the great epic of life. Thus data about the Venusian environment transmitted by the Soviet spacecraft Venus IV, soft landed on that planet for the first time in October 1967, showed a temperature range from an uncomfortable but tolerable 40°C. at an altitude of about fifteen miles to a scorching 240°C. as it drifted lower down during its ninety-minute fall. Earlier data transmitted by the U.S.A. Mariner 2 in 1962 as well as that obtained by radio-telescope observation recorded even higher Venusian surface temperatures (425°C.). These latter-day observations clearly show that Venus is too hot a habitat for the kind of life we know on the earth. They have also debunked the speculation of Venus being a vast shoreless ocean with practically no land surface so that Stapledon's intuition of a varied marine life dominating Venus is very likely to remain a romancer's dream.

His surmise of a similar life on Mars, that darling of the conjurors of planetary populations, is even more wide of the mark. For the earlier interpretation of the thin gossamer lines which indubitably criss-cross the Martian surface as a stupendous canal system has now been generally abandoned as a Lowell fairy tale born of wishful thinking. Nevertheless, the wish to see life of a sort on Mars seems still to persist even in the face of rather doubtful supporting evidence.

Mars would appear to be an arid waterless desert possibly given to frequent volcanic eruptions with the prevailing winds carrying the dust and ash they raise. Under these conditions life could secure a very precarious foothold, if any at all. The scientific data yielded by the U.S.A. spacecraft Mariner IV's fly-by of Mars from as close a range as 6,118 miles in July 1965 seems to confirm the inference. Thus the pictures of the Martian surface transmitted by Mariner's television camera showed craters resembling those on the moon. But the winds, if any, must be extremely tenuous as the atmospheric pressure at the surface is barely 1/200th that on the earth. All in all the Mariner IV observations did indicate the vast differences between the Martian and terrestrial environments even though they did not – as had been expected that they would not – answer the question of the possibility of life on Mars. It is the U.S.A.'s Voyager programme of interplanetary probes, with its first mission scheduled for 1973, that is specifically designed to acquire data relating to extraterrestrial life.

It is the implementation of Voyager-type interplanetary probing missions that will enable us not only to land instruments gently into the crags of the moon and the maria of Mars but even to return them to earth laden with such cargo of life as they may happen to possess. We shall then learn more about the origins of life and its mechanisms by seeing the products of evolution in other celestial climes. We shall then know more positively than we do now whether life can be built on a radically different pattern with, say, silicon in place of carbon, chlorination in lieu of oxidation, controlled disintegration of radioactive atoms instead of photosynthesis and chemical combination, and a nervous system built on the principle of the wireless rather than that of the telephone exchange.

Any of these devices and many more of which we know nothing may yet prove a possible protective armour for life to wear in its eternal struggle against a hostile environment of lethal radioactive fall-outs, cosmic-ray showers, excessive gravity, electromagnetic storms, turbulent hurricanes of stellar winds, bottomless pits of poison gases, and many more of those unknown hazards that our future Odysseuses of space will discover with blood, sweat, tears and travail.

However, casting aside all these speculative possibilities and judging by present indications, our own earth seems to be the only abode of life – at any rate intelligent life – in the solar system. When we consider that barely 10 per cent of the stars in our Milky Way are born single and not every such star has a planetary system, and further that only 10 per cent of its planets may acquire the right blend of mass, axial rotation, distance and other attributes likely to favour the emergence of life and intelligence, we may well appreciate how scarce and lonely life must be wherever in the universe it may have chanced to sprout. Assuming as a rough reckoning that not more than one star out of a million taken at random can possibly have a planet with life on it at some particular stage of development, there may well be hundreds of thousands of inhabited planets in our own Milky Way of some 100 billion stars. In the larger universe of hundreds of millions of galaxies we may expect to find billions of inhabited worlds.

Nevertheless, no matter how numerous the inhabited planets, the universe of galaxies and stars encompasses and swallows them all like an atom, so vastly greater is it in comparison. And yet this frail and fragile atom of life and consciousness whom, as

Pascal said, 'the entire universe need not arm itself to crush', is able to comprehend the blind and uncomprehending hostility of the universe around that knows not what it is doing. This awesome hostility of the universe may frighten it but even a frightened being conscious of its own peril is greater than the entire caravan of bright blind galaxies of stars that, unaware of their own majesty and magnificence, hurry along the intergalactic wastes without knowing why they hurry, whence and whither they hurry, and cannot even ask for

> Another and another cup to drown
> The Memory of this Impertinence!

CHAPTER 18

God and Cosmology

ONE consequence of the creationist theories of cosmology like those of Lemaître, Gamow, Milne, and others is the emergence of a branch of theology which seeks to infer the existence of God by the light of pure reason unaided by the intimations of faith and revelation. Not that it is anything new. In the earlier Middle Ages, St Anselm, for example, worked out an ontological 'proof' of the existence of God by grafting Neoplatonist ideas on Christian orthodoxy. He was followed by St Thomas Aquinas, who took the further step of cutting natural theology adrift of its metaphysical moorings. By attempting to discern in the external universe a foundation for reasoning about God, he made the passage to God a more reasoned affair rather than a mystagogical leap as in St Anselm's ontological proof. But unfortunately for natural theology the further progress of cosmological research tended to make redundant the hypothesis of God even though His disappearance from the scheme of things was slow in coming.

At first Copernicus suggested his new heliocentric theory of planetary motions as a convenient mathematical fiction while still acknowledging the truth of the Biblical stationary-earth theory. Kepler, who discovered the famous three laws of planetary motion that bear his name after a prolonged empirical study of planetary positions, found in them primarily an expression of the beauty and harmony of divine creation. Even Newton, whose celestial mechanics was to provide the basis for the subsequent development of the belligerent anti-religious philosophy of the mechanical materialists, believed that the solar system was 'not explicable by meer [sic] natural causes' so that he was 'forced to ascribe it to the counsel or contrivance of a voluntary agent'. But the divine agent in Newtonian cosmology was apparently required only for the initial arrangement of the planets and for occasionally intervening to prevent them from colliding with one another or falling into the sun under the action of their mutual perturbations. When, therefore, Lagrange and Laplace a century later showed by clever mathematical proofs that the planetary system possessed 'stability' of itself on the basis of Newtonian laws, the classic boast of Laplace that there was no need of the hypothesis of God seemed to be fully justified. From this moment onward there began a conflict between science and religion that reached its apogee towards the close of the nineteenth century when the publication of Haeckel's widely read masterpiece *The Riddle of the Universe* carried to the layman the message of science that the universe around us is potentially explicable by rational inquiry and ordinary observation without invoking God, spirit, *élan*, nisus, entelechy, or some similar mystical or theological principle.

During the last fifty years, however, the newer concepts resulting from the experience of present-day physics and their intellectual interpretation are such a complete reversal of the older picture developed by natural science that in the opinion of some eminent physicists the time has now come for a reappraisal of the relation between science and religion to determine whether the older antireligious belligerent attitude of natural science culminating in Haeckel's time is still valid today. Many of them like Lemaître, Milne and Whittaker, to name only a few, are even more forthright. They do not hesitate to claim that the deeper understanding of the nature of the material universe which has been achieved by modern cosmology has opened up new prospects and possibilities to the advocate of belief in God. But what are they?

Briefly they seem to lie in fresh 'proofs' for the existence of God that modern creationist cosmology is said to furnish now. It is conceded, of course, that these 'proofs' do not possess the imperative persuasiveness of mathematical deductions like, for example, Pythagoras's proof that the square root of 2 cannot be a rational fraction, that is, a quotient of two whole numbers. Such mathematical proofs have the certainty of logical necessity and are therefore universally accepted as the necessary presuppositions of reasoning. But besides them there are also those proofs which are grounded in scientific experiments, like the statement that water is composed of hydrogen and oxygen based on Cavendish's ocular demonstration of it by sparking hydrogen and oxygen into steam.

They too have considerable compelling powers to conviction, though not as great as those of the deductive proofs of pure mathematics. Cosmological 'proofs' of the existence of God do not belong to either of these two types. Lacking the absolute certainty of logical and the comparative security of experimental verification, they are frankly intended to serve only the purpose of practical apologetics, namely to present arguments of such a character that a man already favourably inclined but given to doubt and wavering may find some justification in acting on them.

Consider, for example, Whittaker's modernized version of St Thomas's cosmological argument in favour of God. It opens with the consideration that in the world of sense the connexion between the past and the future is intelligible only on the principle of efficient causation. That is, a thing cannot be the efficient cause of itself: so everything has an efficient cause which is distinct from itself. This in turn has its own cause, and so on, so that there is a sequence of efficient causes. Now Whittaker shows that such a sequence cannot form a closed chain. Nor can it lead to an infinite regress in view of the recent insertion of a creation (as, for instance, in Lemaître-Gamow theory) into the scientific picture of the cosmos. Consequently the sequence must terminate in an ultimate efficient cause which is itself uncaused, and this ultimate cause is God.

In a nutshell, since modern cosmology seems to show that the universe has a beginning in time at the epoch of its creation, it must have been created by an Almighty Creator. It is true that both Milne and Lemaître, the authors of these creationist cosmologies, do not mention God in their technical expositions. But they do require their readers to insert the first cause of the universe without which they consider their cosmological schemes incomplete. No wonder then that even Pope Pius XII could not resist the lure of the argument and chose to cite it as 'scientific' evidence of the existence of God and His 'creation' of the universe some ten billion years ago.

Nevertheless, it is doubtful if even such a modernized version of the old cosmological argument of St Thomas would make any impression on an agnostic mind today. For one may, without being irreverent, inquire what God was doing before He decided to create the universe. It is no answer to say as Milne does that the Creation was the sole supreme irrationality by an otherwise rational Creator. For this is merely a confession of failure that it is futile for rational science to probe into the 'miracle' of creation.

All these attempts to deduce a god from cosmological premises inevitably come to grief because, as Weizsäcker has justly remarked, it is impossible to understand rationally a god in whom one did not believe already. This is why it is no longer possible to rationalize religion and find God in the mathematical laws of the cosmos, as, for example, Kepler could in his day. In fact, the progress of science had already begun to make the rationalization of theological belief so increasingly difficult that a tortured Pascal could no longer experience God in the mathematical laws of the cosmos as Kepler did barely a generation earlier. Within a century it became clear that henceforward there were to be only two choices. Either like Pascal one abandoned mathematics and rational thought and found God where one could, or like Laplace one pursued mathematics and science, leaving God alone.

Why then do their present-day successors insist on discovering rational props for their religious beliefs where none can be found? It seems to me that the reason is not so much the altered content of present-day science as the insecurity of our times. In critical times when man feels threatened by cataclysmic disasters that seem too vast to be averted by human means, he either invokes a superhuman god (or gods) to help him out or, if his means allow it, gives in to the despair of riotous living because 'tomorrow we die'. In the past it was the ignorant masses living a life nasty, brutish and short who invoked God to mitigate their gloom. Surrounded by millions of unshaken faith, a few men of learning could without terror face the agnosticism implicit in any thoroughgoing rational picture of the cosmos.

But today, when the technological means of the advanced industrial nations allow the masses a far richer life than their forefathers of even a century ago could ever dream, they, threatened by wars and alarms of wars, have abandoned themselves to calypso and candy. In the face of such a debacle of religious values and in the midst of an advancing avalanche of scepticism it is the intellectual who is frightened now. This is perhaps one reason why an influential section of contemporary savants has begun to strike an ever-deepening note of mystagoguism born of despair. They no longer endeavour to face squarely the complex problems posed by modern science in a truly scientific spirit but seek to evade them by various sorts of subterfuges such as appeals to religious, mystical, organicist ideas, and sometimes even frankly obscurantist cults.

Lacking the optimistic faith of their illustrious forebears in the intellectual powers of man to comprehend the laws of the universe around him, they have regretfully come to the defeatist conclusion that the laws of the universe are fundamentally unknowable and no true understanding of the ways of the world is really possible without invoking God, spirit, or some similar mystical principle. That this tendency is really, at least in part, due to the intellectual's fright at the widespread diffusion among the masses of agnosticism and scepticism born of riotous living is corroborated by the fact that it is not merely confined to natural science. It is a tendency that is wellnigh universal, being found in present-day literature, art, history, archaeology, sociology and philosophy.

In art and literature, the dominant theme is the problem of death, which gives rise to various kinds of mysticisms in which the writer seeks the significance of transient life in some sort of mystical union with God, as, for instance, in the works of T. S. Eliot, who has drawn inimitable pictures of human society rotten at heart and crumbling as in 'The Waste Land' and 'The Hollow Men'.

In history, the prevailing idea is that western civilization is doomed. Some of these prophets of doom, like Oswald Spengler, almost gloat over their gloomy prognostications. Others, like Arnold Toynbee, Gerald Heard or Pitrim Sorokin, slightly temper their pessimism with vague visions of a new culture, an 'ethereal' or 'ideational' state of immaterial bliss that is bound to arise mysteriously on the ruins of our present 'materialist' civilization.

In philosophy, either the claims of rational thought and reasoning are being superseded by those of mystical intuition of a sort or if they are conceded at all they are used only to limit our thinking activity to a barren formalization of language devoid of all content or meaning.

In natural science, the most 'scientific' and 'up-to-date' view is that the laws of universe are unknowable and we can only construct more or less 'simple', 'economical' or 'elegant' descriptions of phenomena. What 'really' happens around us can never be known. Hence the need for grasping the 'true' essence of the universe by mystical intuition or spiritual second sight. Nor has cosmology been able to ward off the infection, as the insertion of God in some cosmological schemes to perform the 'miracle' of creation or 'guide' 'integrated evolution' clearly shows.

But it does seem futile to try to make scientific cosmology carry grist to the religious mill. It is possible that religion has something important to convey to those who have the faculty to understand its message. But if so, it should try to justify itself by its own criteria of values like art or literature. No one, for instance, would dream of defending surrealism and cubism by an appeal to the tensor calculus or quantum theory and it is as illogical to invoke scientific cosmology in support of God.

For the practice of rationalism is an irreversible process. If once one loses the innocence of naïve belief by venturing to stray into rational thought, there can be no honest way of recovering it. When one has cut oneself off from God by a first sip from the cup of knowledge, one will not rediscover Him by drinking its dregs, no matter how hard they may be boiled. Like Voltaire's Good Brahmin one has no choice but to prefer knowledge in spite of its misery to ignorance with all its accompanying bliss. This is why those intimations of God – they could hardly be called proofs – which are being teased out of new cosmology with almost Munchausen-like abandon convince only those who, like Pope Pius XII, are already convinced, but no others. For they are nothing but coloured phantoms of autistic souls intent on wishful dreaming.

A cosmologist may perhaps, as Jordan suggests, be justified in dreaming about his results provided he has not allowed any human hopes and desires to bias his research. But the innate indeterminateness of the cosmological problem, driving as it does cosmological research in diverse directions, allows so much scope to the play of aesthetic, emotional, religious and metaphysical prejudices that a cosmologist is hard put to it to prevent human motivation from influencing his cosmological judgements. Small wonder that many cosmologists have succumbed to the temptation of gazing at their own cosmological results with such narcissistic persistence that one may say of the God they derive from their cosmology what Zarathustra said of his own:

> Ah, ye brethren, that God whom I created
> was human work and human madness like all Gods!

Appendix

APPENDIX

On How We See the Universe

THERE is a well-known story of six blind men set to observe an elephant. The moral usually drawn is that their disagreements were due to their inability to see that the totality of the partial aspects that each described constituted the whole animal. But we may also draw the further moral that their failure to see the whole elephant arose because lacking vision they could not employ our primary mode of apprehending external objects through radiant light, that eternal coeternal beam which literally wins to us our 'rising world of waters dark and deep' from 'the void and formless infinite'.

Although blind men, whose eyes roll in vain to find a piercing ray, can nevertheless gain some knowledge of the external world by various means such as 'feeling' elephants, stick prodding, hearing sounds or smelling scents and stenches, how horribly mutilated, narrow and limited a view it is may be judged from the laments of men who have spent their light, or, more poetically, from Milton's invocation to holy light.

The reason why blind men's ways of 'feeling' the universe of external objects are so much more cramped than seeing them is that unlike radiant light they involve the transport of some material from the object to the observer. Such, for instance, is the case when we actually carry our hands or stick to an object to touch it or when we smell it by the carriage of material effluvia exhaled by it to our nostrils. For the same reason, even hearing sounds is no match for 'seeing' as a means of gaining knowledge of the external universe around us. Although in this case it is only a state of disturbance that is transported from an object to a listener, such a disturbance in turn needs a material medium like air for its propagation. Because of the need to transport materials to external objects or to have intervening material media to carry disturbances like sound waves in order to produce interactions between them and sentient beings, such means of apprehending external objects suffer from several crippling handicaps.

To mention only two, the materials may remain trapped within the gravitational field of their environment or their motion may be dissipated within a short range like ripples on a lake. If therefore we are to obtain from the depths of the formless void our universe of scintillating stars and glittering galaxies, we have to rely almost wholly on radiant light which alone is free from these drawbacks.

Although we depend on vision for our knowledge of the universe around us, philosophers have never ceased to wonder what aspects of it continue to elude us, limited as we are by our gift of only five senses. In one of his inimitable satires Voltaire describes the chagrin of Micromégas, that intrepid voyager of interstellar space, because in spite of his thousand-odd senses he did not have enough of them to apprehend all that there was in the universe to see. Had the findings of modern physics been known in his day, he would have realized that a plethora of extra senses was not necessarily an advantage, nor, per contra, a lack of them a serious difficulty. For any such extra sense that depended on material transport or material media for producing the interaction required to register a phenomenon on the consciousness of an observer would be of no use in astronomy. What is of value to an astronomer is the means of interpreting only those messages which like radiant light need no material transport.

Now modern physics has shown that radiant light, the small band of colours from violet to red that we see, is but a small segment of an unbroken range of radiations. This range extends far beyond what our eyes can see, with wavelengths both longer and shorter than those of visible light. It begins with red light at one end of the spectrum and continues through infra-red to wireless waves thousands of metres long. At the other (violet) end it proceeds through ultra-violet down to X-rays and gamma rays accompanying nuclear changes. What we need to apprehend completely the universe around us is not a gift of extra senses but the better use of those we do have so that we may learn to see not only with our eyes but with all our senses, somewhat like Zarathustra teaching his listeners to *hear* with their *eyes*.

This is precisely what the physicists of today enable us to do by means of their wondrous apparatus in spite of Nietzsche's gibe at that 'laborious race of machinists' who have nothing but *rough* work to perform. By devising apparatus that can record radiation in the complete range of the extended spectrum of both visible and invisible light in a manner perceptible to some one or other of our five senses, they enable us to 'see' invisible light (cosmic radio waves) with our ears and really teach us how to find a 'higher triumph' in remaining masters of what Nietzsche himself described as the motley whirl of senses. This motley or, as Plato said, mob of senses is reduced to order not by the chimerical conceptual 'networks' the Platonists tried to cast on the world of sensuous perception but by the wide diversity of apparatus designed to detect all kinds of radiations included in the extended spectrum of electromagnetic radiation.

How great the diversity is of apparatus required to observe the entire gamut of electromagnetic radiations may be judged from the fact that visible light is just about one octave of the sixty-four discovered to span the whole range. For if visible light is taken to represent a single octave of wavelengths lying within about 2^{-13} to 2^{-14} cm., the full spectrum of electromagnetic radiations covers about twenty-seven octaves below it down to gamma and cosmic rays at one extreme and about thirty-seven octaves above it to long radio waves at the other as shown in Figure 55. To attempt to see the universe by means of visible light alone is therefore like listening to music with one's ears tuned to hearing only a few notes of a single octave. This is why we have to wear a wide diversity of 'hearing aids' to appreciate fully the austere symphony of the spheres.

The construction of these aids is possible because despite the basic identity in nature of all kinds of electromagnetic emanations, the waves of different length or colour interact rather differently with matter. Thus while light rays are scattered or intercepted by material fogs, radio waves can actually pass through them. It is this diversity of interactions between matter and radiation that enables us to build up inferences about the nature of the objects with which radiation interacts.

Some of these modes of interactions are particularly important

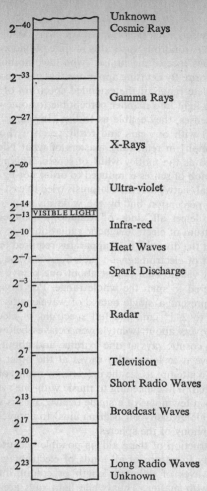

Figure 55. The extended electromagnetic spectrum. The wavelengths are shown as powers of 2 in centimetres. Since the wavelength of visible light extends from 2^{-13} to 2^{-14} cm., the wavelength of cosmic rays (2^{-40} cm.) is 2^{-27} times that of light, that is, twenty-seven octaves lower than that of light. Likewise the wavelength of long radio waves (2^{23}) is 2^{37} that of light, that is, thirty-seven octaves higher.

in astronomy. For example, an object may reflect the light it receives and thus show itself. It is thus that we observe the planets that shine with borrowed beams. Another object may scatter incident light of one or more particular colours or bend it out of its path by refraction or polarize it. Then again there are in outer space sources of radio waves whose existence is revealed to us by the interaction produced by radio waves with external objects which modern equipment enables us to detect. This is the basis of the new science of radio-astronomy. Even more important for the future is the rapid development of better instrumental techniques such as rocket-borne instruments and space observatories that will enable us to 'see' the universe more fully by filling the gap between the infra-red and microwaves on one side and into the far ultra-violet on the other. All these more specialized forms of interactions with light and electromagnetic radiations all the way from long radio waves to extremely short gamma and cosmic rays falling on an object from without can give additional information about the object to an observer placed so as to be able to intercept the appropriate messages and interpret them.

Our interpretation of these messages – the notes of the celestial symphony – exploits a metrical feature that is common to them all. For like ordinary musical notes they too are classified according to their wavelengths. The wavelength in turn is correlated with two others. The correlation is rooted in the fact that they partake of the nature of sea waves which are a manifestation of the periodic lifting and falling of masses of water relative to the earth's centre. Here the separation between successive wave crests is the wavelength λ, and if the wave travels through water with velocity v, it is not difficult to see that wave crests at any point will repeat themselves with a frequency ν given by the equation:

$$v = \lambda\nu.$$

Another way of viewing the phenomenon is that the waves arise on account of the sea water's alternately increasing its potential and kinetic energy in a regular rhythm. In an exactly analogous way electromagnetic radiation including light is the outcome of periodic alternations of the intensity of electric and magnetic fields with regular frequency. If the frequency of this alternation is v, then since all electromagnetic radiations travel with the same speed c as visible light, the wavelength λ of any particular colour is related to the corresponding frequency v by the relation:

$$c = \lambda v.$$

As we mentioned in Chapter 2, it is the wavelength λ or alternatively its correlate frequency v of the radiation in question that fixes its main characteristic that for visible light goes under the name colour so that the wavelength also serves as its distinguishing hallmark. But besides colour electromagnetic radiation has another characteristic feature, namely intensity. Thus radiation of any particular colour may be stronger or intenser than that of the same colour from another source if it carries more energy than the other. For every beam of light travelling through space carries more or less energy with it. It is this energy that darkens our photographic plates and leaves an indelible imprint of the source from which it emanates. By concentrating its energy or fire power by means of an ordinary lens it is possible to produce intense heat which may even burn paper and flesh. What is true of radiant light holds equally for all electromagnetic radiations. The radio waves broadcast by a wireless station, for instance, may carry with them as much as several hundred kilowatts of energy. The amount of energy per unit volume of the space traversed by radiation is defined as its intensity. It is by measuring the intensity of radiation of each wavelength or 'colour' that we compute the temperature of the source by means of Planck's formula quoted in Chapter 2.

Although we have pictured light like other electromagnetic radiations as spreading out in waves, it may also for certain purposes be regarded as a stream of bullets fired from a machine gun. If we choose so to view it, its most appropriate metrical attribute is the energy content E of each bullet or packet of radiation. Since the phenomenon of light radiation can be described in two ways, there must evidently be some underlying connexion between them. This connexion is expressed in the relation between the energy content E of a light packet or quantum and the equivalent wavelength λ:

$$E\lambda = \text{constant.}$$

The relation is of fundamental importance in cosmology. For on it depends the proper correction that must be applied to the measured apparent luminosities of galaxies before their distances can be estimated. As we saw in Chapter 5, Hubble showed that light from the distant galaxies is shifted towards the red end of the spectrum. In other words, the wavelength of the light radiation of each colour is found on arrival here to have increased by the same amount. But an increase in the wavelength λ implies a reduction in the energy content E of the quanta radiation by the galaxy, as otherwise their product would not remain constant. Now a diminution of the energy content of the quanta received from a galaxy would make the galaxy look dimmer than it would in the absence of the red shift. It therefore follows that the measured luminosity of a galaxy must be corrected for the *energy effect* before it can be used to estimate the distance of the galaxy.

This correction is required whether the red shift is interpreted as due to the recession of the galaxy or not. But in case it *is* due to recession, yet another correction must be applied to the measured luminosity. For the recession of the galaxy reduces the rate at which light quanta emitted by it reaches the observer and thus makes it appear dimmer than it would otherwise seem. To understand the reason for it, suppose in one second a stationary galaxy radiates a certain number n of quanta towards the observer. Since all quanta travel with the speed of light, c miles per second, the first quantum to leave the galaxy travels 186,000 miles before the last of the n quanta leaves the galaxy. Thus the n quanta leaving the galaxy during one second are scattered uniformly over a path of length c.

Now consider an exactly similar galaxy receding from the observer with a velocity of v miles per second. As before, it will radiate n quanta per second but during the interval between the emission of the first and the last of these quanta the galaxy will have receded v miles. It therefore results that the same stream of n quanta is now spread over a distance of $v + c$ miles instead of only c miles. Consequently, the quanta stream is diluted and the observer receives fewer quanta per second than previously, so that a receding galaxy appears dimmer than a stationary galaxy would at the same momentary distance. This is why besides the

energy effect a second correction – known as the recession factor*
– must be applied to the observed luminosity of a galaxy in order
to determine its distance.

If we could divine the distances of galaxies independently
without having to depend on their observed luminosities, we
could no doubt use the recession factor to determine whether the
red shift is a genuine recession effect or not. But as there is no
independent measure of the distances of galaxies apart from their
apparent luminosities, the dilution effects of distance and recession
are telescoped into each other with no possibility of their segregation.

This entanglement complicates the correlation between the
magnitudes of the red shifts and the corresponding galactic
distances that Hubble's law seeks to formulate. For there are
two distinct distance scales according to whether the red shifts
are genuine recessional effects or not. Normally one would have
no hesitation in accepting the familiar interpretation of the red
shifts as simple velocity shifts. But the recessional velocities
attributed to the distant galaxies at the fringe of visibility are so
fantastically high as to lead some authorities to doubt the validity
of the recessional interpretation even though they are unable to
suggest any acceptable alternative.

If the recessional interpretation is indeed true, the receding
galaxies, as we mentioned earlier, would fade out of each other's
ken in about twenty billion years. Although cosmologically
speaking this is quite a short period, being no more than about
twice the present 'age' of the universe, in human terms the
recession will take a long while sensibly to affect the quasi-
permanence of the starry heavens that we see. We owe their
comparative preservation once again to the electromagnetic
nature of the messages that the galaxies keep on broadcasting to
us. For unlike their more material counterparts the objects
radiating electromagnetic messages continue to do so automatically without any external prompting and without *any
diminution* of their material content.

If we had to stir a star or a landscape before we could see it (as
we have to strike a tuning fork before we can hear it), we should

* See also page 159.

never have observed any one of them. Further, if their electromagnetic emanations involved any appreciable wastage of their material content, the objects of the observable world would have spent themselves into extinction in a trice in merely allowing themselves to be observed just as a scent exhausts itself in a short while by continual exhalation of its substance in simply announcing its existence.

In all this we have tacitly assumed that an object registers its existence in our consciousness by the electromagnetic radiations that it emits. To be precise, it is not the radiation emitted by an object that enables it to be seen but the *difference* between what it *receives and emits*. For an object not only emits radiation but also receives it from others. It is just the difference between radiation emitted and absorbed that makes its observation at all possible. If all objects in an evironment radiated as much as they received from their neighbours, nothing could be observed in it. For there would be no difference in the radiation received and emitted by objects on which alone their observation depends. Such an unobservable environment may be synthesized (in imagination) by an experiment suggested by A. R. Ubbelohde. Consider a closed furnace in which it is proposed to bake a row of pots of different kinds. If we heat its walls, heat is radiated from the walls to the pots. It is then possible for an observer watching through a minute hole in the furnace wall to distinguish the pots on account of the illumination of the pots by heat radiation. But when the radiation in the enclosure reaches a state of equilibrium so that the pots radiate as much to the walls as they receive from the walls, all of them simply fade out of the observer's perception as he has then no means of distinguishing them.

The question that then arises is this: Would not the continuous enormous radiation into empty space by the stars and galaxies lead the universe to a state of complete radiation equilibrium wherein it would become impossible to observe anything? In other words, would the universe then reach a state of dead evenness or heat death when not only *nothing* could happen therein but also nothing be *seen* at all?

Indices

Subject Index

Acceleration of free particle,
  according to Dirac, 225
  according to Milne, 179, 180,
    181, and
  according to Newton, 123, 124
Acceleration process, of Alfvén, 340
Accretion process, 197
Action, 250
AE Aurigae, 90
Age of Universe, 22, 94, 153, 154,
  155, 222, 224
Ambiplasma, 273, 274, 275, 276
Amino acids, 377, 378
Andromeda, 38, 78, 102
Antares, 33
Anticosmon, 279
Anticosmos, 279
Antimatter, 267, 272, 273
Antiparticles, 267, 272
Antistars and antigalaxies, 267, 279
Arietis, 53, 90
Associations
  Stellar, 82
  O-, 82, 85
  T-, 84
Asteroids, 325, 343
Atmosphere, formation of earth's,
  369
Atom, explosion of primeval,
  161–7
Atomic Number, 311
Atomic theory, 36, 267–71, 309–12

Benzene ring structure, 376
β-processes, 311
Betelgeuse, 33
Bode's law, 325, 335, 348, 351

Cambrian period, 368
Carbon,
  chemistry of, 374–8
  -hydrogen cycle, 44, 45, 54,
    313–14

Centrifugal force, 342
Cepheid variables, 34, 78
Chandrasekhar's limit, 65
Clocks,
  equivalent, 172–5
  graduated, 171–4
Clouds, A-, B-, C-, D- of Alfvén,
  338, 339
Combustion, 43
Comet, 327
Coronium, 36
Cosmic repulsion, 148–9, 154
Cosmical Number, 223, 247–9
Cosmological constant, 145, 154,
  155
Cosmological models, 19, 148–55
  de Sitter's, 148
  Einstein's, 148
  Friedmann's, 148
  Lemaitre – Gamow's, 162–7
  Milne's, 178–80, 184–91
  Oscillating and expanding, 152,
    153, 155–8
  of Klein, 273–7
  and Quasars, 257, 258, 259
Cosmological principle,
  Bondi-Gold's (perfect), 192–3,
    203
  Dirac's, 223–4, 226–8, 232–5
  Milne's (Narrow or Imperfect),
    168, 178, 192, 203
Cosmology,
  Big-bang, 162–7, 216, 277, 306
  Brans-Dicke, 237, 238
  Dirac's, 219–28
  Eddington's, 239–53
  Hoyle-Narlikar, 203–18
  Jordan's, 229–37
  Klein's, 273–7
  Milne's, 168–91
  primitive, 15
Cosmon, 278
Coulomb's law, 221, 240, 271

Covalent bond, 374, 375
Creation, continual, 22, 194–7
Creation field, 206–10, 215
Creation of Universe, 162–7, 214
Creation process, Jordan's, 233
Creation, rate of, 195, 210, 214, 215
Curvature of space, 135–8, 146, 147, 224, 230, 249, 250
Curvature of space-time, 142, 143, 264
Cygnus A, 110, 111, 118

Degenerate state, 53
Density of matter in the Universe, 157, 167, 210
Dimensional analysis, 219, 220
Dimensionless Numbers, 220
Dirac's Numbers, 221, 222
Doppler shift, 37

Earth, 40, 52, 325, 339
Eddies, secondary or roller bearing, 348, 349
Electromagnetic spectrum, 398, 399, 400
Electron, 36, 269, 310
  charge of, 239
  radius of, 249
  shells, 373, 378
Element abundance, 306, 307
Entropy, 285, 287, 288
E-Numbers, 245
Enzymes, 366, 371
Epoch of point-event, 169
e-Process, 315
Equation of continuity, 145
  Laplace's 145, 147
  of Einstein's generalized relativity, 143–6
  Poisson's, 145, 147
Equation of state, 259, 260
  Harrison-Wheeler, 65, 66
Equilibrium, Convective, 54
Equilibrium, radiative, 54;
Equivalence, 175
Equivalence of gravitation and acceleration, 141
Escape velocity, 157
Exclusion principle, Pauli's, 373
Existence symbol, 241–3

Fine structure constant, 239, 250, 251
F-Numbers, 245
Forbidden lines, 36, 73
Force constant, 221, 240, 250
Frequency of wave, 401
Fu Orionis, 87, 88, 89

Galaxies, 27, 28, 29
  attributes of, 75
  classification of (Hubble's), 78, 79
  classification of (revised), 115, 116
  clusters of, 76
  distances of, 78
  distribution of, 77
  elliptical, 78, 81, 101
  evolution of, 79–80 98–102, 105–9, 115–19, 217, 218
  peculiar, 110
  radio, 110, 277
  recession of, 19, 92–5, 160, 221, 224, 403, 404
  Seyfert, 110, 113
  Spiral, 78, 81, 100
Gamma rays, 310
Gasses, inert or noble, 369, 373
Gauss, 96
Geoid, 241
Geometry,
  Euclidean, 20, 131–8
  non-Euclidean, 131–8
  of space, 20, 127–30, 224
Globules, of Bok, 83
God, 15, 16, 130, 372
  and cosmology, 15, 16, 387–93
  ontological proof of, according to Anselm, 387
  ontological proof of, modernized version of Whittaker, 389
Gravitation, 18, 41, 43
  constant of, 22, 123
  according to Brans-Dicke, 237, 238
  according to Dirac, 225, 226
  according to Einstein, 139–46
  according to Milne, 183
  according to Newton, 123
Gravitational collapse, 259–65, 322
Gravitational instability, 80, 98, 99
Grid Number, 247

Heat death of Universe, 22, 300, 301, 405
Helium, 33, 36, 42, 59, 60
Helium abundance, 165, 321
Herbig-Haro objects, 83, 84
H.R. diagram, 51, 56
Hubble's constant, 93, 210
Hydrocarbons, 374–8
Hyperons, 66, 252

Inertial frames, 125
Information, 287, 288, 379
Interstellar gas, 29, 90
Interval, 140, 205
Ionization, 36, 43, 338, 339
Isometric substances, 375
Isotopes, 309

Jupiter, 325, 342, 352, 383

Kinetic theory, 268
Kramer's law, 46

Leidenfrost curtains, 276, 277
Life, 363–8, 370
    abodes of, 381
    origin from inanimate matter, 363–70, 378–80
    origin from other life, 365, 366
    possible abodes of, in solar system, 382–6
    spontaneous origin of, 363
    temperature range of, 381
Light, nature of, 398–402
    polarization of, 96
    quantum of, 402
    radiation pressure of, 41, 42
    velocity of, 17, 126, 239
    wave theory of, 398–401
Light year, 17
Local group, 76, 102
Lorentz transformation, 140
Luminosity, absolute and apparent, 33–4
    mass law, 46
    spectral class diagram, 51, 52
    temperature law, 46

Mach's principle, 141, 149, 205, 208, 240
Magnetic field of sun, 344

Magneto-hydrodynamical effects, 340, 341
Magneto-hydrodynamics, 95, 96, 97
Mars, 325, 339, 383, 384
Martian canals, 384
    maria, 384
Mass, critical, 260
Mass, inertial, 138
Mass-energy equation of Einstein, 44
Mass-luminosity law, 46
Mass Number, 306
Mass ratio of proton and electron, 239, 251
Matter, composition of cosmic, 164, 306, 307
Matter-antimatter cosmology, 273–7
Matter – cold catalysed, 65, 66, 260
Maxwell's demon, 288
Maxwell's equation, 281, 299
Measurement, natural limit of, 248, 249
Measurement, of irreversibility, 285
Mercury, 52, 147, 325, 339, 352, 383
Mesons, 271
Metrical groundform, 134, 136
Microstates and macrostates of crowds, 282–5
Microwave radiation, 161, 214, 216, 217, 274
Milky Way, 27–9, 75, 76, 102
Monomer, 376
Moon, 382
Motion, Newton's laws of, 123–4
Mu Columbae, 90
Mysticism, 391–2

Nebula,
    Coal sack, 30
    Crab, 66, 68–72, 110, 112
    Network of Cygnus, 30, 72
    Orion, 83, 90, 104
    Planetary, 73, 74
    Ring in Lyra, 30
    Rosette, 83
Nebulae (galactic),
    diffuse, 30
    planetary, 30, 73, 74
Nebular hypothesis, 79, 326, 337
Nebulium, 36
Neon, 62
Neptune, 30, 325

Neucleon, 270
Neutrino, 66, 276, 310–12
Neutrino Sink, 66
Neutron, 270
  capture, 163, 164, 313, 318–20
  capture r-process, 313, 318–20
  capture s-process, 313, 318–20
Neutron Star, 21, 53, 65, 67, 70
Non-thermal law of radiation, 111,
  112
Nuclear force 270, 271
Nucleogenesis, 59–63, 164, 308,
  309, 313–20
Nucleon, 309

Olbers's paradox, 129, 130, 160,
  194, 230
Olbers's Sink, 160
Origin of elements,
  astrophysical, 59–63, 308, 309,
    313–20
  cosmological, 162–7, 306, 307

Pair creation, 206, 290–92
  annihilation, 206, 290–92
Parallax, 34
Parallel postulate, 130–32
Parity, 293–6
Particle α-, 44, 309
  β-, 310
Periodicity objection, 289
Perseus, 57, 58
Planck's radiation formula, 37,
  104, 402
Planetary worlds, frequency of,
  354, 385
  Origin of
    Alfvén's theory, 337–45
    Kuiper's theory, 350–54
    Schmidt's theory, 331–6, 353
    Tidal theory, 327–30
    Weizsäcker's theory, 346–9
    Whipple's theory, 345, 346
Pluto, 27, 325
Polymer, 376, 377
Polypeptide, 377
Potentials, retarded and advanced,
  281, 299
Processes, reversible and
  irreversible, 280–82
Proteins, 377, 378
Proton, 270, 309

Proton capture, 313, 319, 320
Proton-proton chain, 44, 50, 54,
  313, 314
Pulsars, 68, 69

Quantum mechanics, 36, 46, 268,
  288, 373
Quantum of light, 402
Quasi stellar galaxies (see Quasars)
Quasars, 16, 21, 22, 31, 32, 53, 92,
  102, 113–15, 116, 118, 213–15,
  254–66, 277
  red shifts of, 254–7
  sources of energy of, 259, 265
Quasar 3C273, 31
Quasar 3C91, 114, 254

Radiation explosion (of Klein), 277
  pressure, 41, 42
Radioactive decays, 309, 310
Radio galaxy M82, 31, 113, 115
Radio galaxy M87, 113, 115, 116
Radio galaxy NGC 1068, 116
Radio sources,
  extragalactic, 31, 92, 102, 213,
    214
  counts of, 210–13, 217–18
Red shift of galaxies, see galaxies
Relativity theory,
  special, 125–7
  general, 139–46
Reversibility and irreversibility,
  280–85, 299
  reversibility objection, 289
Reynolds Number, 103, 105, 347
Rigel, 42, 43
Ring nebula, NGC 6720 in Lyra,
  30
Roche density, 351
  limit, 343
R.R. Lyrae, 62
Running down of Universe, 300,
  301

Satellites, 324
Saturation Number, 373
Saturn, 325, 339, 343
Scale factor, 151, 152
Schwarzchild radius, 260–64
  density, 87, 262, 263
  singularity, 65, 262–5
Sirius, 33

Singular state of the Universe, 153, 154
Space,
    spherical, 230, 249
    -time continuum, 139–42
    -time, curvature of, 142, 143
Spectral index of Radio sources, 111
Spectral lines, 35–8
Spectrum, 35
Stars,
    ages of, 54–7
    Ba, 318
    birth of, 53–6
    blue giants, 45, 81
    central temperature of, 40
    chemical composition of, 39, 42, 43, 44
    cool, 38
    diversity of attributes of, 33
    energy generation in, 44, 45
    evolution of, 48, 53–74
    halo of, 29
    internal constitution of, 39–50
    magnitude of, 34, 35
    main sequence, 51, 54, 55
    models of, 44–50
    nova, 63
    population I and II, 61, 81, 82, 101
    RR Lyrae, 62
    red giants, 52, 55, 60
    spectral class, of, 35, 51
    supergiants, 52
    supernova, 66–73, 235, 330
    Surface temperature of, 38, 39
    T-Tauri, 83–9, 320, 354, 357, 359, 360
    white dwarfs, 52, 53, 55, 65
Statistical laws, 282
Statistical system (Milne's), 182–5
Steady state theory,
    Bondi-Gold's 192–203, 309, 321
    Hoyle-Narlikar, 203–18, 299
    Observational tests of, 210–14
Stereoisomers, 375
Stress-energy-momentum tensor, 143–5
Substratum, 169, 175, 177, 184–5
Sun, 27, 33, 40–42, 50, 325

Symmetry C-, 296, 297
    P-, 293, 296, 297
    T, 296, 297
Synchrotron radiation, 68, 71, 112, 113

Thermodynamic probability, 285, 287
Thermodynamics, second law of, 22, 285–7
Three-halves law, 211
Time
    Einstein's Cosmic, 150, 151
    Einstein's theory of, 126, 127, 140, 141
    Milne's theory of, 169–77
    τ- and t-, 176, 189, 226
    reversal, 290–2, 296
Time's arrow, 287, 289, 298–300
Turbulence, 97, 103–7, 347–9, 350
    Kolmogorov's spectral law of, 104, 350

Universe,
    Mass of, 184, 231, 250
    Radius of, 230, 249
Universon, 278
Uranöid, 241
Uranus, 30, 325

Valence law, 372–3
Venus, 325, 339, 383
Viscosity, 103
Vortices, of Kuiper, 349–54
Vortices of Weizsäcker, 346–9

Wavelength, 37, 398–401
Wheeler-Feynman theory of electromagnetic radiation, 298, 299

X-rays, 21, 32

Ylem, 163

Zeeman effect, 273
Zeta Persei, 82, 90

Name Index

Adams, J. C., 331
Alfvén, H., 322, 337–45, 356
Aller, L. H., 74
Anderson, C. D., 206, 272
Anselm, St, 186, 387
Aquinas, St Thomas, 364, 387, 389, 390
Augustine, St, 169

Baade, W., 19, 72, 202, 224, 252
Babcock, H. W., 49
Baker, N., 62
Batchelor, G. K., 96
Bechlin, E. E., 89
Bergson, H., 364
Bethe, H. A., 44
Bode, J. E., 325, 348, 351
Bok, B. J., 83
Boltzmann, L., 268, 269, 285–7, 288, 289, 299
Bolyai, J., 130–32, 137
Bondi, H., 192, 194, 202, 227, 233, 235, 236, 309
Born, Max, 188
Brahe, Tycho, 18, 73
Brans, 237, 238
Bridgman, P. W., 18
Brown, Harrison, 369
Buckingham, 219, 220
Buffon, G. L. L., 324, 327
Burbidges, G. R., 112, 312, 315, 320, 322

Cameron, A. G. W., 322
Cannon, R. D., 114
Cavendish, Hl, 389
Chandrasekhar, S., 48, 57, 65, 95, 97, 104
Christension, James H., 297
Chui, H. Y., 67
Claus, G., 365
Copernicus, N., 387
Coulomb, C. A., 221, 240

Cowling, T. G., 45, 344
Cronin, James W., 297

Dante, 16
Democritus, 267
Descartes, R., 186, 323
de Sitter, 148, 196, 197, 212, 213
de Vaucouleurs, Gérard, 76
Dicke, R., 161, 162, 237, 238
Dingle, H., 20
Dirac, P. A. M., 22, 206–9, 219, 221–8, 239 267, 272, 291
Doppler, C. J., 37
Drake, F., 70
Dunne, J. W., 169

Eddington, A. S., 19, 22, 82, 144, 149, 226, 239–53, 267, 270, 271, 299, 301, 378
Einstein, A., 44, 123, 126–7, 139–43, 145–50, 154, 169, 188, 189, 203, 204, 206, 209, 210, 212, 213, 231, 235, 236, 237, 252, 261, 281
Eliot, T. S., 391
Eötvös, R., 209
Euclid, 128, 130, 131, 132, 137

Fermi, E., 97
Ferraro, V. C. A., 340
Feynman, R. P., 290, 291, 292, 298
Fitch, Val L., 297
Fowler, W. A., 256, 312, 315, 320, 322
Friedman, A., 148
Friedman, H., 68

Galileo, G., 123, 180
Gamow, G., 162–67, 202, 277, 306, 308, 387, 389
Gauss, C. F., 132
Gold, T., 71, 192, 194, 202, 227, 236, 309

414

Goldhaber, 278, 279
Gurdjieff, 200
Gurevich, L. E., 333

Haeckel, E. H., 23, 388
Haldane, J. B. S., 364, 370, 371
Hall, J. S., 96
Härm, R., 49
Haro, G., 83
Harrison, B. Kent, 65
Hawking, S. W., 165
Heard, Gerald, 391
Heisenberg, W., 96
Herbig, G. H., 83, 84, 86
Herschel, Sir William, 364
Hertzsprung, E., 51
Hiltner, W. A., 96
Hobbes, T., 130
Hoerner, 104
Hoffleit, D., 83
Horowitz, N. H., 379
Howard, R., 49
Hoyle, F., 22, 101, 192, 197, 202,
    203–10, 213–18, 233, 235, 299,
    309, 312, 315, 320, 322, 324, 329,
    337, 345, 354–7, 359
Hubble, E. P., 75, 79, 80, 92, 93,
    94, 105, 109, 150, 155, 177, 186,
    187, 193, 196, 202, 222, 229, 230,
    240, 252, 254, 256, 299, 404
Huygens, C., 196

Iben, Icko Jr., 56

Jeans, J. H., 79, 80, 81, 97, 109,
    235, 332
Jeffreys, H., 328
Johnson, H. L., 85
Jordan, P., 229–35, 239, 392

Kant, I., 20, 79, 169, 324, 326, 327
Kapp, R. O., 22
Kármán, T. von, 104
Kepler, J., 18, 73, 387, 390
Khayyam, Omar, 380
Klein, Oskar, 273–7
Kolomogorov, A., 104, 350
Kraft, Robert, P., 63
Kramers, H., 46
Kuiper, G. P., 332n₄, 349–54

Lagrange, J. L., 235, 388
La Mettrie, J. O. de, 23

Landau, L. D., 260
Laplace, P. S., 79, 145, 147, 248,
    326, 327, 337, 354, 356, 388, 390
Lavoisier, A. L., 195
Lebedinsky, 333
Lee, T. D., 295
Leibniz, G. W., 196
Leidenfrost, 275
Lemaître, Abbé G., 149, 161, 202,
    221, 277, 346, 387, 388, 389, 390
Leverrier, V. J., 331
Levin, B., 333
Lewis, G. N., 287, 289
Lobachevski, N. I., 130–32, 137
Loschmidt, J., 287, 289
Lowell, P., 384
Lucretius, 43, 128
Lundquist, 340
Lyttleton, R. A., 22, 197, 233, 329

McCord, H., 266
MacCrea, W. H., 114, 209
Mach, E., 141, 149, 205, 208
Mann, Thomas, 41
Mariner, R., 370
Mayer, M. G., 308
Maxwell, J. C., 204, 206, 208, 209,
    252, 281, 288, 299, 372
Menzel, D. H., 42
Michelson, A. A., 188
Miller, S. L., 370
Milne, E. A., 150, 168–71, 175–91,
    202, 221, 225, 226, 252, 387, 388,
    389, 390
Milton, J., 397
Minkowski, R., 72
Morgan, Lloyd, 364
Morley, E. W., 188
Morrison, Phillip, 67
Moses, 323
Mössbauer, R., 209
Mumford, L., 323
Munitz, Milton K., 203

Nagy, B., 365
Narlikar, J. V., 203–10, 214, 215, 299
Neugebauer, C., 89
Newton, I., 98, 123, 126, 127, 128,
    129, 137, 138, 139, 141, 145, 147,
    149, 168, 181, 182, 184, 196, 205,
    207, 210, 235, 237, 280, 372, 387
Nietzsche, F., 399

Nölke, F., 328
Noüy, Lecomte de, 371–2, 378

Olbers, H. W. M., 129–30, 160, 194, 195
Oort, J., 29, 73, 74
Ouspensky, 200

Pannekoek, A., 42
Pascal, B., 386, 390
Pauli, W., 67, 310, 373
Peierls, R., 308
Penrose, R., 264, 265
Penzias, A. A., 161, 214, 274
Perutz, M., 367
Peston, M. V., 114
Pius XII, Pope, 389, 392
Planck, M., 37, 38, 239, 250, 402
Plato, 130, 241, 399
Poincaré, H., 189, 289
Poisson, S. D., 145, 147
Ponnamperuma, C., 370
Poole, H. W., 186
Pooley, G. G., 218
Popper, K. R., 288
Pythagoras, 388

Ramsay, W. H., 71, 72
Rayleigh, Lord, 107
Reichenbach, H., 293
Reynolds, J. H., 359
Reynolds, Osborne, 103, 347
Riemann, G. F. B., 132, 136
Robb, 169
Russell, H. N., 42, 51, 329
Rutherford, Lord, 270
Ryle, M., 210, 211, 214

Sagan, C., 370
Saha, M. N., 43
Sandage, A., 19, 31, 32, 48, 58, 59, 60, 157, 214
Schmidt, O., 331–6, 353, 359, 369
Schonberg, 48, 57
Schrödinger, E., 149
Schwarzschild, K., 261
Schwarzschild, M., 48, 49
Sciama, D. W., 197, 198
Seeliger, H., 129
Shklovsky, I. S., 73, 112
Smouluchowski, M. V., 289
Sorokin, Pitrim, 391
Spengler, Oswald, 391

Spinoza, B., 130
Spitzer, I., 328–9, 330
Stapledon, Olaf, 380, 384
Strömgren, E., 42, 48
Struve, O., 336
Suess, Hans, 306, 320, 321, 369

Taylor, R. J., 165
Teller, Edward, 226, 227, 308
Tennyson, A., 193
Ter Haar, D., 345, 354
Thorne, Kip S., 65, 261
Titius, 325
Turlay, René, 297
Tolman, R. C., 22, 301
Toynbee, Arnold, 391
Truran, T. W., 321

Ubbelohde, A. R., 405
Unsöld, A., 42
Urey, M. C., 306, 332, 336, 353, 358, 370

Van de Kamp, Peter, 331
Vernon, Phillipe, 213
Voltaire, 18, 126, 392, 398

Wakano, Masani, 65
Walker, Merle F., 84, 85
Weizsäcker, C. F. von, 98, 104, 105, 107, 108, 109, 300, 324, 332, 337, 345, 346–9, 350, 351, 390
Wells, H. G., 141
Weyl, H., 149
Wheeler, J. A., 65, 260, 298
Whipple, F. L., 337, 345
Whitehead, A. N., 19, 246
Whitrow, G. J., 282
Whittaker, E. T., 191, 252, 253, 388, 389
Whyte, L. L., 228
Wilson, R. W., 161, 214, 274
Woodger, J. H., 130, 364
Wu, Madame, 295

Yang, C. N., 295
Yukawa, H., 271

Zahlank, A. B., 379
Zanstra, H., 73
Zeldovich, Ya. B., 308
Zermelo, E., 289
Zeeman, P., 273